STUDIES IN APPLIED MECHANICS 20

Micromechanics of Granular Materials

STUDIES IN APPLIED MECHANICS

1. Mechanics and Strength of Materials (Skalmierski)
2. Nonlinear Differential Equations (Fučík and Kufner)
3. Mathematical Theory of Elastic and Elastico-Plastic Bodies
   An Introduction (Nečas and Hlaváček)
4. Variational, Incremental and Energy Methods in Solid Mechanics and Shell Theory (Mason)
5. Mechanics of Structured Media, Parts A and B (Selvadurai, Editor)
6. Mechanics of Material Behavior (Dvorak and Shield, Editors)
7. Mechanics of Granular Materials: New Models and Constitutive Relations (Jenkins and Satake, Editors)
8. Probabilistic Approach to Mechanisms (Sandler)
9. Methods of Functional Analysis for Application in Solid Mechanics (Mason)
10. Boundary Integral Equation Methods in Eigenvalue Problems of Elastodynamics and Thin Plates (Kitahara)
11. Mechanics of Material Interfaces (Selvadurai and Voyiadjis, Editors)
12. Local Effects in the Analysis of Structures (Ladevèze, Editor)
13. Ordinary Differential Equations; Introduction to the Theory of Ordinary Differential Equations in the Real Domain (Kurzweil)
14. Random Vibration — Status and Recent Developments (Elishakoff and Lyon, Editors)
15. Computational Methods for Predicting Material Processing Defects (Predeleanu, Editor)
16. Developments in Engineering Mechanics (Selvadurai, Editor)
17. The Mechanics of Vibrations of Cylindrical Shells (Markuš)
18. Theory of Plasticity and Limit Design of Plates (Sobotka)
19. Buckling of Structures — Theory and Experiment (Elishakoff et al., Editors)
20. Micromechanics of Granular Materials (Satake and Jenkins, Editors)

STUDIES IN APPLIED MECHANICS 20

Micromechanics of Granular Materials

Proceedings of the U.S./Japan Seminar on the Micromechanics of Granular Materials, Sendai-Zao, Japan, October 26–30, 1987

Edited by

Masao Satake

Department of Civil Engineering, Tohoku University, Sendai, Japan

and

James T. Jenkins

Department of Theoretical and Applied Mechanics, Cornell University, Ithaca, NY, U.S.A.

Sponsored by the Japan Society for Promotion of Science and the United States National Science Foundation

ELSEVIER

Amsterdam — Oxford — New York — Tokyo 1988

ELSEVIER SCIENCE PUBLISHERS B.V.
Sara Burgerhartstraat 25
P.O. Box 211, 1000 AE Amsterdam, The Netherlands

Distributors for the United States and Canada:

ELSEVIER SCIENCE PUBLISHING COMPANY INC.
52, Vanderbilt Avenue
New York, NY 10017, U.S.A.

Library of Congress Cataloging-in-Publication Data

US/Japan Seminar on the Micromechanics of Granular Materials (1987 : Sendai-shi, Miyagi-ken, Japan and Zaō-machi, Japan)
Micromechanics of granular materials.

(Studies in applied mechanics ; 20)
"Sponsored by the Japan Society for Promotion of Science and the United States National Science Foundation."
Includes bibliographies.
1. Granular materials--Congresses. I. Satake, Masao. II. Jenkins, James T., 1942- . III. Nihon Gakujutsu Shinkōkai. IV. National Science Foundation (U.S.) V. Title. VI. Series.
TA418.78.U18 1987 620.1'9 88-24367
ISBN 0-444-70523-6 (U.S.)

ISBN 0-444-70523-6 (Vol. 20)
ISBN 0-444-41758-3 (Series)

Printed in The Netherlands

Dedicated

to Professor Takeo Mogami

and

to his pioneering work

PREFACE

This volume contains the proceedings of the United States-Japan Seminar on the Micromechanics of Granular Materials held in Sendai and Zao, Japan, from October 26 through October 30, 1987.

With the success of the last two Seminars held in Sendai in 1978 and at Cornell University in 1982, we were again able to invite researchers who had made fundamental contributions to the micromechanics of granular materials from Japan, the United States and England. The purpose of this seminar was to facilitate an exchange of ideas between scientists working with statistical and continuum theories, computer simulations and experiments on both static and dynamic behavior. The papers collected here indicate that a promotion of a good understanding of the mechanics of granular materials has been and will continue to be valued in a variety of technical disciplines.

Since the first seminar it has been unanimously agreed that the microstructure of the granular materials plays a crucial role in their mechanical behavior. The fabric tensor introduced after the first meeting has been well accepted and applied to investigate not only granular materials but also other discontinuous media. The crack tensor employed in the field of rock mechanics is an example.

In describing the solid like behavior of granular materials, many new ideas on the constitutive relations were introduced. As an application of the analysis, the mechanism of liquefaction was also discussed. Computer simulations have become a vital tool in analyzing the micromechanical behavior which otherwise would not be experimentally tested. In numerical simulations and theoretical analyses of rapid granular flow, various modifications on the nature of materials and boundaries were made. At this seminar possible applications of the techniques of the stereology and analysis based on geometrical statistics were also stressed.

The seminar was sponsored by the Japan Society for the Promotion of Science and the United States National Science Foundation under the United States-Japan Cooperative Science Program. The social programs were supported by the Tohoku Branch of the Japanese Society of Soil Mechanics and Foundation Engineering. We are also greateful for their financial support by some enthusiastic Japanese corporations.

With great regret we note that Professor Takeo Mogami passed away on December, 1987. He, a pioneer in the field of mechanics of granular materials, delivered a special lecture at the first Seminar. We would like to dedicate this volume to Prof. Mogami. Finally, we would also like to thank the contributors for their care in the preparation of their manuscripts.

Masao Satake and James T. Jenkins

TABLE OF CONTENTS

PARTICIPANTS

J. Robin F. Arthur
Professor
Department of Civil and Municipal Engineering
University College London
London WC1E 6BT, England

Christopher E. Brennen
Professor
Division of Engineering and Applied Science
California Institute of Technology
Pasadena, California 91125

Charles S. Campbell
Assistant Professor
Department of Mechanical Engineering
University of Southern California
Los Angeles, California 90089

Stephen C. Cowin
Professor
Department of Biomedical Engineering
Tulane University
New Orleans, Louisiana 70118

Peter A. Cundall
Senior Research Associate
Department of Civil and Mineral Engineering
University of Minnesota
Minneapolis, Minnesota 55455

Harold J. Frost
Professor
Thayer School of Engineering
Dartmouth College
Hanover, New Hampshire 03755

Kouichi Hashiguchi
Associate Professor
Department of Agricultural Engineering
Kyushu University
Fukuoka 812

Kenji Ishihara
Professor
Department of Civil Engineering
University of Tokyo
Bunkyo-ku, Tokyo 113

James T. Jenkins
Professor
Department of Theoretical and Applied Mechanics
Cornell University
Ithaca, New York 14853

Ken-ichi Kanatani
Associate Professor
Department of Computer Science
Gunma University
Kiryu 376

Tadahiko Kawai
Professor
Department of Electrical Engineering
Science University of Tokyo
Shinjyuku-ku, Tokyo 162

Yuji Kishino
Associate Professor
Department of Civil Engineering
Tohoku University
Sendai 980

Ryosuke Kitamura
Associate Professor
Department of Ocean Civil Engineering
Kagoshima University
Kagoshima 890

Kazutaka Makino
Professor
Department of Chemical Engineering for Resources
Akita University
Akita 010

Hajime Matsuoka
Professor
Department of Civil Engineering
Nagoya Institute of Technology
Showa-ku, Nagoya 466

Takaaki Nagao
Professor
Department of Foundation Engineering for Production
University of Tokyo
Bunkyo-ku, Tokyo 113

Siavouche Nemat-Nasser
Professor
Department of Applied Mechanics and
Engineering Sciences
University of California, San Diego,
La Jolla, California 92093

Masanobu Oda
Associate Professor
Department of Foundation Engineering
Saitama University
Urawa 338

Nobunori Oshima
Professor
College of Industrial Technology
Nihon University
Narashino 275

Masao Satake
Professor
Department of Civil Engineering
Tohoku University
Sendai 980

Hayley H. Shen
Associate Professor
Department of Civil and Environmental
Engineering
Clarkson University
Potsdam, New York 13676

Anthony J. M. Spencer
Professor
Department of Theoretical Mechanics
University of Nottingham
Nottingham, NG7 2RD, England

Colin Thornton
Reader
Department of Civil Engineering
Aston University
Birmingham, B4 7ET, England

Otis R. Walton
Lawrence Livermore National Laboratory
Livermore,
California 94550

OBSERVERS

Norbert L. Ackermann
Professor
Department of Civil and Environmental Engineering
Clarkson University
Potsdam, New York 13676

John K. Beddow
Professor
Chemical and Materials Engineering
University of Iowa
Iowa City, Iowa 52242

Colin B. Brown
Professor
Department of Civil Engineering
University of Washington
Seattle, Washington 98105

Ching S. Chang
Associate Professor
Department of Civil Engineering
University of Massachusetts
Amherst, Massachusetts 01003

Motohiko Hakuno
Professor
Earthquake Reserch Institute
University of Tokyo
Bunkyo-ku, Tokyo 113

Yotaro Hatamura
Professor
Department of Mechanical Engineering for Production
University of Tokyo
Bunkyo-ku, Tokyo 113

Hiroyoshi Hirai
Lecturer
Department of Civil Engineering
Kumamoto University
Kumamoto 860

Hideyuki Horii
Associate Professor
Department of Civil Engineering
University of Tokyo
Bunkyo-ku, Tokyo 113

Isao Ishibashi
Professor
Department of Civil Engineering
Old Dominion University
Norfolk, Virginia 23529

Genji Jimbo
Professor
Department of Chemical Engineering
Nagoya University
Chigusa-ku, Nagoya 464

Masataka Kamata
Technical Research Institute
Fujita-Kogyo Co.
Kohoku-ku, Yokohama 223

Kouji Kawasaki
Technical Research Institute
Taisei Co.
Totsuka-ku, Yokohama 245

Jun-ichi Konishi
Associate Professor
Department of Civil Engineering
Shinshu University
Nagano 380

Hironobu Matsui
Department of Mechanical Engineering for Production
University of Tokyo
Bunkyo-ku, Tokyo 113

Morteza M. Mehrabadi
Associate Professor
Department of Mechanical Engineering
Tulane University
New Orleans, Louisiana 70118

Toshihiro Mikami
Department of Civil Engineering for Production
University of Tokyo
Bunkyo-ku, Tokyo 113

Nobuchika Moroto
Professor
Department of Civil Engineering
Hachinohe Institute of Technology
Hachinohe 031

Masami Nakagawa
Research Associate
Department of Civil Engineering
Tohoku University
Sendai 980

Teruo Nakai
Associate Professor
Department of Civil Engineering
Nagoya Institute of Technology
Showa-ku, Nagoya 466

Susumu Nakamura
Engineering Research Institute
Sato Kogyo Co.LTD.
Sanda 47-3, Atsugi 243-02

Shigeru Niiseki
Research Associate
Department of Civil Engineering
Tohoku University
Sendai 980

Fusao Oka
Associate Professor
Department of Civil Engineering
Gifu University
Gifu 501-11

Norio Takeuchi
Lecturer
Department of Civil Engineering
Meisei University
Hino, Tokyo 191

Yoshio Tobita
Lecturer
Department of Civil Engineering
Hachinohe Institute of Technology
Hachinohe 031

Ikuo Towhata
Associate Professor
Department of Civil Engineering
University of Tokyo
Bunkyo-ku, Tokyo 113

Yasuo Yamada
Associate Professor
Institute of Engineering Mechanics
University of Tsukuba
Tsukuba 305

Eiji Yanagisawa
Professor
Department of Civil Engineering
Tohoku University
Sendai 980

Micromechanics of Granular Materials, edited by M. Satake and J.T. Jenkins
Elsevier Science Publishers B.V., Amsterdam, 1988 — Printed in The Netherlands

STEREOLOGICAL ESTIMATION OF MICROSTRUCTURES IN MATERIALS

K. KANATANI

Department of Computer Science, Gunma University, Kiryu, Gunma 376 (Japan)

SUMMARY

Typical types of stereological procedures are classified according to the type of estimation, and their mutual relationships are discussed. First, principles of estimating average quantities are reviewed. Then, approaches for estimating sphere size distribution from observation of material cross-sections are briefly discussed. Finally, methods of estimating structural anisotropy due to internal distribution of line tissues and surfaces are studied in mathematical terms. The structural anisotropy is characterized by the distribution density, which is expressed in terms of the "fabric tensors." The distribution density is related to observed data by the "Buffon transform," and estimation is done by computing its inversion. Related applications are also presented.

INTRODUCTION

"Stereology" is the study of methods to estimate "geometrical characteristics" of the space (of any dimensionality - mostly 2 and 3) from "partial observations" such as observations of material cross-sections or thin slices. By definition, definite solutions cannot be attained; the problem is statistical in its nature. Hence, statistical models are employed with appropriately assigned probabilities, and the solution has a meaning only in the statistical sense. In the following, we classify and summarize typical techniques studied in the past.

AVERAGE DENSITY ESTIMATION

The mathematical structure of stereological estimation depends on what is to be estimated. The quantities that can be estimated most easily are such "densities" as V_V, the volume of a specific phase per unit volume of the material , A_V, the area of a specific internal surface structure per unit volume, and L_V, the length of specific internal tissues per unit volume. Various techniques have been developed in the past. Usually, observations are made on a material surface after cutting the material randomly or on a probe line randomly placed on it. For example, we observe the material surface and measure such quantities as A_A, the area of a specific phase per unit area (Fig. 1(a)), L_A, the length of intersections with the internal surface structure per unit area (Fig. 1(b)) and N_A, the number of intersections with internal line

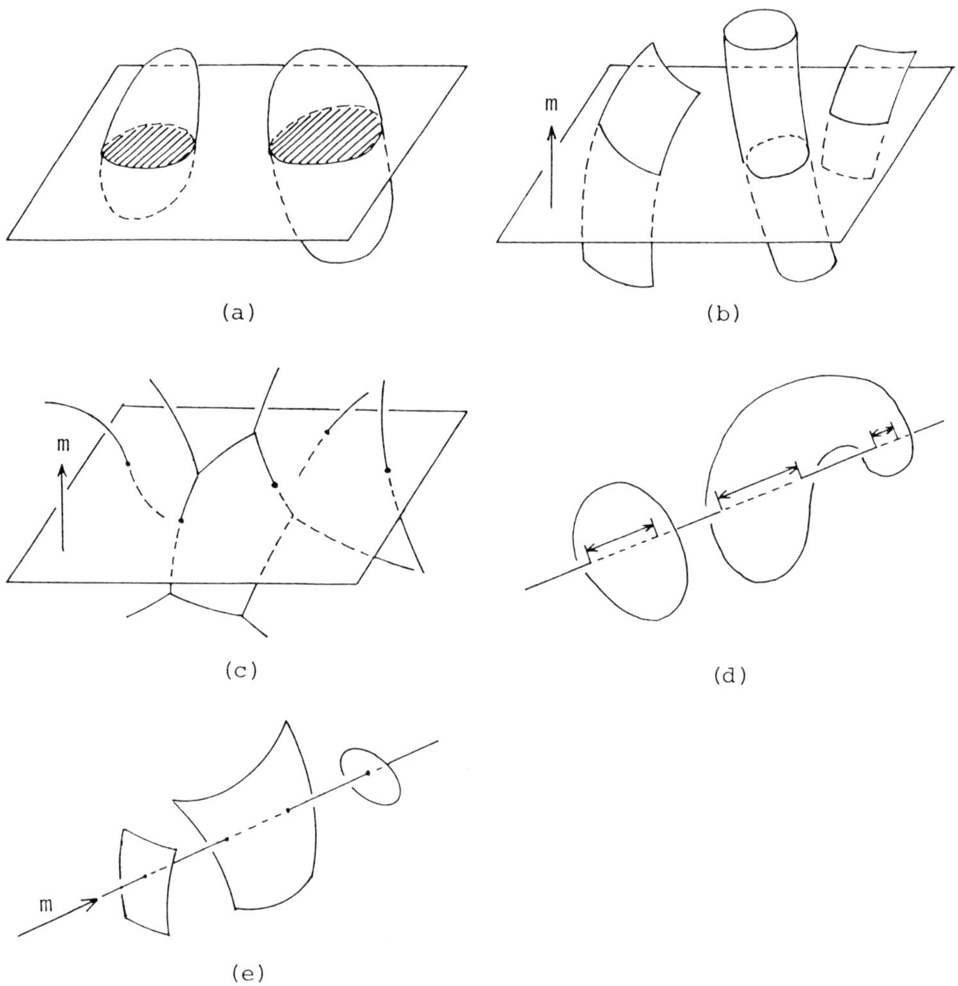

Fig. 1. (a) A_A, (b) L_A, (c) N_A, (d) L_L, (e) N_L. For (b), (c) and (e), N(m) is observed if m is fixed.

tissues per unit area (Fig. 1(c)). Alternatively, we place a probe line and measure suh quantities as L_L, the length of intercepts made by a specific phase per unit length (Fig. 1(d)) and N_L, the number of intersections with the internal surface structure per unit length (Fig. 1(e)).

Densities such as V_V, A_V, L_V, A_A, L_A, N_A, L_L and N_L all have meanings in the sense of "average" - per unit volume, per unit area, or per unit length. This type of estimation has been extensively studied in biology, medicine, metallurgy and geology in order to quantify the internal biological tissues, inclusions in metals, underground minerals, etc. from microscope observations (or areal

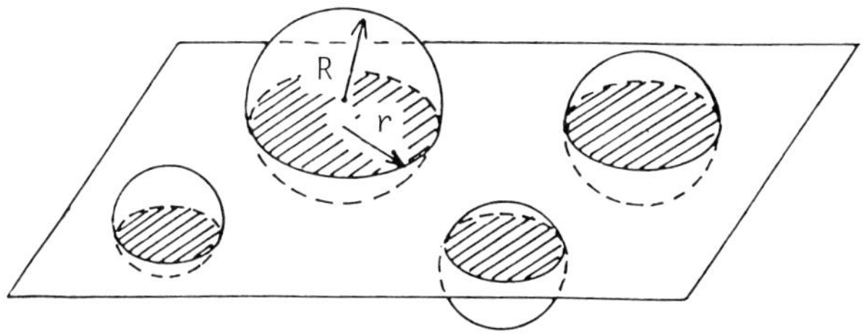

Fig. 2. Distribution of spheres in the material and their corss-sections on a cutting plane.

photographs) of material surfaces and cross-sections. Mathematical relationships among those quantities, such as $L_V = 2N_A$ and $A_V = 4L_A/\pi = 2N_L$, are listed in many books (refs. 1-3).

SIZE DISTRIBUTION ESTIMATION

At times, we want to know not noly average densities by also their "distributions." For example, if many objects of the same shape but different sizes are distributed in a material, we often need to know the distribution of their sizes. Most extensively studied are spherical inclusions. Although other shapes such as ellipsoids and cubes are also important, the assumption of sphere makes the theory very simple. Moreover, spherical particles often appear and play an important role in biology, medicine, metallurgy and geology (refs. 1-3).

Let the "distribution density" F(R) be defined so that F(R)dR is the number of spheres of radii between R and R + dR in unit volume of the material. The spheres are assumed to be distributed randomly. Suppose we place a cutting plane randomly in the material and observe particle cross-sections on the cutting plane (Fig. 2). Let the observed distribution density f(r) be defined so that f(r)dr is the number of circular cross-sections of radii between r and r + dr in unit area of the cutting plane. Then, F(R) and f(r) are shown to be related by the integral equation of the form

$$f(r) = 2r\int_r^{R_{max}} \frac{F(R)dR}{\sqrt{R^2 - r^2}}, \tag{1}$$

where R_{max} is the maximum particle radius. A small change of variables reduces this equation to an integral equation of Abel type, and it can be analytically inverted to express F(R) in terms of f(r) in the form

$$f(R) = \frac{R}{\pi}\int_R^{R_{max}} \frac{d(f(r)/r)}{\sqrt{r^2 - R^2}} . \tag{2}$$

Many numerical schemes are proposed to compute this process by using only a finite number of data. Various other effects are also incorporated - the use of a thin slice instead of a cutting plane (the "Homes effect" or "overprojection"), the "resolution threshold" and the "capping effect" (refs. 1-7).

Thus, the problem is completely solved in principle. However, there remain several issues. First, most numerical schemes were devised on the basis of some discrete distribution models. For example, each particle is assumed to belong to one of prefixed classes of the particle size. The effectiveness of these schemes were often evaluated only by the numerical outcomes of real or synthetic examples. However, we should recall that the exact integral equations (1) and (2) are already obtained. Hence, every numerical scheme must be a discrete approximation of (1) or (2) in some sense, and hence it should be evaluated by the criterion of how well it approximates the exact integral equation. This viewpoint is strongly emphasized by Kanatani and Ishikawa (ref. 7).

Another issue is the use of the (cumulative) "distribution function" $\Phi(R)$, the number of spheres of radii smaller than R in unit volume of the material, and $\phi(r)$, the number of cross-sections of radii smaller than r in unit area of the cutting plane, rather than the "distribution densities" F(R) and f(r). This is because $\Phi(R)$ and $\phi(r)$ are well-defined functions from both a practical and a mathematical points of view, while densities F(R) and f(r) are their "differentiations" involving infinitesimals dR and dr. Making a discrete histogram is nothing but one process of "numerical differentiation", which is very susceptible to noise.

In many books, not only F(R) and f(r) are used but also they are interpreted as "probability densities" normalized so that the total becomes unity. However, there is no physical sense to do so, because they describe distributed quantities "per unit volume" and "per unit area". Normalization only complicates mathematical treatment.

In terms of the distribution functions $\Phi(R)$ and $\phi(r)$, equations (1) and (2) are integrated into the form

$$\phi(r) = 2R\overline{N} - 2\int_r^{R_{max}} \sqrt{R^2 - r^2}\, d\Phi(R), \tag{3}$$

$$\Phi(R) = N - \frac{1}{\pi}\int_R^{R_{max}} \frac{d\phi(r)}{\sqrt{r^2 - R^2}}, \tag{4}$$

where N is the "numerical density", i.e., the number of particles per unit volume, and $\overline{R}$ is the "mean radius" of the particles. Integration is done in the Stieltjes sense. From these relations various discrete numerical schemes are derived (ref. 7).

Another factor to keep in mind is the distinction of error sources. The reliability of the estimation increases as the number of observations increases and as the accuracy of each observation is improved. This is because "sampling errors" are reduced. On the other hand, the use of accurate formulae also improves the accuracy. This is because "computational errors" are reduced.

These two error sources are different in their nature and should not be mixed up. In the pase, many estimation schemes were tested experimentally, and their effectiveness was evaluated according to the overall accuracy. However, good overall accuracy does not imply that the formula itself is accurate. Different error sources should be examined separately, and the computational accuracy should be studied from the viewpoint of numerical analysis. For instance, Kanatani and Ishikawa (ref. 7) pointed out that the computational accuracy depends dominantly on the treatment of "singularities" involved in the discrete approximation of original integral equations. They also studied the numerical stability in terms of the "condition number" of the matrices involved in the numerical schemes.

ESTIMATION OF STRUCTURAL ANISOTROPY

Characterization of a material which has internal structures is not limited to computing average densities and their size distributions. Sometimes, the degree of "anisotropy" of the distribution becomes important. If the distribution is not isotropic, the same observation process yields different results depending on the way the material is placed. In the past, the material anisotropy has been characterized directly by these dependences of particular measurement values on the material orientation. Many "indices of anisotropy" are devised for this purpose (refs. 1-3).

Although this type of approach is common in biology, medicine, metallurgy and geology, it involves many ad-hoc assumptions which makes theoretical treatment very difficult. In many cases, interrelationships between different indices cannot be established because these indices are defined without any regard to the "internal mechanism" which produces the anisotropy. The right approach should be as follows: We first define a definite "model" of anisotropy in mathematical terms, and determine the parameters which characterize it irrespective of whether or not these parameters can be measured by available means. Then, we devise some means for measuring these parameters. If they cannot be measured directly, we try to devise some schemes to estimage their approximate values.

Let us consider two types of material - materials which have internal line tissues and materials which have internal surface structures. First, consider a material in which line tissues are distributed. These line tissues may be straight or curved. They may be dissected into separate "needles" or linked to form a connected "network." Let the line tissues be hypothetically dissected into infinitesimally small line segments. Let each line segment be given a unit vector n indicating its orientation. Since n and -n both indicate the same orientation, let us choose one randomly with a probability of 1/2 out of these two possibilities. (Theoretically, we could also think of "polarized" tissues whose orientations are uniquely determined.) Let the "distribution density" $f(n)$ be defined so that $f(n)d\Omega(n)$ is the total length of those line segments, in unit volume of the material, whose orientations are inside the differential solid angle $d\Omega(n)$ around n. By definition, $\int f(n)d\Omega(n)$ equals L_V, the length of the line tissues per unit volume. If the distribution is isotropic, $f(n)$ = const.

Next, consider a material in which surfaces are distributed. These surfaces may be planar or curved, separate "disks" or connected "cell walls." Let these surfaces be hypothetically dissected into infinitesimally small surface segments. Let each surface segment be given a unit normal vector n indicating its orientation. Since n and -n indicates the same orientation, choose one randomly with a probability 1/2 out of these two possibilities. (Again, we could imagine surfaces which have "polarity" telling us which side is front or back.) Let the "distribution densities" $f(n)$ be defined so that $f(n)d\Omega(n)$ is the total area of those surface segments, in unit volume of the material, whose normals are inside the differential solid angle $d\Omega(n)$ around n. By definition, $\int f(n)d\Omega(n)$ equals A_V, the area of the surface per unit volume. If the distribution is isotropic, $f(n)$ = const.

In either case, our aim is now "well-defined" in mathematical terms as estimation of the distribution density $f(n)$. Hence, the first stage is done. Next, let us turn to the second stage of "parameterizing" the distribution density $f(n)$. Since it is a function of unit vector n, it can be expanded in "spherical harmonics" (ref. 8). In the Cartesian tensor notation, the expansion takes the form

$$f(n) = \frac{c}{4\pi}[1 + D_{ij}n_in_j + D_{ijkl}n_in_jn_kn_l + \ldots]. \tag{5}$$

Here, we adopt the Einstein summation convention over repeated indices. Terms of odd degrees do not appear because $f(n)$ is "symmetric" with respect to the origin, i.e., $f(n) = f(-n)$. (Terms of odd degrees appear if we consider distributions of "polarized" line tissues or surfaces.) In the above equation, $c = \int f(n)d\Omega(n)$ is the "total density", and D_{ij}, D_{ijkl}, ... are "deviator tensors" for which contraction of any two indices yields 0 - they are completely "traceless" over any pair of indices. Define the "moment tensors" by

$$N_{ij} = \frac{1}{c}\int n_in_jf(n)d\Omega(n), \qquad N_{ijkl} = \frac{1}{c}\int n_in_jn_kn_lf(n)d\Omega(n), \quad \ldots . \tag{6}$$

Then, tensors D_{ij}, D_{ijkl}, ... and given by taking their "deviator parts":

$$D_{ij} = \frac{15}{2}N_{\{ij\}}, \quad D_{ijkl} = \frac{315}{8}N_{\{ijkl\}}, \quad \ldots, \quad D_{i_1\ldots i_n} = \frac{2n+1}{2^n}\binom{2n}{n}N_{\{i_1\ldots i_n\}}. \tag{7}$$

Here, the brackets { } over indices designate the following operations. Let $A_{i_1\ldots i_n}$ be a symmetric tensor. We define

$$A_{\{ij\}} = A_{ij} - \frac{1}{3}\delta_{ij}, \qquad A_{\{ijkl\}} = A_{ijkl} - \frac{6}{7}\delta_{(ij}A_{kl)mm} + \frac{3}{35}\delta_{(ij}\delta_{kl)}A_{mmnn}, \ldots,$$

$$A_{\{i_1\ldots i_n\}} = c_0^n A_{i_1\ldots i_n} + c_1^n \delta_{(i_1 i_2}A_{i_3\ldots i_n)j_1 j_1} + \cdots$$

$$+ c_n^n \delta_{(i_1 i_2}\delta_{i_3 i_4}\cdots\delta_{i_{n-1} i_n)}A_{j_1 j_1 j_3 j_3\ldots j_{n-1} j_{n-1}}. \qquad (8)$$

where the parentheses ( ) designate the symmetrization of the indices between them.

Tensors D_{ij}, D_{ijkl}, ... are are called the "fabric tensors" of the distribution f(n). This expansion is the same as the "multiple moment expansion" of an electric potential well known in electromagnetics. Since spherical harmonics of each degree form a basis of an irreducible representation of the three-dimensional rotation group SO(3), this expansion corresponds to resolution of the function space over unit sphere to the irreducible representation spaces of SO(3) (ref. 8).

Let a cutting plane with unit surface normal m be placed randomly in the material, and let N(m) be the number of intersections with line tissues per unit area of the cutting plane (Fig. 1(c)). It can be shown that the expectation value of N(m) is related to the distribution density f(n) of the line tissues in the form

$$N(m) = \int |m \cdot n| f(n) d\Omega(n). \qquad (9)$$

This relation was first suggested by Hilliard (ref. 9) and was called the "Buffon transform" by Kanatani (ref. 10). Let a probe line of orientation m be placed randomly in the material, and let N(m) be the number of intersections with surfaces per unit length of the probe line (Fig. 1(e)). The expectation value of N(m) is also shown to be related to the distribution density f(n) of the surfaces by the Buffon transform (9) (cf. ref. 10). Let a cutting plane with unit surface normal m be placed randomly in the material, and let N(m) be the length of intersections with surfaces per unit area of the cutting plane (Fig. 1(b)). The expectation value of N(m) is shown to be related to the distribution density f(n) of the surfaces in the form

$$N(m) = \int |m \times n| f(n) d\Omega(n), \qquad (10)$$

which was also called the "Buffon transform" by Kanatani (ref. 10).

Thus, the estimation of the distribution density f(n) reduces to inverting the Buffon transform (9) or (10). Let the right-hand side of (9) or (10) be written as Bf(m), and call the operator B the "Buffon operator." Let the rotation operator R be defined by $Rf(m) = f(R^{-1}m)$. Then, the Buffon operator B "commutes" with the rotation operator R, and hence is an "invariant" operator: BRf(m) = RBf(n) or $B = R^{-1}BR$. As is well known in harmonic analysis, the spherical harmonics generate "eigenspaces" of the operator B, and the inversion of (9) or (10) is computed if observed $f(n)$ is also expanded in spherical harmonics:

$$N(m) = \frac{C}{4\pi}[1 + F_{ij}m_i m_j + F_{ijkl}m_i m_j m_k m_l + \ldots]. \tag{11}$$

The coefficients C, F_{ij}, F_{ijkl}, ... are obtained from $N(m)$ just in the same way that c, D_{ij}, D_{ijkl}, ... are obtained from $f(n)$ (cf. (5), (6), (7)). If we substitute this expression into the Buffon transform (9) or (10), we find that c and the fabric tensors D_{ij}, D_{ijkl}, ... are given as follows (ref. 10):

$$D_{i_1 i_2 \ldots i_n} = \lambda_n F_{i_1 i_2 \ldots i_n}, \tag{12}$$

$$c = C/2\pi, \quad \lambda_n = (-1)^{n/2-1}2^{n-1}(n-1)(n+2)/\binom{n}{n/2} \quad \text{for (9)}, \tag{13}$$

$$c = C/\pi^2, \quad \lambda_n = -\,2^{2n-1}(n-1)/\binom{n}{n/2}^2 \quad \text{for (10)}. \tag{14}$$

Thus, the problem is "theoretically" solved. However, this does not mean that the problem is solved in "practical" terms. For practical purposes, we must estimate the coefficients of the expansion (11) from a "finite" number of observed data $N(m^{(1)})$, $N(m^{(2)})$, ..., $N(m^{(N)})$. The most straightforward method is the "Monte Carlo method," choosing cutting orientations $m^{(1)}$, $m^{(2)}$, ..., $m^{(N)}$ randomly from the entire solid angle, i.e., uniformly over the unit sphere (ref. 11).

To do so, however, we must prepare a large number of material samples having the same statistical characteristics and cut these samples from various different orientations. This is often very difficult. On the other hand, if the anisotropy is "weak", i.e., if the distribution is nearly isotropic, higher order terms of (9) or (10) can be neglected. Then, there exist methods which use only three types of surfaces - planes parallel to the xy-, the yz- and the zx-planes. Errors involved in this approximation are also estimated (ref. 11).

This principle leads to a variety of applications in many different fields. Our approach can be used for characterization of particles in powder technology. For example, the orientation of maximum alignment of the needle-like particles of Fig. 3(a) is estimated to be 125° from the x-axis (ref. 12). In relation to geology, the internal crack distribution in a rock can also be estimated from observations of its surfaces (ref. 13). Applying the same procedure, we can also measure the strain of a material by observing the structural anisotropy if the internal structure is assumed to be isotropic in the original state. For example, the orientations of maximum extension and compression of the material shown in Fig. 3(b) are estimated to be 2° and 92° from the x-axis respectively and the shearing strain to be 0.18 in the xy-plane (ref. 10). The same principle is also used in computer vision and artificial intelligence to reconstruct three dimensional information from two dimensional camera images (ref. 14). For example, applying this stereological technique, we can compute

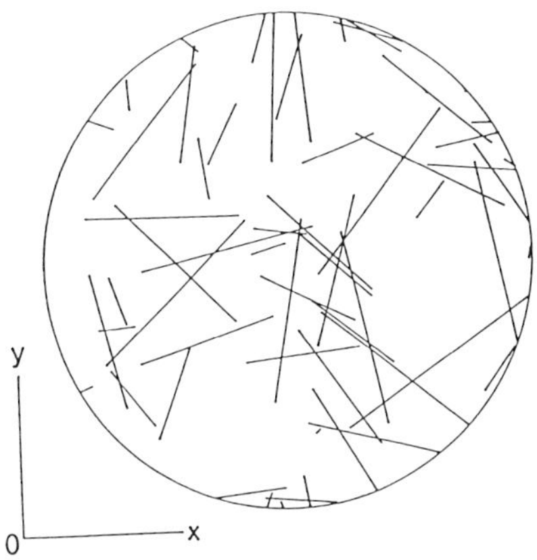

(a)

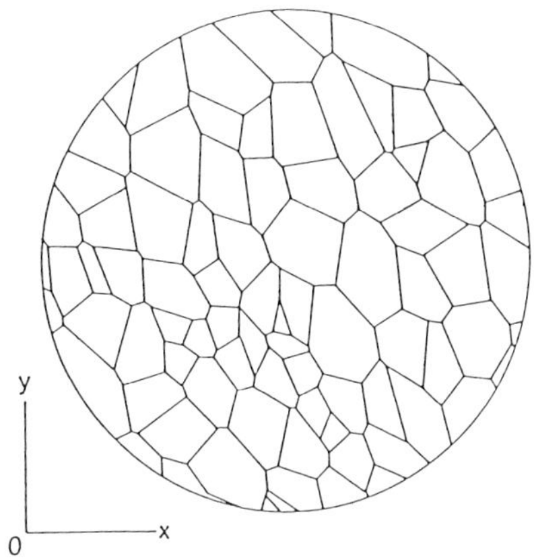

(b)

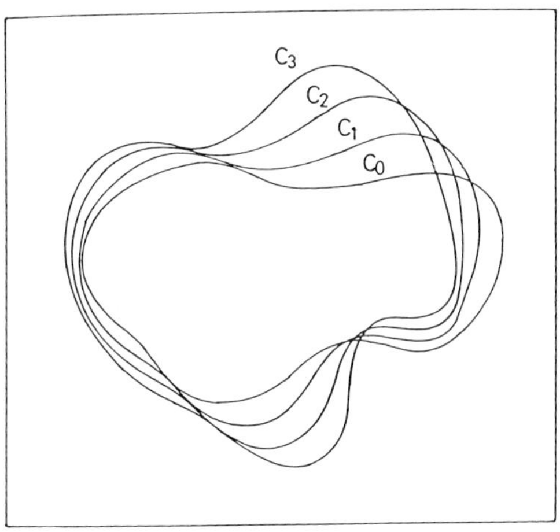

(c)

Fig. 3. The stereological principle of estimating anisotropy has applications in many different areas such as powder technology, metallurgy, geology, artificial intelligence and computer vision.

the three-dimensional motion of a moving plane shown in Fig. 3(c) (ref. 15). On the other hand, the fabric tensors can become a powerful tool in analysis and characterization of various material state - the packing structure of granular materials, for instance.

CONCLUDING REMARKS

Stereology is an interdisciplinary area, and the same principle can be used in many different areas for many different purposes. Hence, results in one particular area could bring about new applications in other areas. However, this is possible only if the results are described in "precise mathematical terms", because only mathematics is the common "language" of all branches of science and technology. In the past, researches in a particular area seem to have been addressing only researchers in the same area. As a result, the same principle has been rediscovered many times separately in different areas.

The approach given here is (I) first formulating the problem exactly in precise mathematical terms and (II) then devising necessary experimental means that are easy to perform. We have also presented analytical techniques wich can be useful in many different areas. In particular, the characterization of a distribution density and the introduction of the fabric tensor discussed have many applications in various different problems.

REFERENCES

1 R.T. DeHoff and F.N. Rhines, Quantitative Microscopy, McGraw-Hill, New York, 1968.
2 E.E. Underwood, Quantitative Stereology, Addison-Wesely, Reading, MA, 1970.
3 E.R. Weibel, Stereological Methods, Vols. 1, 2, Academic Press, New York, 1979, 1980.
4 L.M. Cruz-Orive, Distribution-free estimation of sphere size distribution from slabs showing overprojection and truncation, with a review of previous methods, J. Microscopy, 131(3) (1983) 265 - 290.
5 R. Coleman, The size of spheres from profiles in a thin slice I. Opaque spheres, Biometric J.,24(3) (1982) 273 - 286.
6 R. Coleman, The size of spheres from profiles in a thin slice II. Transparent spheres, Biometric J., 25(8) (1983) 745 - 755.
7 K. Kanatani and O. Ishikawa, Error analysis for the stereological estimation of sphere size distribution: Abel type integral equation, Journal of Computational Phys., 57(2) (1985) 229 - 250.
8 K. Kanatani, Distribution of directional data and fabric tensors, Int. J. Engineering Sci., 22(5) (1984) 149 - 164.
9 J.E. Hilliard, (1967): Determination of structural anisotropy, in: H. Elias (Ed.), Stereology - Proc. 2nd Int. Congress for Stereology, Chicago, IL, April 8-13, 1967, Springer, Berlin, 1967, pp. 219 - 227.
10 K. Kanatani, Stereological determination of structural anisotropy, Int. J. Engineering Sci., 22(5) (1984) 531 - 546.
11 K. Kanatani, Procedures for stereological estimation of structural anisotropy, Int. J. Engineering Sci., 23(5) (1985) 587 - 598.
12 K. Kanatani, Measurement of particle orientation distribution by a stereological method, Particle Characterization, 2 (1985) 31 - 35.
13 K. Kanatani, Measurement of crack distribution in a rock mass from observation of its surfaces, Soils and Foundations, 25(1) (1985) 77 - 83.
14 K. Kanatani, Detection of surface orientation and motion from texture by a stereological technique, Artificial Intelligence, 23(2) (1984) 213 - 237.
15 K. Kanatani, Tracing planar surface motion from projection without knowing the correspondence. Computer Vision, Graphics, and Image Processing, 29(1) (1984) 1 - 12.

Micromechanics of Granular Materials, edited by M. Satake and J.T. Jenkins
Elsevier Science Publishers B.V., Amsterdam, 1988 — Printed in The Netherlands

Morphological Analysis of Particulate Materials

John Keith Beddow
The University of Iowa
Department of Chemical and Materials Engineering

Introduction

Morphological Analysis is the science of the analysis of points in space. In defining the relationship between particle characteristics and their microstructures we seek to relate particle morphologies to the microstructure. It has been shown that the particle morphology is represented by the (R,Θ) Fourier method.[1] The microstructure or morphological texture is represented by the Bessel-Fourier method.[2]

Morphological Analysis

The essential steps in morphological analysis are:

- determination of the mathematical of the particle representation.
- obtain data that measures the positions of pioints in space
- convert the data to an analytical relationship
- transform to invariant form--these are the features.
- extraction of the features from the representations
- formulation and ranking of the significant features in order to charaterize the particle (or particle set) or the texture (microstructure)

In general, the mathematical form chosen depends upon the parameterization to be represented as shown in the Luerkens Equations of Table 1. Those representations are derived from basic principles by means of the calculus of variations in the form of the **morphological variational principle**:

> **A mathematical representation of the Boundary Function for surfaces may be derived by finding a "Normalized Function" which causes the surfaces area integral to take on stationary values.**

Particle Characterization

The representation for particle profile expounded in Table 1 shows the profile expressed in polar coordinates has the form:

$$R\Theta = a_0 + \sum^{\infty} (a_n \cos n\Theta + b_n \sin n\Theta) \qquad (1)$$

The initial urge is to consider (a_n, b_n) as shape features. This is unwise because (a_n, b_n) vary as the particle is moved by rotation, for example. This means

Table 1

The Luerkens Equations[3-8]

Parameterization	Representation
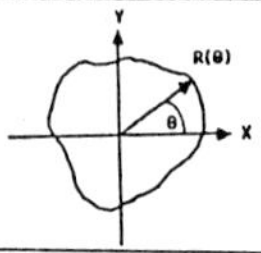	Profile $R(\theta) = a_0 + \sum a_n \cos n\theta + b_n \sin n\theta$
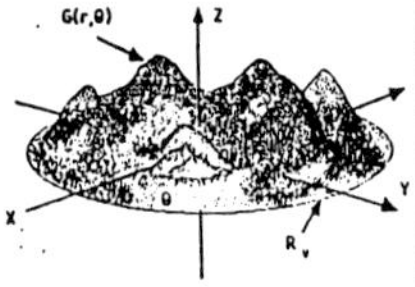	Extended Surface $G(r,\theta) = \sum_{n=0}^{\infty}\sum_{m=0}^{\infty} (a_{m,n}\cos n\theta + b_{m,n}\sin n\theta) J_n(\gamma_{m,n}\frac{r}{R_v})$
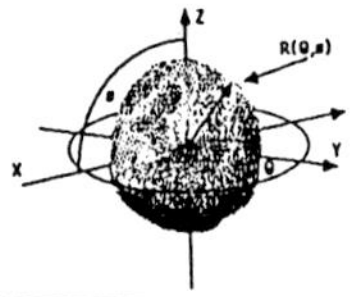	3-D Bulky Particle $R(\theta,\varphi) = \sum\sum (a_{m,n}\cos m\theta + b_{m,n}\sin m\theta) P_n^m(\cos\varphi)$
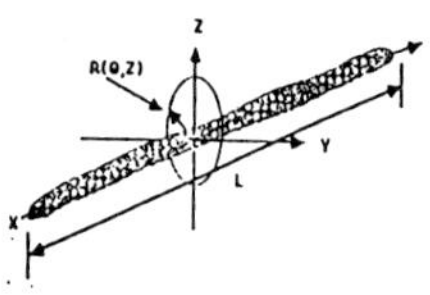	Finite Fiber $R(\theta,z) = \sum_{m=0}^{\infty}\sum_{n=0}^{\infty}(a_{m,n}\cos m\theta \cos(\frac{\pi n z}{L}) + b_{m,n}\cos m\theta \sin(\frac{\pi n z}{L}) + (c_{m,n}\sin m\theta \cos(\frac{\pi n z}{L}) + d_{m,n}\sin m\theta \sin(\frac{\pi n z}{L})$
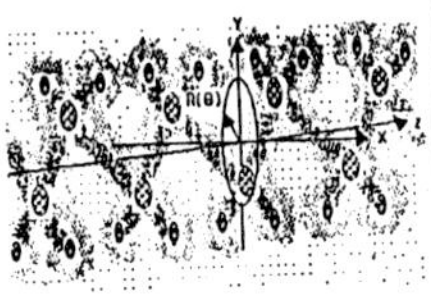	Infinite Fiber or Thread $R(\theta,z) = \sum_{m=1}^{\infty} \cos m\theta \int_0^{\infty}[A(w)\cos wz + B(w)\sin wz]\,dw + \sin m\theta \int_0^{\infty}[C(w)\cos wz + D(w)\sin wz]\,dw$
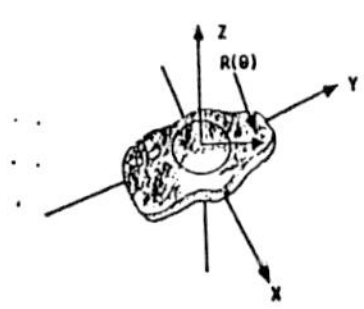	Flake $R(\theta) = a_0 + \sum_{n=1}^{\infty} a_n \cos n\theta + b_n \sin n\theta$

that (a_n, b_n) are definitely not invariants. They have to be transformed in order that the invariants might be obtained.

In general, feaatures are moments of distributions. The deep significance of statistics in morphological analysis is shown by the **fundamental morphological principle** for size, R_0, in which R_0, the equivalent radius is defined

$$R_0^2 = a_0^2 + {}^1/_2 \sum_{n=1}^{\infty} a_n^2 + b_n^2 \tag{2}$$

In which πR_0^2 is the area of the profile

a_0 is the mean radius (the mean of the radial distribution)

a_n, b_n, are Fourier coefficients.

In words equation 2 reads

$$\text{(The Size)}^2 = \text{(The mean radius)}^2 + \text{(The Shape Information)} \tag{3}$$

μ_2 is the second moment about the mean of the radial distribution, μ_2 is identical to ${}^1/_2 \sum a_n^2 + b_n^2$. Therefore,

$$R_0^2 = a_0^2 + \mu_2 \tag{4}$$

In words, equation 4 reads,

(The Size)^2 = (The Mean)^2 + The second moment about the mean

The key features obtained from morphological analysis are shown in Table 2. Also included in Table 2 are the so-called **classical terms**. Classical terms are empirical and ad hoc in origin. They have no underlying theoretical basis.

Characterization

Particle characterization results from a non-simple process of information structuring. Particle characterization is defined as the communication of the description of the essential multidimensional properties of a particle or a collection of particles. Each multidimensional property constitutes a characterization vector which is determined by a measurement vector operating on a basis set of the property features.[9] The resultant characterization comprises a set of mathematical descriptors and a corresponding set of verbal descriptors.

Both individual particle characteristics and the characteristics of bulk lots of particles are numerous and diverse as is demonstrated in Table 3. Therefore, inherent in particle characterization is the prerequisite that both theory and practice deal with sets--sets of measurments and key features alike. Characterization of the size and also the shape of a simple particle or of a batch

Table 2 - Examples of Key Features[10]

Distribution or Series/Set	Features
Single Particle	
Radial Distribution (L_{2n})	Moments $\mu_0, \mu_1, \mu_2, \mu_3 \dots R_0$ Rotational Symmetries $C_1, C_2, C_3, C_4, \dots$ Reflection Symmetries $\sigma_1, \sigma_2, \sigma_3, \sigma_4, \dots$
Diameter Distribution	Moments
Aspect Ratio Distribution	Mean & Standard Deviation, Maximum Value
Roughness	
Classical Terms	Perimeter area, $P^2/2\pi A$, Centroid Aspect Ratio, General Aspect Ratio, Martin's Diameter, Peret Diameter, $^{Dm}/_{Df}$, etc.
Fractal Measurment	Perimeter Measurment Fractal Measurment
Particle Sets	
Grand Population Radial Distribution	Moments $\mu_{0g}, \mu_{1g}, \mu_{2g}, \dots$
R_0 Distribution	Moments $u_0, u_1, u_2, \dots$
Roughness Distribution	
Shape Distribution	Moments $\mu_0 2, \mu_1 2, \mu_2 2, \dots$
Symmetry Distribution	Moments $\mu_0 C_n, \mu_1 C_n, \mu_2 C_n, \dots$ $\mu_0 \beta_n, \mu_1 \beta_n, \mu_2 \beta_n, \dots$
Distribution of Classical Terms	Moments of Distribution for Area, Perimeter, Martin's Diameter, Peret Diameter, Sphericity, etc.
Fractal Population Measurment	Perimeter Fractal Distribution, Area Fractal Distribution

Table 3 - The Characterization Vector

$$\begin{Bmatrix} a_0 \\ a_1 \\ \cdot \\ \cdot \\ \cdot \\ \cdot \end{Bmatrix}$$

Characterization Vector

a_0 = Circularity
a_1 = Cardiodicity
a_2 = Ellipicity
a_3 = Triangularity
a_4 = Squareness
.
.
.

Each Coefficient Represents a Symmetry Vector

of particles has been indicated in Table 2. In more detail these procedures are as follows:

Size: In the case of a set of particles let the size measurments produce a size distribution with moments of $u_0, u_1, \ldots, u_n$. These may be written in the form of the moment vector

$$\begin{pmatrix} u_0 \\ u_1 \\ . \\ . \\ . \\ u_n \end{pmatrix}$$

The moment vector uniquely determines the characteristic function Φ such that

$$\Phi = a_0\Phi_0 + a_1\Phi_1 + a_2\Phi_2 + \ldots + a_n\Phi_n \qquad (5)$$

In which Φ is the (unknown) characteristic function determined by the measurements.

By appropriate substitution

$$\begin{pmatrix} u_0 \\ u_1 \\ u_2 \\ . \\ . \\ . \\ u_n \end{pmatrix} = a_0 \begin{pmatrix} u_0 \\ u_1 \\ u_2 \\ . \\ . \\ . \\ u_n \end{pmatrix}_0 + a_1 \begin{pmatrix} u_0 \\ u_1 \\ u_2 \\ . \\ . \\ . \\ u_n \end{pmatrix}_1 + \ldots a_n \begin{pmatrix} u_0 \\ u_1 \\ u_2 \\ . \\ . \\ . \\ u_n \end{pmatrix}_n \qquad (6)$$

And So,

$$\begin{pmatrix} u_0 \\ u_1 \\ u_2 \\ . \\ . \\ . \\ u_n \end{pmatrix} = \begin{pmatrix} u_{0,0} & u_{0,1} & \ldots & u_{0,n} \\ u_{1,0} & u_{1,1} & \ldots & u_{1,n} \\ u_{2,0} & u_{2,1} & \ldots & u_{2,n} \\ . & . & \ldots & . \\ . & . & \ldots & . \\ . & . & \ldots & . \\ u_{n,0} & u_{n,1} & \ldots & u_{n,n} \end{pmatrix} \begin{pmatrix} a_0 \\ a_1 \\ a_2 \\ . \\ . \\ . \\ a_n \end{pmatrix} \qquad (7)$$

Measurment Vector Basis Set Characterization Vector

The basis set is the standard; the measurment vector is obtained from the experimental data; and the characterization vector is generated by Equation 7.

Shape In this discussion the rotational symmetries C_n and the reflection symmetries σ_n will be combined. The final version of the relationship is similar to that for size in equation 7.

$$\begin{Bmatrix} C_1 \\ C_2 \\ \cdot \\ \cdot \\ \cdot \\ \cdot \\ C_1 \\ \sigma_1 \\ \sigma_2 \\ \cdot \\ \cdot \\ \cdot \\ \sigma_n \end{Bmatrix} \quad \begin{Bmatrix} C_{10} & C_{11} & C_{12} & \ldots & C_{1n} \\ C_{20} & C_{21} & C_{22} & \ldots & C_{2n} \\ \cdot & \cdot & \cdot & \ldots & \cdot \\ \cdot & \cdot & \cdot & \ldots & \cdot \\ \cdot & \cdot & \cdot & \ldots & \cdot \\ \cdot & \cdot & \cdot & \ldots & \cdot \\ C_{n0} & C_{n1} & C_{n2} & \ldots & C_{nn} \\ \sigma_{10} & \sigma_{11} & \sigma_{12} & \ldots & \sigma_{1n} \\ \sigma_{20} & \sigma_{21} & \sigma_{22} & \ldots & \sigma_{2n} \\ \cdot & \cdot & \cdot & \ldots & \cdot \\ \cdot & \cdot & \cdot & \ldots & \cdot \\ \cdot & \cdot & \cdot & \ldots & \cdot \\ \sigma_{n0} & \sigma_{n1} & \sigma_{n2} & \ldots & \sigma_{nn} \end{Bmatrix} \quad \begin{Bmatrix} a_0 \\ a_1 \\ a_2 \\ \cdot \\ \cdot \\ \cdot \\ \cdot \\ \cdot \\ \cdot \\ \cdot \\ \cdot \\ \cdot \\ a_n \end{Bmatrix} \tag{8}$$

Measurment Vector Basis Set Characterization Vector

Again, the basis set is the standard against which all other shapes of profile are compared. The measurment vector is obtained experimentally using (for a profile) the R-θ method. The characterization vector is derived from equation 8.

A word about the basis set: the standard figures for compilation of the basis set are:

circle
cardiod
lemniscate
equilateral triangle
square
pentagon
hexagon, etc.

Thus, the first column of the basis set corresponds to the circle, the second column corresponds to the cardiod and so on.

In the case of the characterization vector, each "a" term describes the proportion to which the profile being analyzed corresponds to the symmetry of a circle (a_0), a cardiod (a_1), and so on. Each "a" term therefore describes an appropriate symmetry vector. This is summarized in Table 3.

In general all of the various methods of fine particle representation discussed above can be summarized as follows.

Let P = Feature measurment vector.
A = Basis set
B = Feature characterization vectors.

Then,

$$B = A^{-1}P \tag{9}$$

Table 4
Some Morphological Textural Features[2]

A. Statistical Features: mean gray level, standard deviation, moments of an image gray level distribution.

$$G(r,\Theta) = \sum_{m=0}^{\infty} \sum_{n=0}^{\infty} (A_{m,n} \cos m\Theta + B_{m,n} \sin m\Theta)\, J_m(\lambda_{m,n} \frac{r}{R_v})$$

$$\bar{G}_T = \sum_{i=1}^{L} i \quad P_i$$

$$\sigma 2 = \sum_{i=1}^{L} (i - \bar{G}_T) 2\, P_i$$

$$M_n = \sum_{i=1}^{L} (i - \bar{G}_T) n \quad P_i$$

B. Coefficients of the Bessel-Fourier Function: $A_{m,n}$, $B_{m,n}$ and frequency of an image which are computer based on Equation (1).

C. Symmetry Indices:

i. rotational symmetry

$$C_R = \frac{\int_0^{R_v} \int_0^{2\pi} G(r,\Theta)\, G(r,\Theta + \Delta\Theta) r dr d\Theta}{\int_0^{R_v} \int_0^{2\pi} G2(r,\Theta)\, r dr d\Theta}$$

Table 4 (Continued) - Some Morphological Textural Features

ii. translational symmetry

$$C_R = \frac{\sum_{m=0}^{\infty} \sum_{n=0}^{\infty} (H_{m,n})_R 2 \cos m \, 2\pi/_R \, J'_m 2 (\lambda_{m,n})}{\sum_{m=0}^{\infty} \sum_{n=0}^{\infty} (H_{m,n}) 2 \, J'_m 2 (\lambda_{m,n})}$$

iii. rotational coefficients
iv. translational coefficients.

D. Descriptors:

i. coarseness

$$4 - 2\,[C_R + C_T] = F_{COR}$$

ii. contrast

$$\alpha_4 = \mu_4/\sigma_4 = F_{CON}$$

iii. roughness

$$F_{ROU} + F_{CRS} = F_{CON}$$

iv. regularity

$$F_{REG} = \sum_{T=1}^{n} C_T + \sum_{R=1}^{m} C_R$$

Microstructure: Morphological Texture Analysis

In morphological texture analysis the gray levels of the image are transformed via statistical and mathematical models to a set of invariants called Morphological Textural Features (MTFs). The statistical model is the gray level distribution which yields the moments as MTFs. The mathematical model is a Bessel-Fourier representation of the image which yields Bessel-Fourier Coefficients, and rotational and translational symmetry coefficients. These and other MTFs are shown in Table 4.

Conclusion

The objective of relating particle morphology to microstructure will be pursued in the form of relating the corresponding basis set equations.

$$B_{R\Theta} = A_{R\Theta}^{-1} P_{R\Theta} \text{ and } B_{BF} = A_{BF}^{-1} P_{BF} \tag{10}$$

References

1. Luerkens, D.W., Beddow, J.K. and Vetter A.F., "Invariant Fourier Descriptors," Powder Technology, 31, 2, 217-220, 1982.
2. Hsyung, N.B., Luerkens, D. W. and Beddow, J.K. "Morphological Textural Features", paper submitted to IEEE TRANS ON PATTERN RECOGNITION AND MACHINE INTELLIGENCE, 1987.
3. Luerkens, D.W. "The R-Theta Method Derived from A Variational Principle" Particulate Science and Technology, Volume 4, Number 4, p. 353-361, 1986
4. Luerkens D.W. "Particle Surface Representations Derived from A Variational Principle"--1. The Grey Level Function" Particulate Science and Technology, Volume 4, Number 4, p. 361-371, 1986
5. Luerkens D.W. "Particle Surface Morphologies--The Three-Dimensional Particle" Particulate Science and Technology, Volume 4, Number 4, p. 371-378, 1986
6. Luerkens D.W. "Particle Surface Morphologies--Finite Fibers, Part I Mathematical Description", Particulate Science and Technology, Volume 4, Number 2, p. 151-160, 1986
7. Luerkens D.W. "Particle Surface Morphologies--Flakes", Particulate Science and Technology, In Press, Volume 5, Number 3, 1987
8. Luerkens D.W. "Infinite Fibers and Threads", Particulate Science and Technology, In Press, Volume 5, Number 3, 1987
9. Luerkens, D.W., Beddow, J.K., and Vetter, A.F., "Structure and Morphology, The Science of Form Applied to Particle Characterization", Powder Technology, 50, 93-101, 1987.
10. Courtesy of the Shapespeare Corporation, Iowa City, Iowa, 1986.

THE AUTHORS ARE GRATEFUL TO DR. MORRIS OJALVO, PROGRAM DIRECTOR, PARTICULATE PROCESSING AND MULTIPHASE FLOW PROGRAM OF THE NATIONAL SCIENCE FOUNDATION FOR PARTIAL SUPPORT UNDER GRANT NUMBER CPE8414748.

Micromechanics of Granular Materials, edited by M. Satake and J.T. Jenkins
Elsevier Science Publishers B.V., Amsterdam, 1988 — Printed in The Netherlands

THE MICROGEOMETRY OF DENSE RANDOM PACKINGS

H. J. Frost

Thayer School of Engineering, Dartmouth College, Hanover N.H 03755
U.S.A.

SUMMARY

There are a variety of different approaches to the characterization of the local geometry in a random packing of particles. This paper gives a brief review of several different methods, with concentration on those methods which describe the geometry of groups of particles which surround the cavities between the particles.

INTRODUCTION

A granular material is usually an irregular packing of particles, which are often approximately spherical. In the general case, the particles will have various sizes and shapes. The structure of such an irregular packing has often been used as a model of the structure of liquid metals and metallic glasses. (Bernal, refs. 1,2) The simplest case is that of a dense random packing (DRP) of identical spheres.

A complete description of an irregular packing is given by specifying the location, size, shape and orientation of each particle in the packing. For a large array of particles, the information required is unmanagable. In order to extract a more useful description for practical applications, the geometric structure of an irregular packing can be characterized according to the statistical distribution of some geometric feature. No particular statistical description contains all the information required to specify the structure. Different features may be more or less useful for modelling particular aspects of the behavior. In this paper, we briefly review the various methods that have been used to characterize the local geometric structure of dense random packings, drawing primarily on the literature relating to the atomic structure of liquid metals and metallic glasses. This review is certainly incomplete because it does not properly consider the characterization approach of the fabric tensor, which has been developed for the description of granular materials.

CHARACTERIZATION BY THE AVERAGE SURROUNDINGS OF A PARTICLE

The most common characterization method is by radial distribution functions which give the number of pairs at various spacings. (Reviewed, for example, by Cargill (ref. 3) or Finney (ref. 4). This method immediately gives the average number of contacts between neighboring spheres, and the number of near contacts. The coordination number of a particle depends on the cut-off distance beyond which two particles are not considered to be neighbors. Bennett (ref. 5) argued that the average number of contacts should be exactly six, based on the number of degrees of freedom and the requirement of stability. This average number of contacts appears to apply to both loose random packings and dense random packings. The difference between various packings must therefore be reflected in the non-contacting neighbors. Radial distribution functions for different models differ most notably in their second peaks, which may be split. (ref. 4) The sub-peaks of the second peak have been related to particular geometric arrangements, such as the distance through two tetrahedra, the distance through two equilateral triangles, and twice the sphere diameter (corresponding to three particles in a line). It is difficult, however, to extract any information from the radial distribution function that describes the geometric arrangements of groups of particles.

There are several other methods which follow the general approach of characterizing the average surroundings of the particles. For example, Scott and Mader (ref. 6) introduced a pole figure which showed the average angular distribution of first shell neighbors. The fabric tensor, which is not covered here, also follows the general averaging approach.

DIFFERENT SURROUNDINGS OF DIFFERENT PARTICLES

A second approach for characterizing the structure is to describe the arrangements of spheres about a particular sphere, in such a way as to illustrate the different surroundings of different spheres. This can be done by reporting the shapes of the Voronoi polygons that surround each particle. (ref. 7) The Voronoi polygon of a particle is that region of space which is closer to that particular particle center than to any other particle center. The Voronoi polygon is bounded by planes which are perpendicular bisectors between neighboring particle centers. The edges of the Voronoi polygon are lines which are equidistant from three particle centers, and the vertices of a Voronoi polygon are points which are equidistant from four particle centers. The number of faces on a Voronoi polygon gives the number of first shell neighbors, including neighbors which are not touching. (For a packing of equal-sized spheres, any two particles with centers closer than $\sqrt{2}$ diameters must share a Voronoi polygon face.) A description of the shapes of the Voronoi polygons contains information that gives the radial distribution function for the first shell of neighbors. It also contains some information on the geometric shapes formed by groups of neighboring particles, although not in a convenient form.

A similar description has been introduced by Gellatly and Finney (ref. 8) to describe packings with different sizes of spheres, based on radical plane polyhedra. The radical plane is the locus of points from which tangents to the two spheres are equal. It is an improvement over the Voronoi approach for multiple-sized models because the radical plane always lies between spheres of different sizes; the Voronoi plane does not always.

There are several other methods which follow the general approach of characterizing the surroundings of particular particles. For example, the local environment for models of metallic glasses has been described using atomic-level internal stresses and local symmetry coefficients. (See for example, Egami et al.(ref. 9) and Srolovitz et al. (ref. 10)) This approach is related to the fabric tensor approach, but is designed to report on the variations in local stress tensor for different local atomic arrangements.

SURROUNDINGS OF THE VOIDS BETWEEN PARTICLES

A third approach is to characterize the arrangement of spheres around the cavities which lie between the particles. This is usually done by listing the various polyhedra formed with the particle centers as corners (such as tetrahedra, octahedra, etc.). Bernal (ref. 2) used this approach to describe the DRP in terms of five canonical holes, which are shown in Figure 1. Whittaker (ref. 11) and Frost (ref. 12) elaborated on this approach, using some additional shapes not mentioned by Bernal. There are some difficulties with this approach, because it implicitly relies on identifying the various polyhedral shapes by using a network of nearest-neighbor connections. If only touching particles are considered to be neighbors, then there are not enough connections to divide up the DRP into convenient polyhedra. On the other hand, if the nearest-neighbor distance is taken to be too large, then the volume may be divided in an ambiguous way. For equal sized spheres

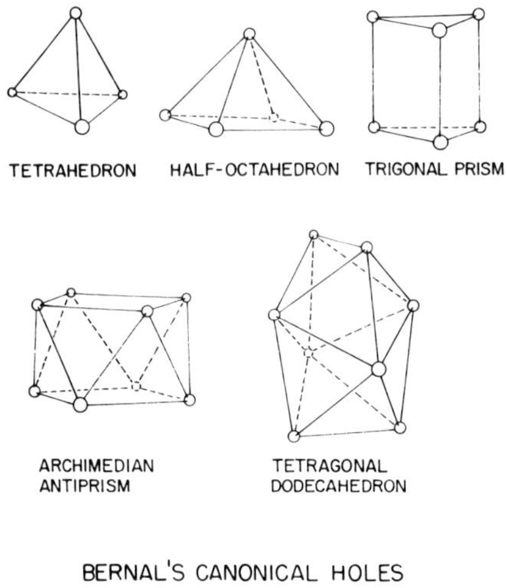

Fig. 1. Bernal's Five Canonical Holes.

this ambiguous description becomes possible when the nearest neighbor distance exceeds $\sqrt{(12/7)}$ sphere diameters. This arrangement is shown in Figure 2; it can be divided into two tetrahedra top and bottom, or three tetrahedra around a vertical axis. The polyhedral description of any particular region depends on the exact choice of the nearest-neighbor distance. Bernal evidently used a flexible nearest neighbor distance which was locally extended until the volume was appropriately divided.

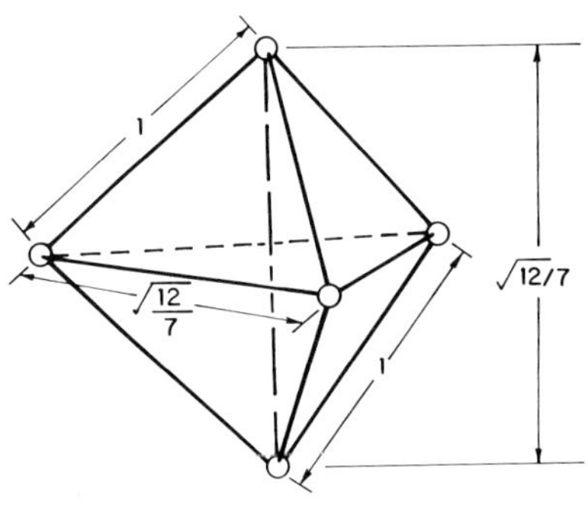

Fig. 2. Arrangement showing the minimum nearest neighbor distance that allows for interpenetrating tetrahedra.

The characterization in terms of the polyhedra which surround the voids between particles must include some criterion for deciding exactly how to divide up the volume. This corresponds to selecting a list of the different polyhedra which may be identified. One simple approach is to use deltahedra, which are polyhedra with all triangular faces. The volume can then be divided in a straightforward manner: Any two particle centers within a chosen nearest-neighbor distance form an edge. Any three edges which join as a triangle form a face. And finally, each volume that is totally enclosed by faces describes a deltahedron. Although this approach is logical and unambiguous, it results in the identification of an unmanageably large list of deltahedra. Figure 3 shows those

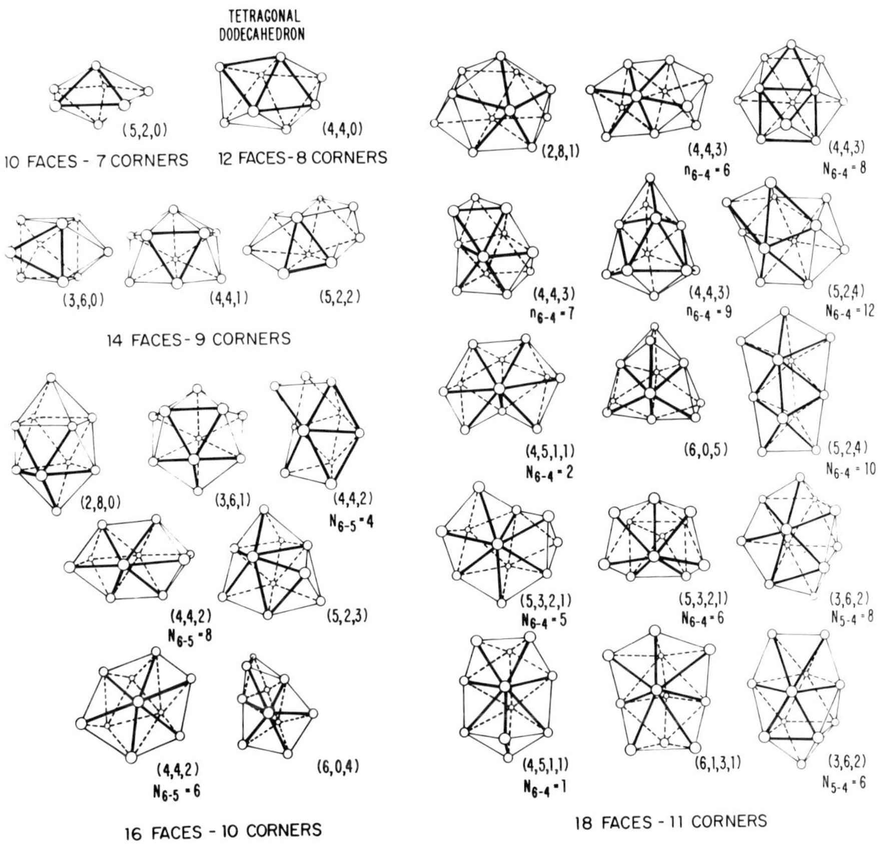

Fig. 3. Deltahedra with seven to eleven vertices. Integers give the number of vertices with four, five, six and sometimes seven edges.

deltahedra with between seven and eleven vertices which can be identified in the DRP. Even for a nearest neighbor distance approaching $\sqrt{(12/7)}$, a large fraction of the volume must be assigned to even larger and more complex deltahedra.

In order to provide a more convenient description of the local geometry than deltahedra, we may simply allow the volume to be divided by four-edged faces in addition to triangular faces. The term "four-edged face" is used because the edges are generally not co-planar, and do not form a planar quadralateral. Figure 4 shows a catalogue of such shapes which cannot be divided up further by four-edged faces; hence the name "irreducible shapes". By this

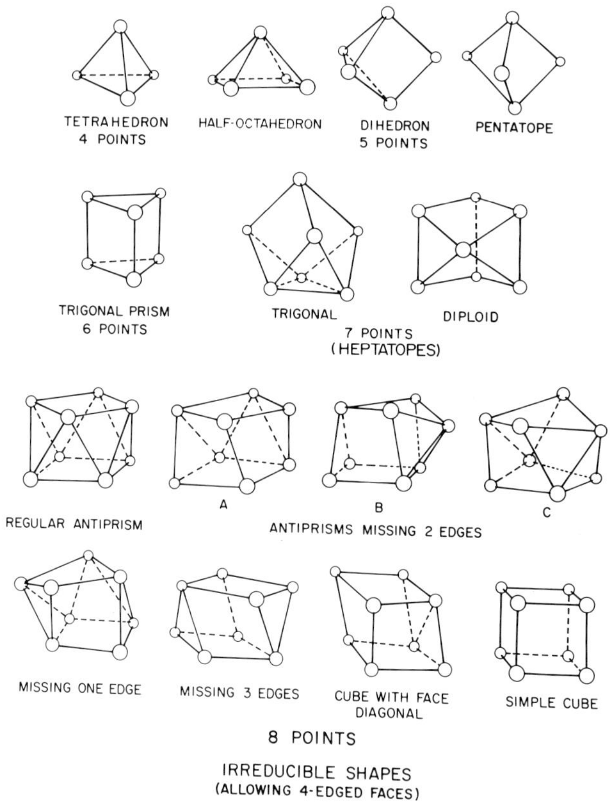

Fig. 4. All the irreducible shapes, allowing both triangular and four-edged faces, with up to eight points.

approach, almost the entire volume of the DRP may be assigned to shapes with nine or fewer points. This approach does leave some room for ambiguity: for example, a regular octahedron may be divided into two half-octahedra in three different ways. Nevertheless, the result is a manageable list of shapes which do represent the local geometry of the neighboring particles. A more detailed explanation of the descriptions according to deltahedra or to "irreducible shapes" has been given by Frost (ref. 12). It should be remembered that this type of description may occasionally give an inaccurate identification of a particular region. For example, Figure 5 shows two distorted shapes intermediate between an octahedron and a trigonal prism. With a short neighbor distance the distorted octahedron could be identified as a trigonal prism; with a long neighbor distance the distorted trigonal prism could be identified as an octahedron. Even with these difficulties, the polyhedral approach can be useful for describing local clusters of atoms in a metallic glass.

Another related characterization approach is to characterize the size of the interstitial sites between the particles where an interstitial sphere could be inserted. (refs. 12, 13) The centers of the interstitial sites are actually the vertices of the Voronoi

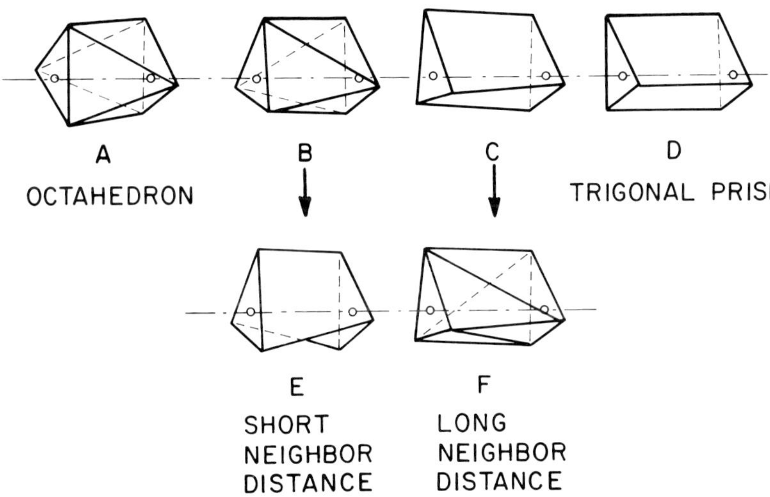

Fig. 5 Demonstration that of the possibility of misidentification of shapes at short or long neighbor distances.

polyhedra discussed above. Figure 6 shows the interstitial site size distribution for the Finney DRP model. Other packing models have different distributions (ref. 13). The statistics of these distributions must be gathered carefully. Each interstitial site is defined by four particles which the small interstitial sphere would touch. It is possible to arrange the particles so that one small interstitial sphere would overlap a neighboring small interstitial sphere. It is unreasonable to count both such interstitial sites. If only two overlap, we may count the larger and neglect the smaller. It would be possible, though more difficult, to redefine the smaller site for a sphere that just touches three particles and the larger interstitial sphere. This has not been attempted. If three sites overlap, with the center the largest, but the other two not overlapping each other (as in Figure 7A), then we must choose between the maximum number of sites (using the two smaller) and the maximum size of sites (the largest one). Figure 6 has chosen the latter. Figure 7B shows that chosing the largest site first and neglecting any that overlap can sometimes result in more sites than the alternative. It is also possible for the overlap elimination to

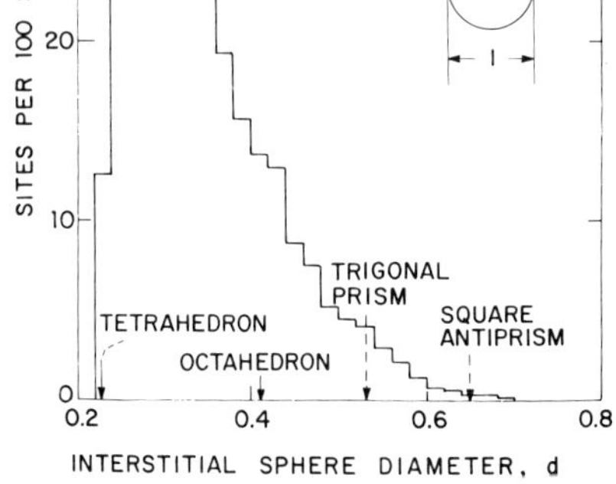

Fig. 6. Distributions of interstitial-sphere sizes in the Finney dense random packing model.

leave surviving sites that do not accurately reflect the cavity size, as shown in Figure 7C. The smaller site is not constrained by its three neighboring particles, and could be replaced by a slightly larger interstitial sphere with a slightly shifted center. An extension of this approach is the characterization of the correlation between nearby interstitial sites, which was introduced by Finney and Wallace (ref. 13) Ahmadzadeh and Cantor (ref. 14). Approaches using interstitial site sizes and correlations are particularly useful for describing the solubility or diffusion of small atoms in a metallic glass.

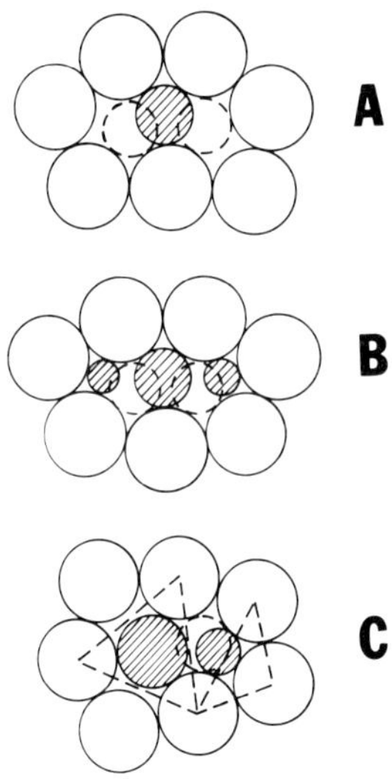

Fig. 7. Possible arrangements for overlapping interstitial sphere sites, in a two dimensional analogue.

CONCLUSIONS

None of the approaches discussed above are exactly suited to the description of the static mechanical state of a granular material, in which forces are only transmitted through touching neighbors. They may, however, find application in consideration of the pattern of changes in particle contacts which must occur in a deforming granular material.

REFERENCES

1. J. D. Bernal, Geometry of the structure of Monatomic Liquids, Nature 185, (1960), 68-70.
2. J. D. Bernal, The Structure of Liquids, Proc. Roy. Soc. 280A, (1964), 299-322.
3. G. S. Cargill III, Radial distribution functions and microgeometry of dense random packings of hard spheres, in Physics and Chemistry of Porous Media, edited by D. L. Johnson and P. N. Sen, AIP, New York,(1984), pp. 20-36.
4. J. L. Finney, Modelling the Structures of Amorphous Metals and Alloys, Nature 266, (1977), 309-314.
5. C. H. Bennett, Serially Deposited Amorphous Aggregates of Hard Spheres, J. Appl. Phys. 43, (1972), 2727-2734.
6. G. D. Scott and D. L. Mader, Angular Distribution of Random Close-Packed Equal Spheres, Nature 201,(1964), 382-383.
7. J. L. Finney, Random Packings and the Structure of Simple Liquids I, Proc. Roy. Soc. 319A, (1970), 479-493.
8. B. J. Gellatly and J. L. Finney, Characterisation of Models of Multicompoment Amorphous Metals, J. Non-Crystalline Solids 50, (1982), 313-329.
9. T. Egami, K. Maeda and V. Vitek, Structural Defects in Amorphous Solids: A Computer Simulation Study, Phil. Mag. A41, (1980), 883-901.
10. D. Srolovitz, K. Maeda, V. Vitek and T. Egami, Structural Defects in Amorphous Solids: Statistical Analysis of a Computer Model, Phil. Mag. A44,(1981), 847-866.
11. E. J. W. Whittaker, The Cavities in a Random Close-Packed Structure, J. Non-Crystalline Solids 28, 1978, 293-304.
12. H. J. Frost, Cavities in Dense Random Packings, Acta Metallurgica 30, (1982), 889-904.
13. J. L. Finney and J. Wallace, Interstice Correlation Functions; A New, Sensitive Characterisation of Non-Crystalline Packed Structures, J. Non-Crystalline Solids 43, (1981), 165-187.
14. M. Ahmadzadeh and B. Cantor, Interstitial Structure and Diffusion in DRP Metallic Glasses, J. Non-Crystalline Solids 43, (1981), 189-219.

Micromechanics of Granular Materials, edited by M. Satake and J.T. Jenkins
Elsevier Science Publishers B.V., Amsterdam, 1988 — Printed in The Netherlands

EXPERIMENTS ON DISC ARRAYS: COORDINATION AND VOID ENTROPIES

Colin B. Brown

Department of Civil Engineering, University of Washington, Seattle, Washington (U.S.A.)

SUMMARY

Uncertainties in a theory of granular media have been expressed as probabilities which maximize the information entropy subject to the known statistical and bounding constraints. Implicit in this approach is that a preferred entropy state in a granular medium always exists. This paper treats this implication experimentally and reports tests on like discs on a plane which are disturbed by shear racking from an initial fixed array. These two-dimensional experiments suggested that the use of a maximum entropy approach to generate least-biased probabilities could be justified.

INTRODUCTION

In a previous U.S.-Japan Seminar, the author developed a consistent theory for the kinematics and dynamics of granular media where uncertainties were expressed as probabilities which maximized the information entropy, subject to known statistical and bounding constraints (ref. 1). The consequent theories are approximate in the sense that the least-biased rather than the actual probabilities are employed and that the entropies in the theories are upper bounds on the actual entropies. Implicit in these arguments is that a disturbed granular medium always moves towards definite entropy states. The teleological nature of the theories implicit in the mathematical programs to maximize entropy suggests that invariant minimum entropy measures are the inevitable end states of disturbed granular media. The object of this study was to test whether such teleological entropies exist or are barmecidal.

The paper commences with an argument which constrains the experiments to perturbations on disc arrays, and then discusses two geometric entropy measures obtained from the counting of contacts and bounding surfaces of voids in plane experiments. The experiments are then explained and typical results and analyses provided. The conculsions suggest that these two measures on like disc arrays tend to fixed entropies irrespective of the boundary states and the initial configuration. The experimental work was conducted at the University of Washington by a trio of graduate students, K. Nikzad, M. T. Hanson and E. R. Worden (refs. 2-4).

BACKGROUND

In the previous discussion of the geometry of a granular medium, the porosity appears as the fundamental characteristic (ref. 1) and was included in a mathematical program by means of Filep's 28 feasible microstates of like rigid spheres in a three-dimensional medium (ref. 5). Such microstates are identified as cells wherein each sphere has the same coordination number and each cell a definite porosity. The granular medium is then made up of random arrangements of these 28 microstates where the cells of microstate i occupy a total volume of V_i and, when all V_i (i = 1 to 28) are combined, the total volume, V_o, is obtained as:

$$V_o = \sum_{i=1}^{28} V_i \tag{1}$$

The probability of the cells of microstate i occurring is:

$$p_i = \frac{V_i}{V_o} \tag{2}$$

and the entropy of the system:

$$E = - K \sum_{i=1}^{28} p_i \mathrm{Ln}\, p_i \tag{3}$$

where K is a positive constant. The entropy statistic, E, can be examined experimentally by counting the presence of each microstate which has a definite but not unique coordination number, N_i, and the system mean is μ_N, where

$$\mu_N = \sum_{i=1}^{28} p_i N_i \quad . \tag{4}$$

Each microstate has a definite porosity as a measure of voids of P_i and the system mean porosity, μ_p, is

$$\mu_p = \sum_{i=1}^{28} p_i P_i \tag{5}$$

In two-dimensional arrays of like discs only four microstates exist. These are shown in Fig. 1. Each has a unique coordination number and porosity. Cell 1 is stable in any deviant perturbation under dilatational pressure, whereas cells 2, 3 and 4 collapse whilst maintaining the same coordination numbers and reduced porosities.

Experiments on three-dimensional arrays of spheres provide definite problems in determining the probabilities of each microstate. The array has to be disrupted in order to examine the states and the coordination numbers have to be estimated vicariously, for instance, by the injection of marking fluids. The initial state in a definite system microstate is only possible for those of the 28 cells that are stable under a deviatric perturbation; the others tend to collapse under the ubiquitous gravity effect. Finally, the details of the internal structure are elusive. In two dimensions, these problems are not evident. Photographs show the disc arrangements at any experimental time and direct counts of the coordination numbers and voids are possible. Cells 2, 3 and 4 can be maintained without the collapsing orthogonal force of gravity and the internal grouping and arrangements can be directly discerned.

CELL NUMBER	1	2	3	4
MICROSTRUCTURE				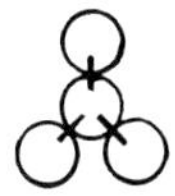
P_i	0.0931	0.1582	0.2344	0.3954
N_i	6	5	4	3

Fig. 1. Microstructure, Porosity (P_i) and Coordination Number (N_i)

EXPERIMENTAL ARRANGEMENTS

The tests involved arrays of like discs in a 19" side frame on an air table with jets at 1" spacing. The initial configuration was in either cell 2, 3 or 4, and the deviatric perturbation was by racking the frame. Measurements of the number of oscillations and the angle of perturbation were recorded. The discs were 3/16" thick hard rubber of 1", 1/2", and 1/4" diameters. Three boundary conditions were explored: (1) simple shear with no isotropic pressure; (2) shear with constant area and accommodating pressure; and (3) shear with constant pressure. The kinematic evidence was recorded by a camera mounted on a frame above the air table with axis normal to the table.

The results were arranged as follows:

(a) The coordination number of each sphere was counted. In Fig. 1 values of 3, 4, 5, and 6 are evident; however, in the tests values of 0, 1, and 2 occurred and therefore the micropicture was for five states with coordination numbers 6, 5, 4, 3, and less than 3. For equiprobable occurrence of each state, the entropy is:

$$S_{uniform} = \mathrm{Ln}\ 5 = 1.6094 \tag{6}$$

and the relative entropy is:

$$V_1 = \frac{\sum_{i=1}^{5} \bar{p}_i \, \mathrm{Ln}\, \bar{p}_i}{1.6094} \tag{7}$$

where $0 \leq V_1 \leq 1$ and $\bar{p}_i$ is the measured frequency of state i.

(b) Professor D. Elms of the University of Canterbury, New Zealand, suggested that a measure of each void could be obtained by counting the number of surrounding discs. This was used as a surrogate for porosity and is related to the states in Fig. 1 where cell 1 has a void number of 3, cell 3 has 4, cell 4 has 6, and cell 2 a mixture of 3 and 4. The frequency of the occurrence of void numbers was counted and the void number entropy determined as:

$$E_2 = \sum_{i=3}^{6} \tilde{p}_i \, \mathrm{Ln}\, \tilde{p}_i \tag{8}$$

where $\tilde{p}_i$ is the void number frequency.

The perturbation by racking left to right was over different angles and for various numbers of cycles.

RESULTS

Fig. 2(a) plots the calculated relative entropy, versus the number of racking oscillations, for simple shear tests with different initial microstates of 2, 3, and 4. Fig. 2(b) shows the same ordinate and the abscissa as the angle of racking in a one-quarter cycle for constant area tests in the three microstates. Fig. 2(c) is for constant pressure tests in microstates 2 and 4. Figs. 3(a), (b) and (c) plot the calculated void number entropy, E_2, versus number of racking oscillations with 1" diameter discs for the above three constraints and microstates.

The effect of disc size is indicated in Fig. 4(a) and (b). In Fig. 4(a) the void number entropy is plotted against cycles for 1/4" diameter discs in the three microstates 2, 3, and 4 under simple shear. Fig. 4(b) has the same plot for microstate 3 with 1", 1/2", and 1/4" diameter discs.

DISCUSSION

The question initially raised concerned the use of information entropy as an invariant measure of state in a granular material. These experiments addressed the way that disc arrays settled down after perturbation from a regular arrangement. Inspection of the photographs showed a collapse from regular arrays to complete disorder. The extent that entropy measures are invariant expressions of this disorder is judged to be related to the appropriateness of the principle of maximum entropy to assess unbiased

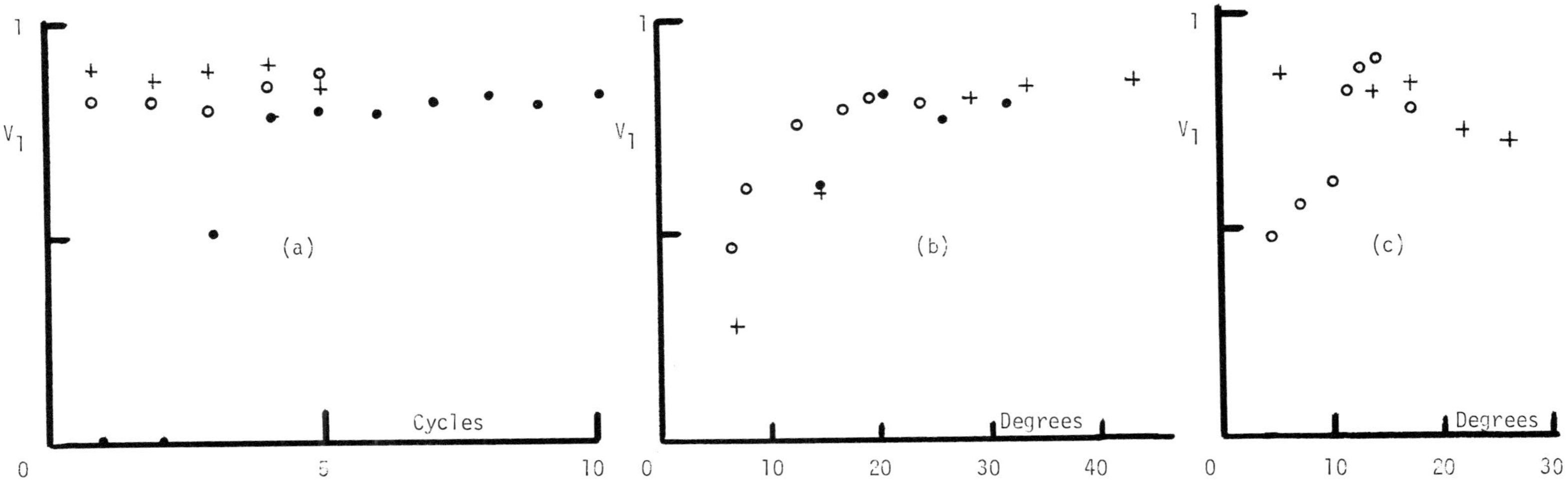

Fig. 2. Relative Coordination Entropy (V_1) versus (a) cycles and (b), (c) angles for various boundary conditions, initial microstates and 1" discs

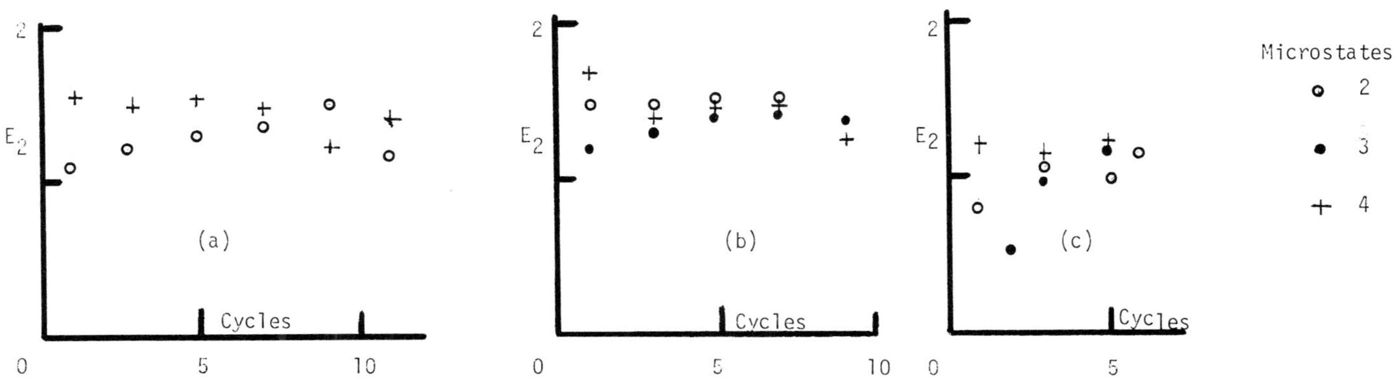

Fig. 3. Void Entropy (E_2) versus Cycles for various boundary conditions, initial microstates and 1" discs

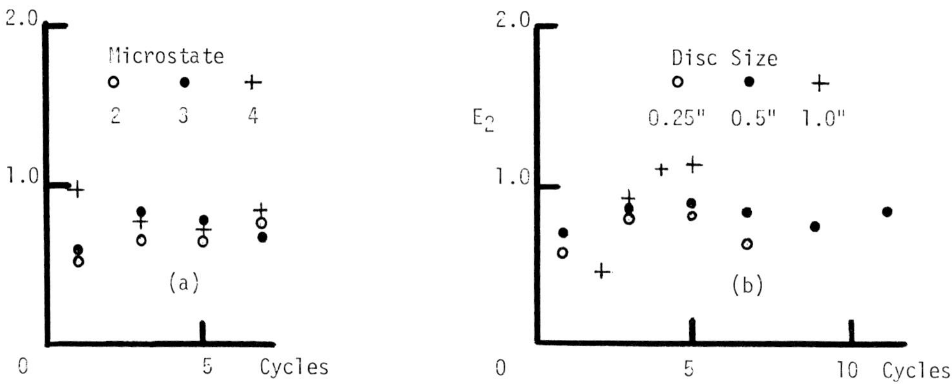

Fig. 4. Void Entropy (E_2) versus Cycles for simple shear (a) with 0.25" discs and various microstates, and (b) with various discs and microstate 3

probabilities in a theory of granular mechanics. If the entropy measures showed purposeless changes, then the invariance was not evident and the theory suspect. If, on the other hand, a teleological effect was displayed in the entropy measure, then the theory is not disproved. All the tests indicated movements of the entropies from zero towards definite values. These are exhibited in Figs. 3 and 4. The manner in which these shifts occurred and the values attained varied according to the initial uniform microstate and the size of the discs relative to the frame side. The tendency to fixed entropies was also shown to be independent of the disc-table friction (ref. 4).

Of the three initial microstates tested it is evident that 4 exhibits an immediate overshooting of the final entropy and then rebounds to the state value. From the other initial microstates, the final entropy levels are attained assymptopically. This suggests that the instability of 4 leads to excessive immediate disorder and then a more structured disc arrangement is subsequentially attained. However, whatever the initial state, the attained entropy appeared fixed for each boundary condition and disc size. In this respect it should be noted that higher values of entropy signify greater disorder or uncertainty in the system. Lower values imply less equiprobable microstates. In the experiments all initial microstates, when perturbed, settled down to the same level of disorder, as expressed through the entropies, for discs of the same size. The larger (1") discs tended to higher values of entropy with all microstates being more equally probable than in the smaller (1/2" and 1/4") discs (Fig. 4(1)). The entropy levels were sensitive to the boundary conditions, and the constant pressure case reduced the voids and hence ordered the system with lowered entropy levels.

CONCLUSIONS

The tests showed that both the coordination entropy and the void entropy settled down to definite values irrespective of the initial microstate, disc size or boundary conditions. This indicates that the use of a maximum entropy theory in the mechanics of granular media is supported by physical evidence. The entropy values were sensitive to the system states; simple shear tests and smaller discs produced more ordered arrays. The entropy values were approached from below in microstates 2 and 3 and from above (after the initial perturbation) in microstate 4.

ACKNOWLEDGMENT

This work was supported by the National Science Foundation under Grant CEE-8211600.

REFERENCES

1 C. B. Brown, The Use of Maximum Entropy in the Characterization of Granular Media, Proc. U.S.-Japan Seminar on Continuum Mechanical and Statistical Approaches in the Mechanics of Granular Material, Sendai, Japan, 1978.

2 K. Nikzad, Some Experimental Evidence on the Usage of the Principle of Maximum Entropy in Granular Media, M.S.C.E. Thesis at the University of Washington, Seattle, 1984.

3 M. T. Hanson, Further Experimental Evidence in the Use of Entropy to Characterize Granular Materials, M.S.C.E. Thesis at the University of Washington, Seattle, 1985.

4 E. Worden, The Effect of Friction on Two-dimensional Experiments to Test the Usage of the Principle of Maximum Entropy in Granular Media, Internal Report, Department of Civil Engineering, University of Washington, Seattle, 1985.

5 P. W. Jowitt and J. Munro, The Influence of Void Distribution and Entropy on the Engineering Properties of Granular Media, Proc. 2nd International Conference on Applications of Statistics and Probability in Soil and Structural Engineering (ICOSSAR), Aachen, Germany, 1975.

Micromechanics of Granular Materials, edited by M. Satake and J.T. Jenkins
Elsevier Science Publishers B.V., Amsterdam, 1988 — Printed in The Netherlands

A NOTE ON FABRIC IN TERMS OF VOIDS

J. KONISHI[1] and F. NARUSE[2]
[1]Department of Civil Engineering, Shinshu University, Nagano 380 (Japan)
[2]Doshitsu Consultant, Co., Sasazuka, Shibuya-ku, Tokyo 151 (Japan)

SUMMARY
In order to characterize the size, shape and orientation of local voids and overall void system, each void is replaced by vector polygon consisting of vectors which connect contact points sucsessively arround the associated void. Then, local void tensors and mean void tensor are defined. They may offer different aspects other than the fabric tensor in terms of contact normals. Some measurements are conducted on two-dimensional assemblies.

INTRODUCTION

The fabric of a granular material may be quantified by several measures as follows:

(1) the orientation of non-spherical particle(Fig.1(a), ref.1,2)

(2) the orientation of contact normal, i.e., the normal at the point of contact between two particles(Fig.1(b), ref.1-4)

(3) the orientation of branch vector which connects the centroid of the first granule to that of the second(Fig.1(c), ref.5,6)

Among these measures, the contact normal has been most widely accepted as a key measure for fabric in connection with the mechanical behavior of the material. The granular material, however, consists of two phases: the solid particles and associated voids. It can, therefore, be expected that the voids might be an alternative or a complementary measure for fabric.

For cellular materials, forms and cancellous bone, "the mean intercept length", which is the average distance between two solid/void interfaces in a given direction, has been used to quantify the fabric(ref.7,8). A grid of

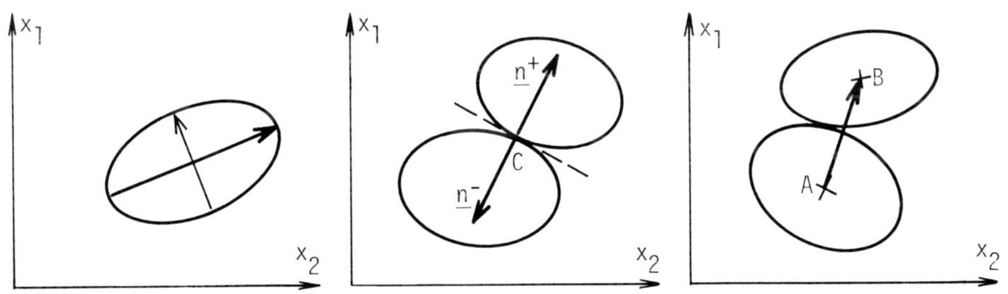

(a) orientation of a particle (b) contact normal (c) branch vector

Fig. 1. Fabric measures associated with solid particles.

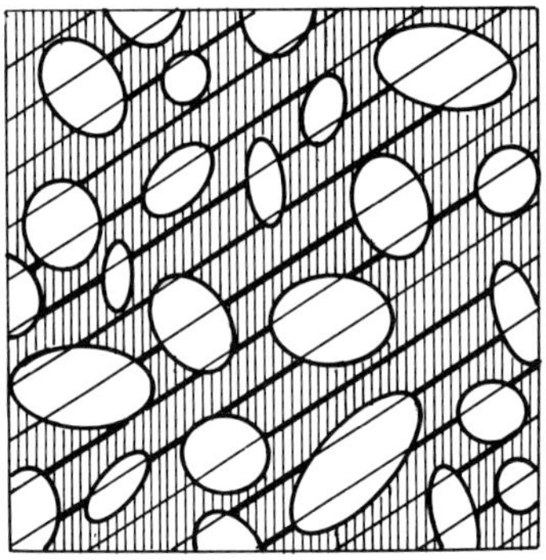

Fig. 2. Test grid of parallel lines superimposed on a polous structure.

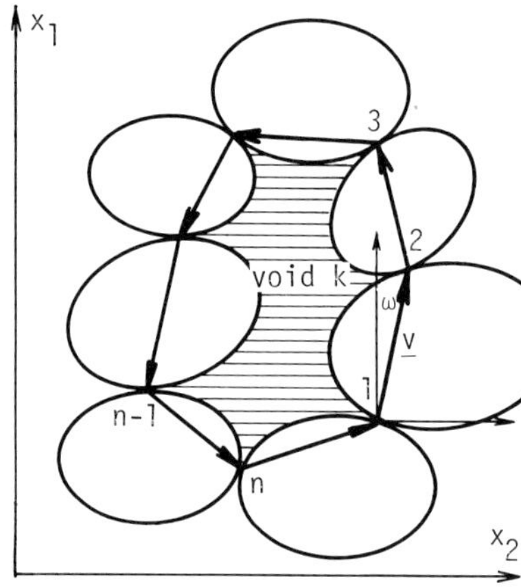

Fig. 3. A void k with the associated n particles and a replaced polygon

parallel test lines is laid over a specimen in a given direction, as shown in Fig. 2, and the length of the test line is divided by the number of intercepts to give mean intercept length. Measurements are repeated for test lines oriented in several directions. Oda et al. applied a similar approach to the voids in granular material to get an overall indication of void shape(ref.2).

We will now introduce an alternative approach to quantfy the shape, size and orientation of individual local voids as well as those of overall assemblies.

A LOCAL VOID TENSOR

Let us imagine a "unit void" surrounded by several contacting particles. In general case, the void in granular material may be continuous space with complex configulation and the void should be divided into a number of "unit voids" in a certain way. But our discussions will be limited within the two-dimentional assemblies in which the void space consists of a lot of discrete units. Fig. 3 shows a unit void k limited by n contacting particles. The boundary of the void consists of n curved segments(arcs) linked each other at contacting points of associated particles. To quantfy the size, shape and orientation of the void, the void with n curved segments may be replaced by a polygon with n vectors along the chords as illustrated in Fig. 3. The vectors of the replaced polygon pass through the associated solid particles. In this context, the polygon may also represent a unit skeleton of the material.

Let us denote a vector on any chord of the polygon by $\underline{v}$ ($\underline{v}$ stands for two vectors $\underline{v}^+$ and $\underline{v}^-$ of opposite sense). We may calculate the sum of the product $\underline{v}\underline{v}$ with respect to a void k to define a following tensor:

$$P^k_{ij} = \frac{1}{2}\sum_{\ell=1}^{2n} v_i v_j \qquad (1)$$

where n is a number of the associated vectors of the polygon k. This tensor may be tentatively called as "local void tensor"(or "local skeleton tensor"). Its principal values are calculated as follows:

$$P_1^k,\ P_2^k = \frac{1}{2}\left\{ \frac{1}{2}\sum_{\ell=1}^{2n} v_\ell^2 \pm \frac{1}{2}\sqrt{\left(\sum_{\ell=1}^{2n} v_\ell^2\cos 2\omega_\ell\right)^2 + \left(\sum_{\ell=1}^{2n} v_\ell^2 \sin 2\omega_\ell\right)^2}\right\} \tag{2}$$

where ω_ℓ is a direction angle of vector v_ℓ. The difference of the principal values is denoted by D^k:

$$D^k = P_1^k - P_2^k \tag{3}$$

The major principal direction with respect to a fixed coordinate is calculated as follows:

$$\zeta^k = \frac{1}{2}\arctan\left\{ \left(\sum_{\ell=1}^{2n} v_\ell^2\sin 2\omega_\ell\right) / \left(\sum_{\ell=1}^{2n} v_\ell^2\cos 2\omega_\ell\right)\right\} \tag{4}$$

Now we may characterize the size, shape and orientation of a local void k by a vector $\underline{p}^k$ whose magnitude is the difference of the principal value D^k of the local void tensor P_{ij}^k and its direction coinsides with the major principal axis of P_{ij}^k. The vector p^k may be called as "local void vector"(see Fig. 4).

MEAN VOID TENSOR

Finally we should characterize a void system(or a skeleton system) consisting of a number of local voids(or local skeletons). We will introduce a tensor, which may be tentatively called "mean void tensor", in the following form:

$$V_{ij} = \frac{1}{2N}\frac{1}{\sum_{k=1}^{2N} D^k}\sum_{k=1}^{2N} D^k h_i h_j \tag{5}$$

where N is the number of local voids in the specified area and $\underline{h}$ is the unit vector along $\underline{p}$. Principal values of V_{ij} are denoted by V_1, V_2 and the direction of the major principal axis is measured by Ψ from the fixed coordinate x_1.

(a) Detailed photograph of an assembly of two-dimensional granules (b) Replaced polygons and local void vectors

Fig. 4. An example of replaced polygons and local void vectors.

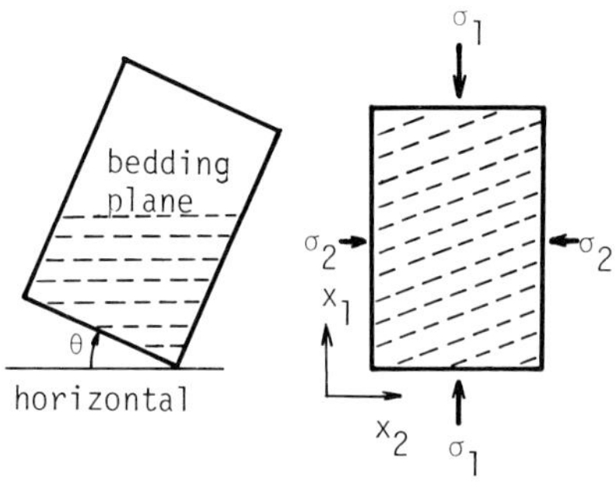

(a) during packing (b) during testing

Fig.5. Definition of bedding angle θ and of reference axes x_1 and x_2.

SEVERAL EXAMPLES OF MEASUREMENTS

Two-dimensional assemblies in biaxial compression

To get a concrete immage of the mean void tensor, several measurements were conducted on photoelastic photographs taken from the biaxial compression tests on two-dimensional assemblies of rod-shaped particles. The tests were performed at Northwestern University and various results have already been reported elsewhere(refs. 2, 9-11).

Rod-shaped particles are made of polyurethane rubber, having three kinds of cross-sections: i) circular, ii) oval with axial ratio of 1.1:1(oval I) and iii) oval with axial ratio of 1.4:1(oval II).

Loading frame can be tilted in its plane and held at a desired angle θ when particles are stacked in the frame to get a sample with desired bedding plane as illustrated in Fig. 5(a). After the assembly is completed the frame is brought back to its normal position. Then the assembly is loaded bi-axially with constant horizontal stress σ_2 and increasing vertical stress σ_1 (Fig.5(b)). Measurements were conducted within the inner one half area of the assembly in order to avoid the boundary effect. Contacts which do not transmit detectable contact force were neglected when the polygons were constructed(see Fig. 4.).

Results

Polar diagrams of Figs. 6-8 illustrate directional distributions of the "local void vectors" $\underline{p}$ divided by the number of local voids N at three stages of the test, i.e., at initial stage, at peak stress stage and at residual stress stage. These diagram may reflect the size, shape and orientation of representative(mean)local void.

In Table 1 calculated values of mean void tensors are summarized for five different assemblies at three different stages.

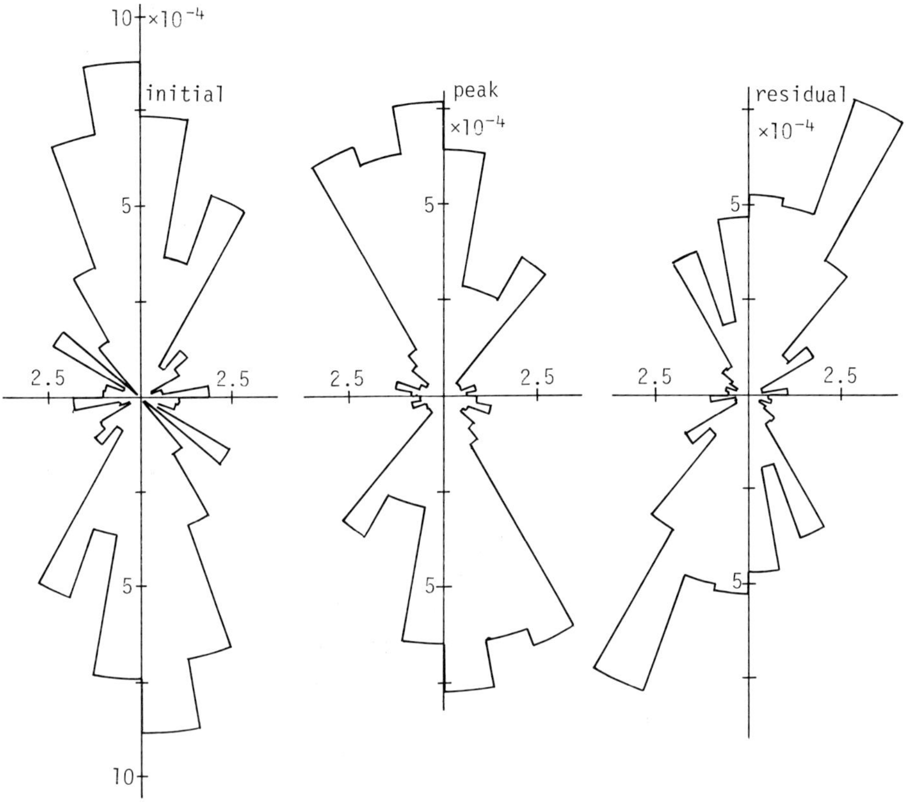

Fig. 6. Distribution of the local void vectors p (circular particles, θ =0°)

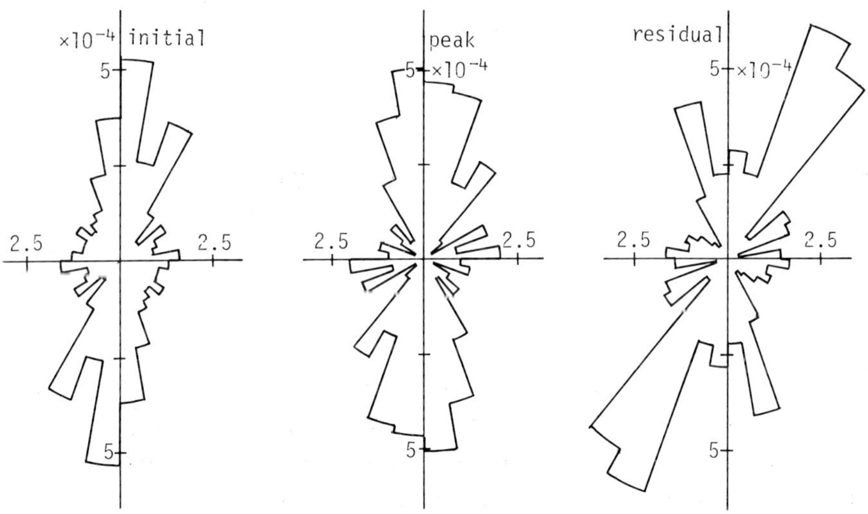

Fig. 7. Distribution of the local void vectors p (oval I particles, θ=0°)

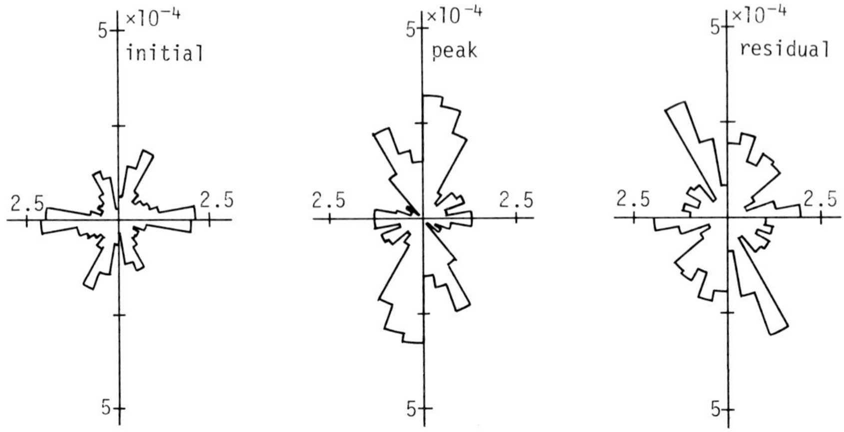

Fig. 8. Distribution of the local void vector p (oval II particles, θ=0°)

Average size of local voids and inherent anisotropy

If we compare the size of the initial distribution of p for circular, oval I and oval II particle assemblies with θ=0°, Figs. 6,7 and 8, respectively, it is clear that the size of the diagram is largest for circular particle assembly and smallest for oval II particle assembly. This means that the average size of local voids is largest for circular particle assembly and is smallest for oval II (flat) particle assembly though the overall voids ratio is nearly the same for all assemblies.

It can also be seen from the shape of the initial distributions that the inherent anisotropy of the fabric in terms of void shape is most significant for circular particle assembly and smallest for oval II particle assembly.

The same these results can also be drawn from comparisons of the magnitudes of mean void tensors V_{ij} and of the ratio of V_{11}/V_{22} at initial stage given in Table 1.

As for the effect of the bedding angle, from a comparison of data for oval II, θ=0°, 60° and 90°, in Table 1, it seems that the larger is the bedding angle, the larger is the average size of local voids and the more significant is the inherent anisotropy.

Induced anisotropy

Fig. 9(a) illustrates the relation between the stress ratio σ_1/σ_2 and the ratio of mean void tensor V_{11}/V_{22} which is a measure of induced anisotropy due to void shape. It can be said that the induced anisotropy due to void shape will develop as the stress ratio increases for each case. This tendency corresponds with the visual observation as follows: as the stress ratio increases, small neiboring voids merge with each other to form elongated voids

Table 1. Observed values of mean void tensor V_{ij}

	σ_1/σ_2	V_{11}*	V_{12}*	V_{22}*	V_{11}/V_{22}	V_1*	V_2*	Ψ+	Φ_{11}/Φ_{22}
circle initial	1.0	3.80	-0.03	1.16	3.28	3.80	1.16	-0.6	1.45
θ= 0° peak	4.1	3.60	-0.17	1.03	3.51	3.61	1.02	-3.8	1.56
residual	1.9	3.20	0.49	1.02	3.13	3.31	0.92	12.0	1.41
oval I initial	1.0	2.19	-0.08	1.13	1.93	2.19	1.13	4.3	1.48
θ= 0° peak	4.1	2.48	0.11	1.09	2.27	2.49	1.08	-4.7	1.54
residual	1.2	2.39	0.47	1.33	1.80	2.56	1.16	20.6	1.38
ovalII (initial)	4.6	0.88	0.07	0.97	0.91	1.02	0.84	60.8	2.39
θ= 0° peak	11.6	1.60	0.08	0.84	1.89	1.61	0.83	6.2	2.77
residual	1.8	1.50	0.02	1.12	1.33	1.50	1.12	3.6	2.57
ovalII initial	#	2.28	0.16	0.55	4.12	2.29	0.53	5.4	1.08
θ=60° peak	4.3	4.17	0.49	0.68	6.17	4.23	0.61	7.8	1.52
residual	2.5	3.59	0.55	0.70	5.12	3.69	0.60	10.5	1.51
ovalII initial	#	1.73	0.09	0.64	2.71	1.74	0.63	-4.0	0.81
θ=90° peak	5.3	4.42	0.11	0.38	11.50	4.42	0.38	-1.0	1.43
residual	3.1	4.10	0.06	0.56	7.35	4.10	0.56	0.9	1.33

*: $\times 10^{-3}$; +: in degree; #: not measured

σ_{ij} : stress tensor ; $\Phi_{ij} = (1/2m) \sum n_i n_j$: fabric tensor in terms of contact normals.

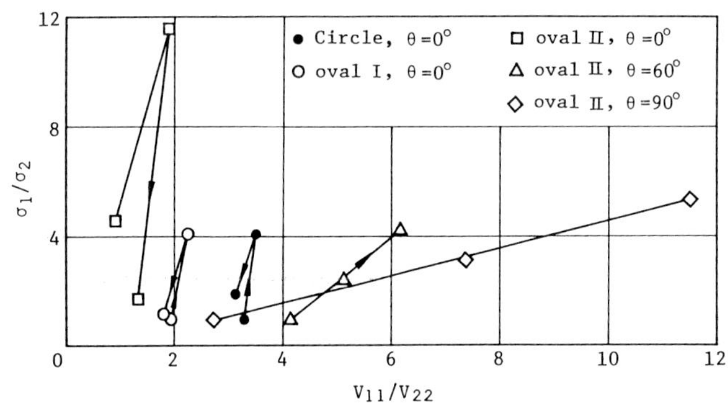

(a) Relation between mean void tensor and stress tensor

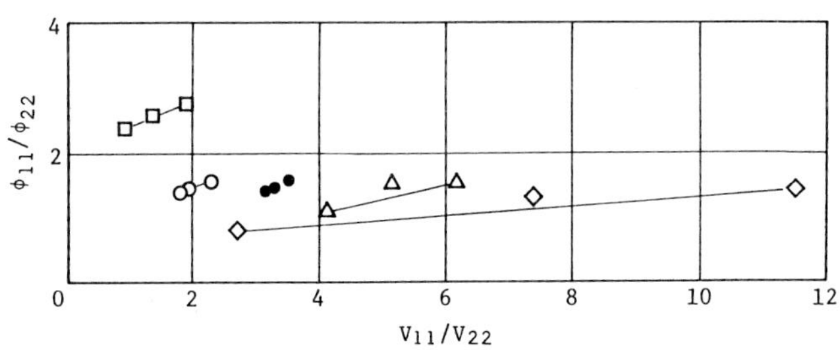

(b) Relation between mean void tensor and fabric tensor in terms of contact normals

Fig. 9. Induced anisotropy due to void shape.

with longer axis oriented to the direction of the major compressive stress.

Relation between the mean void tensor and the fabric tensor due to contact normals is illustrated in Fig. 9(b), where the fabric tensor due to contact normals is defined after Satake(ref. 4.) as $\phi_{ij} = (1/2m)\sum n_i n_j$(m is the number of contacts, $\underline{n}$ is the contact normal). It can be seen that larger ratio of ϕ_{11}/ϕ_{22} may corresponds to smaller ratio of V_{11}/V_{22} though the relation seems less clear.

CONCLUDING REMARKS

The proposed "mean void tensor" can be a measure for the size, shape and orientation of local voids or local skeletons. It may offer different aspects of fabric other than the fabric tensor in terms of contact normals.

ACKOWLEDGEMENTS

The authors should note with sincere gratitude that the photographs used in this study were the results of a series of tests performed at Northwestern University collaborated with Professor S. Nemat-Nasser of the University and Professor M. Oda of Saitama University.

REFERENCES

1 M. Oda, Initial fabrics and their relations to mechanical properties of granular material, Soils and Foundations, 12(1), 1972, 17-36.
2 M. Oda, J. Konishi and S. Nemat-Nasser, Stress-induced anisotropy in granular materials, Soils and Foundations, 25(3), 1985, 85-97.
3 M. Oda, J. Konishi and S. Nemat-Nasser, Some experimentally based fundamental results on the mechanical behaviour of granular materials, Geotechnique, 30, 1980, 479-495.
4 M. Satake, Constitution of mechanics of granular materials through graph representation, in:Theor. Appl. Mech. 26, 1978, Univ. of Tokyo Press, 257-266
5 M.M. Meherabadi, M. Oda and S. Nemat-Nasser, On statistical description of stress and fabric in granular materials, Int. J. Numer. Anal. Meths. Geomech. 6, 1982, 95-108.
6 S. Nemat-Nasser and M. M. Meherabadi, Stress and fabric in granular masses, in: J. T. Jenkins and M. Satake(ed), Mechanics of Granular Materials, Elsevier, 1983, 1-8.
7 T. P. Harrigan and R. W. Mann, Characterization of microstructural anisotropy in orthotropic materials using a second rank tensor, J. Mater. Sci.,19, 1984, 761-767.
8 C. H. Turner and S. C. Cowin, Dependence of elastic constants of an anisotropic porous material upon porosity and fabric, J. Mater. Sci.,22, 1987, 3178-3184.
9 J. Konishi, M. Oda and S. Nemat-Nasser, Inherent anisotropy and shear strength of assembly of oval cross-sectional rods, Deformation and Failure of Granular Materials, A.A.Balkema, 1982,403-412.
10 J. Konishi, M. Oda and S. Nemat-Nasser, Induced anisotropy in assembries of oval cross-sectional rods in biaxial compression, in: J. T. Jenkins and M. Satake(ed), Mechanics of Granular Materials, Elsevier, 1983,31-39.
11.M. Oda, J. Konishi and S. Nemat-Nasser, Experimental micromechanical evaluation of strength of granular materials: effects of particle rolling, Mech. of Materials, 1, 269-283.

Micromechanics of Granular Materials, edited by M. Satake and J.T. Jenkins
Elsevier Science Publishers B.V., Amsterdam, 1988 — Printed in The Netherlands

SOME CONSIDERATIONS ON SHEARING RESISTANCE ANGLES OF SANDS

N.MOROTO
Department of Civil Engineering,Hachinohe Institute of Technology,
Hachinohe 031(Japan)

Abstract

From both a theoretical and practical point of view, it is essential to make a standard of the choice of shearing strength parameters for the stability problem of cohesionless soil masses. In this paper, (1) based on the stress-dilatancy formulae and energy correction equation, the shearing resistance angles of ϕ_r and ϕ_{cv} are thoretically related to the intergraular friction angle of ϕ_μ, (2) a strength equation expressed in terms of anisotropy, remoulding, density and material property parameters is newly introduced, and (3) comparisons of experimental results with calculated ones by the proposed equations are presented.

1. Introduction

From both a theoretical and practical point of view, it is essential to appreciate the fundamental characteristics of shearing strength parameters and their proper use for stability analysis of cohesionless soil masses. Thus, in this paper,

(1) shearing resistance angles and their proper use are described,

(2) concept of shearing strength of granular material is briefly explained,

(3) based on the stress ratio-strain increment ratio relationship (the so called stress -dilatancy relationship) and energy correction equation, the shearing resistance angle of ϕ_r and ϕ_{cv} are theoretically related to the intergranular friction angle of ϕ_μ, and comparisons of experimental results with calculated ones by the proposed equations are presented,

(4) strength parameter λ is newly introduced and a strength equation expressed in terms of anisotropy, remoulding, density and intergranular friction is examined.

The data used in this paper were obtained using the conventional triaxial compression test. The shearing resistance of sand in a very loose state was not taken into consideration.

2. Choice of Shearing Resistance Angles

It is essential to have proper shearing resistance angles for stability analysis of sand and gravel. Shearing resistance of granular materials such as sand depends not only on the friction angle between sliding solids but also on the degree of shearing strain, density and particle characters. Therefore, engineers must choose the proper shearing resistance angle in the design of earthen structure. The angles of shearing resistance are :

(1) ϕ : defined at the peak of the stress-strain curve,

(2) ϕ_{cv} : defined at the critical state,

(3) ϕ_r : obtained by Bishop's energy correction ($\fallingdotseq \phi'$, ex. in reference [1]),

(4) ϕ_μ : particle-to-particle friction angle.

In most problems encountered engineering practic, it is not possible to tolerate large strains within a sand mass. Thus, the shearing resistance angle ϕ is properly used.

When one evaluates the resistance encountered by a tracked vehicle and the friction force between sliding sand and the rough surface of the structures such as concrete retaining wall and pile, it would be appropriate to use ϕ_{cv} to represent the strength of the sand, because large strains occur for such cases. If the surface of the structure is very smooth, as in the case of sand sliding on unrested steel, the shearing resistance angle is most likely equal to ϕ_μ in reference [2].

As almost no volume change can be permitted for granular materials deep in fill type dam, it would be better to choose $\phi_r(=\phi')$ as the shearing resistance angle.

3. Basic Concept of Stress Ratio Law

An ideal model is employed to simply express the complicated phenomena. This is the case in mechanics of granular materials. The shearing resistance angle of face centered cubic array with equall spheres was obtained by Scott [3] as

$$\tan\phi=\frac{\sqrt{3}+4\sqrt{2}f}{2(\sqrt{6}-f)} \tag{1}$$

This was derived on the assumption that the particles have a tendency to slip in the direction of least resistance. Let refer to this assumption as the 'least-stress ratio law'. However, this law cannot explane the work-hardening nature which is attributable to the shearing behavior of granular materials.

Rowe [4] developed a stress ratio theory, the so-called stress dilatancy theory, regarding the shearing behavior of ideal equal sphere packing and also of a random assemblage of irregular particles.

Introducing a new concept for increasing anisotropy for a model granular material, Horne [5, 6] theoretically developed Rowe's theory.

Oda[7] carried out microscopic research in order to directly measure the increasing anisotropy of sand spesimens. His test results showed that Horne's concept of increasing anisotropy during shear was valid. According to his paper, the mobilized stress ratio was written as

$$\frac{\sigma_1}{\sigma_3}=\frac{4}{\pi}\frac{S_1}{S_3}\tan\left(\frac{\pi}{4}+\frac{\phi_\mu}{2}\right) \tag{2}$$

where S_1, S_3 are equal to the summations of the projected area of the contact surfaces of the particles on the principal planes (radial and vertical planes, respectively), and thus, the ratio S_1/S_3 becomes an important parameter showing the degree of anisotropy developed in the sand specimens. Oda's experimental results clarified that the fabric index S_1/S_3 increased with an increase in the shear strain and finally took a certain limitted value as shown in Fig. 1. According to Eq.(2), so did the stress ratio σ_1/σ_3. Considering this fact, one may denote the stress ratio-anisotropy relationship given by Eq.(2) as the "modified least stress ratio law". Using Eq.(2), one is able to obtain

$$\sin\phi=\frac{1-\frac{4}{\pi}\frac{S_1}{S_3}\tan\left(\frac{\pi}{4}+\frac{\phi_\mu}{2}\right)}{1+\frac{4}{\pi}\frac{S_1}{S_3}\tan\left(\frac{\pi}{4}+\frac{\phi_\mu}{2}\right)} \tag{3}$$

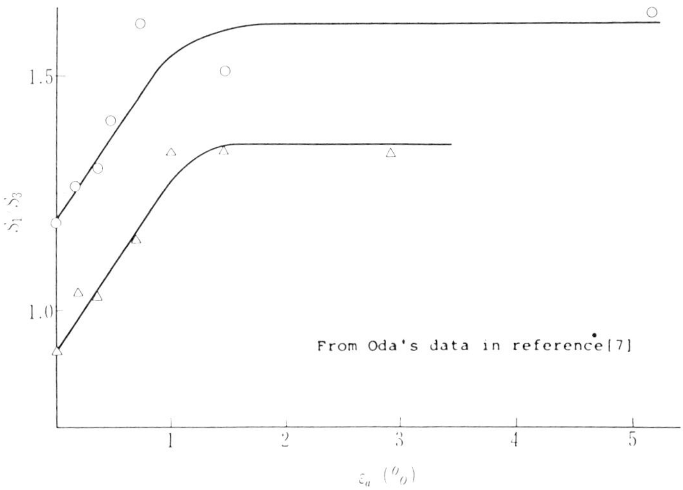

Fig 1. Degree of Anisotropy and Shearing Strain

4. Relationship Between ϕ_r and ϕ_μ

Taylor [8] first remarked that the effect of dilatancy on the friction angle could best be explained by giving consideration to the strain energy. Following Taylor, Bishop [9] defined a portion of the deviator stress which can be associated with the frictional heat absorbed internally as

$$(\sigma_1-\sigma_3)_r=(\sigma_1-\sigma_3)+\sigma_3\frac{dv}{d\varepsilon_1} \qquad (4)$$

where $dv/d\varepsilon_1$ is the tatio of a small increment of volumetric strain to a corresponding small increment of axial strain occurring at failure. In the drained triaxial compression test in which the cell pressure is held constant during the test, Eq.(4) can be rewritten as

$$\left(\frac{\sigma_1}{\sigma_3}\right)_r=\frac{\sigma_1}{\sigma_3}+\frac{dv}{d\varepsilon_1} \qquad (5)$$

or

$$\left(\frac{\sigma_1}{\sigma_3}\right)_r=\frac{\sigma_1}{\sigma_3}+\frac{2d\varepsilon_3}{d\varepsilon_1}+1 \qquad (6)$$

and

$$\sin\phi_r=\frac{\left(\frac{\sigma_1}{\sigma_3}\right)_r-1}{\left(\frac{\sigma_1}{\sigma_3}\right)_r+1} \qquad (7)$$

The value for ϕ_r is the angle of friction which corresponds to the total work absorbed internally after elimination of the external work carried out by diation.

Referring to the stress-dilatancy equations developed by Rowe Horne and Oda, we can write the stress ratio and strain increment ratio as

$$\frac{\sigma_1}{\sigma_3}=A\cdot\tan\left(\frac{\pi}{4}+\frac{\phi_\mu}{2}\right) \qquad (8)$$

$$-\frac{2d\varepsilon_3}{d\varepsilon_1}=A\cdot\tan\left(\frac{\pi}{4}-\frac{\phi_\mu}{2}\right) \qquad (9)$$

$$\tan\phi_f = \chi \cdot \tan\phi_\mu \tag{10}$$

where coefficient χ increases as the degree of remoulding and associated energy loss rises in reference [4]. Coefficient A is a measure of fabric anisotropy which is summarized as

$$A = \tan\alpha \qquad (Rowe) \tag{11}$$

$$A = m_1/m_3 \qquad (Horne) \tag{12}$$

$$A = S_1/S_3 \qquad (Oda) \tag{13}$$

Based on Oda's experimental data, we get

$$\left(\frac{\sigma_1}{\sigma_3}\right)_r = 1.5\left(\frac{S_1}{S_3}\right) + 0.7 \tag{14}$$

where

$$\frac{\sigma_1}{\sigma_3} = k_1\frac{S_1}{S_3} + k_2 \qquad k_1 = 2.8 \qquad k_2 = -0.9$$

$$-\frac{dv}{d\varepsilon_1} = k_3\frac{S_1}{S_3} + k_4 \qquad k_3 = 1.3 \qquad k_4 = -1.6$$

From Eq.(14), we are able to realized that the stress ratio largely depends on the degree of anisotropy induced in sand sample due to shearing action. Substituting Eqs.(8) and (9) into (6), we are able to derive the following equation:

$$\left(\frac{\sigma_1}{\sigma_3}\right)_r = 2A \cdot \tan\phi_f + 1$$

$$= 2\lambda + 1 \tag{15}$$

Where $\lambda = A \cdot \chi \cdot \tan\phi_\mu$ (16)

Rowe[4] reported that the value of χ depends on the initial density, confining pressure, angularity of particles and $\chi = 1.1$-1.2 for angular materials; $\chi = 1.4$-1.5 for rounded materials. He directly observed the value of ϕ_μ and their values were

$\phi_\mu = 26°$ (medium to fine grained sand)

$\phi_\mu = 24°$ (Leighton Buzzard sand)

$\phi_\mu = 23°$ (coarse sand)

Table 1 shows the calculated values of

$$\frac{8}{\pi} \cdot \chi \cdot \tan\phi_\mu$$

with the values of χ and ϕ_μ being varied.

Table 1. The Values of $\frac{8}{\pi} \cdot \chi \cdot \tan\phi_\mu$

	χ				
ϕ_μ	1.1	1.2	1.3	1.4	1.5
22°	1.13	1.24	1.34	1.44	1.54
23°	1.19	1.30	1.41	1.51	1.62
24°	1.24	1.36	1.47	1.58	1.70
25°	1.31	1.43	1.54	1.66	1.78
26°	1.37	1.49	1.61	1.74	1.86

One can sce in Eq.(14) that the coefficient of the term S_1/S_3 is 1.5 and the constant is 0.7, and the value of 1.5 is lying on the middle of the calculated ones shown on Table 1 and the value of 0.7 is in the proximity of unity which is obtained theoretically in Eq.(15). These things lead to the conclusion that the stress ratio-strain increment ratio relationship is valid. For simplicity, Eq.(15) which is derived theoretically will be employed in this paper.

Substituting Eq.(15) into (7), we have

$$\sin\phi_r = \frac{\tan\phi_\mu}{\tan\phi_\mu + \dfrac{1}{A\cdot x}} \tag{17}$$

or

$$\sin\phi_r = \frac{1}{1+\dfrac{1}{\lambda}} \tag{18}$$

Strength parameter λ becomes of interest, because it represents the effects of anisotropy, remoulding and intergranular friction on the shearing strength of granular materials. The value of λ increases with increase in grain friction, degree of anisotropy and degree of remoulding in the sand samples subjected to shear stress. One can easily understand that the angle of shearing resistance ϕ_r given by Eq.(18) increases with the value of λ.

5. Strength Parameter

Rowe[4] reported that the value of x depends on the grain shape and initial density. Oda [7] wrote that coefficient A at failure changed its value depending on grain characteristics and initial fabric. Strength parameter λ, therefore, is related to the arrangement of particles and particle characters.

Konwing the value of ϕ_r from experiment, one is able to obtain the value of $1/\lambda$ in Eq.(18) as

$$\frac{1}{\lambda} = \frac{1-\sin\phi_r}{\sin\phi_r} \tag{19}$$

In order to examine the relationship between $1/\lambda$ and the initial density of the sand sample, the data for conventional compression test conducted by several authors are available. An shown in Fig. 2, for example, we are able to see that there are linear regression relationships between $1/\lambda$ and initial void ratio e_0 which, generally, can be written as

$$\lambda\cdot e_0 = m \quad \text{(constant)} \tag{20}$$

The above relationship is coincidental with the well-known strength formula proposed by Caquot and is as follows:

$$e\cdot\tan\phi = \text{constant} \tag{21}$$

The constant m lies in the proximity of unity.

From Eq.(20), we can write

$$\frac{S_1}{S_3}\cdot x\cdot\tan\phi_\mu\cdot e_0 = \frac{\pi}{4}m \quad \text{(constant)} \tag{22}$$

In Eq.(22), initial void ratio e_0, degree of anisotropy S_1/S_3, degree of remoulding x and intergranular friction coefficient $\tan\phi_\mu$ are interrelated in a multi-product form.

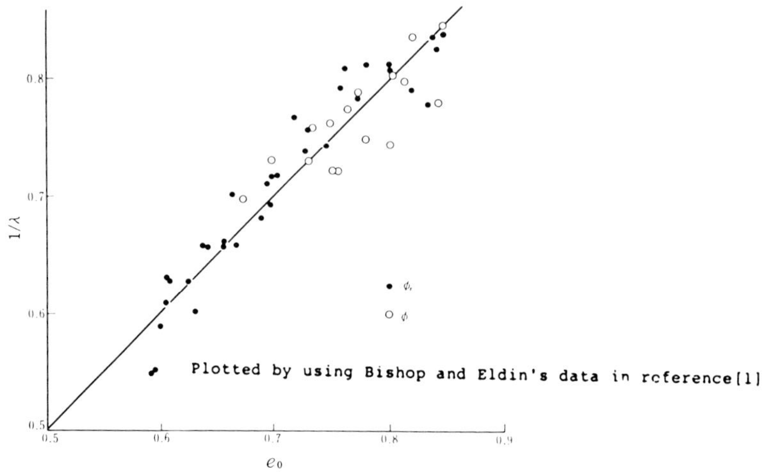

Fig 2. Relationship between $1/\lambda$ and e_0

6. Relationship Between ϕ_{cv} and ϕ_μ

At strains which are considerably greater shearing deformation occurs at constant volume where the corresponding angle of friction is denoted ϕ_{cv}. Roscoe, Schofield and Wroth[10] proved that tha assemblage of particles attain a final single void ratio e_{cr} which may depend on the particle shape and grading. This ultimate condition is reached no matter what the initial void ratio of the samle for a given confining pressure is. The angle of shearing resistance ϕ_{cv} has been correlated with friction angle ϕ_μ by the investigators. Caquot presented the theoretical analysis leading to the formula

$$\tan\phi_{cv}=\frac{\pi}{2}\tan\phi_\mu \tag{23}$$

Bishop[9] developed an approximate expression in the triaxial compression as shown in the form of

$$\tan\phi_{cv}=\frac{15\tan\phi_\mu}{10+3\tan\phi_\mu} \tag{24}$$

At the critical state, the value of ϕ_r approaches to the value of ϕ_{cv}. Then Eq.(17) can be written as

$$\sin\phi_{cv}=\frac{\tan\phi_\mu}{\tan\phi_\mu+a_{cv}} \tag{25}$$

where

$$a_{cv}=\left(\frac{1}{A\cdot\chi}\right)_{cv} \tag{26}$$

Horne[6] plotted the values of ϕ_{cv} and the corresponding values of ϕ_{cv} and the corresponding value of ϕ_μ for different granular materials such as quartz sand, feldspar sand, bronze balls, zircon balls and glass balls. These values are prepared from Horne's graph and are tabulated in Table 2. The listed valuds of a_{cv} are calculated by Eq.(25). One can see that the values of lie in the narrow range from 0.4. After a more detailed investigation, a linear relationship such as the following was able to be found

$$a_{cv}=0.509-0.181\tan\phi_\mu \tag{27}$$

Substituting Eq.(27) into (25), we get

$$\sin \phi_{cv} = \frac{1.22 \tan \phi_\mu}{\tan \phi_\mu + 0.62} \tag{28}$$

The values tabulated in Table 2 and 3 are plotted in Fig. 3. The values of ϕ_μ and the corresponing values of ϕ_{cv} which are calculated from the formulas of Caquot, Bishop and the author are listes on Table 4. From Table 4, the author's formularization of Eq.(28) provides excellent results. Using 0.4 as the value for a_{cv} in Eq.(26), we obtain the following approximate relationship

$$\left(\frac{S_1}{S_3} \cdot \chi \right)_{cv} = 2.0 \tag{29}$$

There is a tendency in Eq.(29) for the degree of anisotropy to be in inverse proportion to the degree of remoulding.

Table 2. The Values of ϕ_μ and ϕ_{cv}

Symbols	Materials	ϕ_μ	ϕ_{cv}	a_{cv}
⊛	Crashed glass	38°	43°	0.364
⊕	Feldspar sand	36°	41°	0.381
▲	Quartz sand	28°	35°	0.395
△	Quartz sand	26°	32°	0.432
□	Zircon	23°	29°	0.450
●	Bronze	20°	27°	0.438
○	Glass ballotini	17°	24°	0.446

Table 3. Additional Data between ϕ_μ and ϕ_{cv}

Symbols	ϕ_μ	ϕ_{cv}	References
▽	27°	32.5°	Brasted sand. [11]
◆	28.8°	36.6°	Stonecourt sand. [12]
△	26.6°	34°	Biddulph sand. [12]

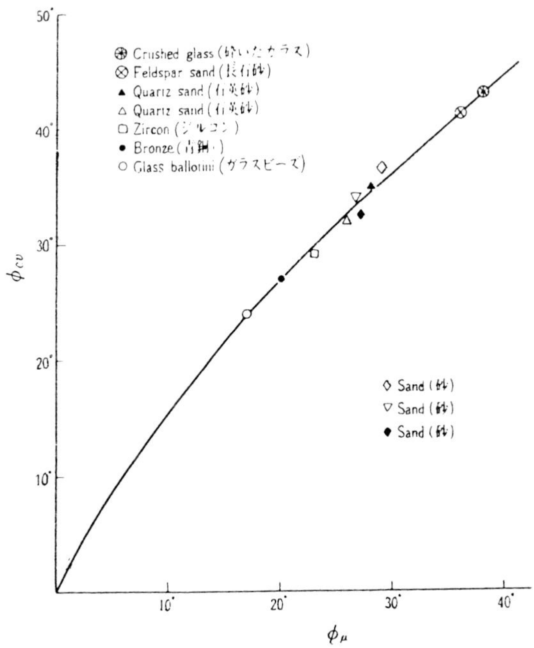

Fig 3. Relationship between ϕ_{cv} and ϕ_μ

Table 4. Experimental and Theoretical Values of ϕ_{cv}

Material	Experimental		Theoretical (ϕ_{cv})		
	ϕ_{μ}	ϕ_{cv}	Caquot	Bishop	The Author
Steel balls	7°	14°	11°	10°	11.6°
Glass balls	17°	24°	26°	25°	23.8°
Quartz sand	26°	32°	38°	40°	32.5°
Feldspar	37°	42°	50°	67°	42.0°

(The values of ϕ_{μ} and ϕ_{cv} are reported in reference [13])

References

1. Bishop, A.W. and A.K. Eldin, "The Effect Stress History on the Relation between ϕ and Porosity in Sand", Proc. 3rd ICSMFE., Vol. I 1953, pp. 126-130.
2. Lambe, T.W. and R.V. Whiteman, Soil Mechanics, John Wiley, 1969.
3. Scott, R.F., Principle of Soil Mechanics, Addison-Wesley, 1963.
4. Rowe, P.W., "The Stress-Dilatancy Relation for Static Equilibriun in Contact, Proc. Roy. Soc. London, A 269, 1962, pp. 500-527.
5. Horne, M.R., "The Behaviour of an Assembly of Rotund, Rigid, Cohessionless Particles, Part I and Part II", Proc. Roy. Soc. London, A 286, 1965, pp. 62-78 and pp. 79-97.
6. Horne, M.R., "The Behaviour of an Assembly of Rotund, Rigid, Cohessionless Particles, Part III, Proc. Roy. Soc. London, A 310, 1969, pp. 21-34.
7. Oda, M., "The Mechanics of Fablic Changes during Compressional Deformation of Sand, "Soils and Foundations", Vol. 12, No. 2, 1972, pp. 1-17.
8. Talor, D.W., Fundamentals of Soil Mechanics, Jone Wiley, 1948.
9. Bishop, A.W., Discussion of A.D.M. Penman (1953), Geotechnique, Vol. 4, No. 1, 1954, pp. 43-45.
10. Roscoe, K.H., A, N, Schofield and C.P. Wroth, "On the Yielding of Soils", Geotechnique, Vol. 8, No. 1, 1958, pp. 22-53.
11. Rowe, P.W., Correspondence of Conforth (1964), Geotechnique, Vol. 14, No, 2, 1964, pp. 143-168.
12. King, J.W. and E.A. Dickin, "Comparison of Stress-Dilatancy Theories", proc. ASCE, SM5, 1970, pp. 1697-1714.
13. Lee, K.I., "Stress-Dilatancy Performance of Feldspar", Proc. ASCE SM 2, 1966, pp. 79-103.

Micromechanics of Granular Materials, edited by M. Satake and J.T. Jenkins
Elsevier Science Publishers B.V., Amsterdam, 1988 — Printed in The Netherlands

MEASURMENT OF THE BULK DENSITY STRUCTURE OF GRANULAR MATERIALS IN A POWDER VESSEL

K.Makino and K.Kuramitsu

Dept. of Chem.Eng. for Resources,Mining College,Akita Univ.,
AKITA 010 (JAPAN)

SUMMARY

In this paper,X-ray radiograph method is applied to the measurement of the bulk density distribution of powder by use of the analyzing method presented by us.As a result of it,the measuring system is pointed out to have the possibility to measure the bulk density distribution and the constitution relation of granular materials indirectly.

INTRODUCTION

In industial particulate processes,the development of powder handling is urgently needed.First of all,it is considered to be important to have stability when feeding the liquified powder into the reactor;hence,no choking must occur.However,as of yet a continuous stable powder feeding technique has not yet been sufficiently established. Therefore,in this paper,from the standpoint of developing a design method for stable no choking powder feeding from a powder vessel,the bulk density structures of the following three cases are qualitatively and quantitatively measured by X-ray radiograph system.

i) Initial bulk density distribution in a powder vessel
ii) Increase in bulk density after closing the discharge valve
iii) The change of bulk density near shearing plane before and after shearing powder in shear test

1 INITIAL BULK DENSITY DISTRIBUTION IN A POWDER VESSEL

The heterogeneity of the bulk density distribution for powder in a powder vessel is predominantly affected by the powder packing method into a powder vessel. However,the packing mechanism for powder into a powder vessel has not yet been elucidated fundamentally or experimentally. Therefore,in this paper, an X-ray radiograph system was constructed as shown in Fig.1. The system is composed of an X-ray generator, a test cell(140 x 180 x 30), an image intensifier, a monitor and an X-ray shield box. With this equipment,the initial powder density distribution in powder vessel can be qualitatively and quantitatively observed corresponding to the powder feeding method. Fig.2 shows a conceptional diagram for an experimental powder feeding apparatus. Some typical results for the initial bulk density distribution pictured and analyzed by X-ray radiograph,are exibited in Figs.3 to 4. The black and white parts in the picture relate to the dense and loose packing zones, respectively. Also,the numbers seen in the right picture(Fig.3) indicate the porosity of the powder transformed from the bulk density and three dimensional representation (Fig.4)means the bulk density distribution analyzed by use of the method presented by us(1).

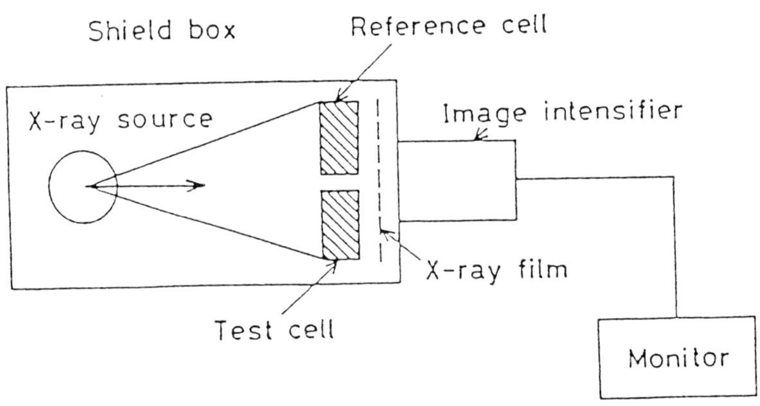

Fig.1 Conceptional diagram of X-ray radiograph system

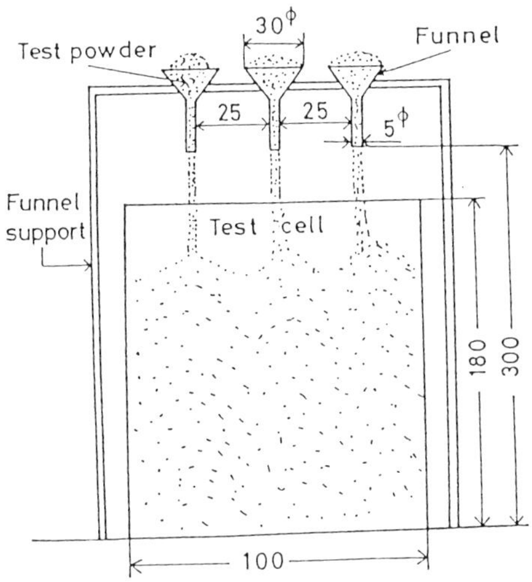

Fig.2 Conceptional diagram of a powder feeding method (Funnel metod)

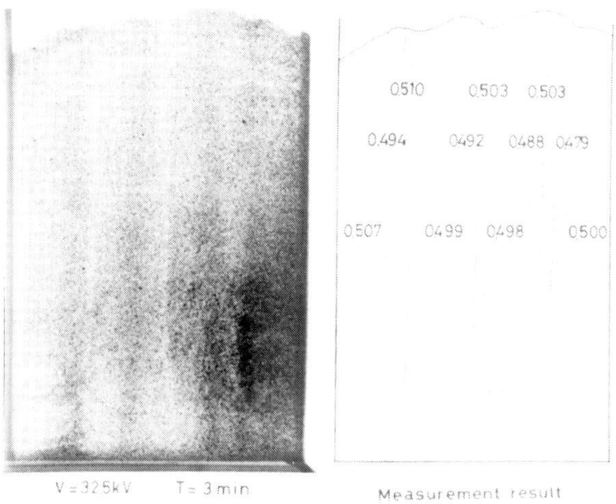

Fig.3 A typical example of bulk density distribution in a powder vessel (Funnel method)

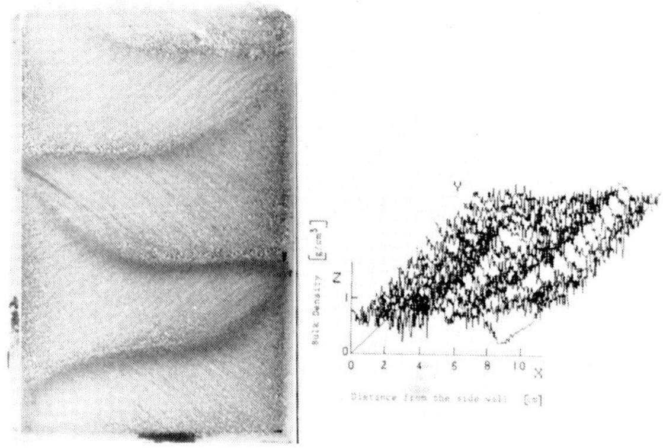

Fig.4 A typical example of bulk density distribution in a powder vessel (Horizontally reprocating method)

From these results,it is concluded that each powder feeding system shows its own peculiar distribution mode for the powder bulk desity.

2 INCREASE IN BULK DENSITY AFTER CLOSING THE DISCHARGE VALVE OF POWDER VESSEL

Bulk density distribution around a discharge valve was observed before and after closing the valve by use of X-ray radiograph system.

Test powder was initially fed into a vessel (to the bottom of which a slide valve was attached),through a funnel.An X-ray radiograph was taken at this initial state in the porosity distribution of the powder. Next,a fixed quantity of test powder was discharged by openning the valve for a given period of time ; whereby subsequent to its reclosing ,an X-ray radiograph was taken.This procedure was repeated until the powder was completely discharged.

Fig.5 shows a series of typical bulk density distribution around the discharge valve subsequent to being closed(the valve angle is 0°).According to this picture,we can observe the increase in bulk density near the valve. However,in case of the valve angle 45°,no increase in bulk density near the valve appears as shown in Fig.6. The difference between them can be considered to be a new experimental finding.

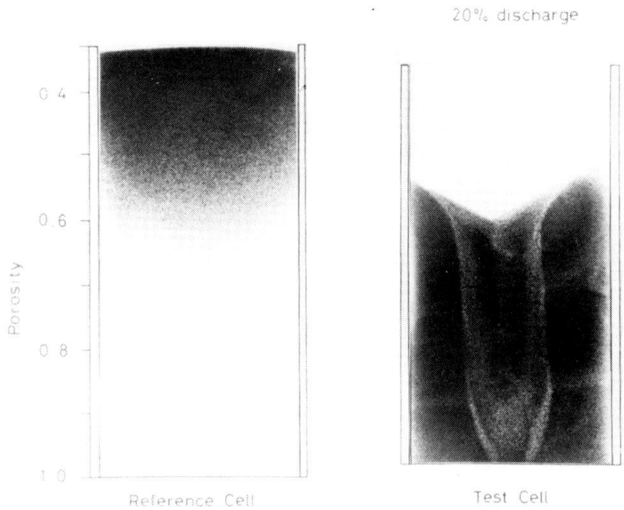

Fig.5 A typical example of bulk denesity distribution near the valve after closing it (The valve angle is 0°)

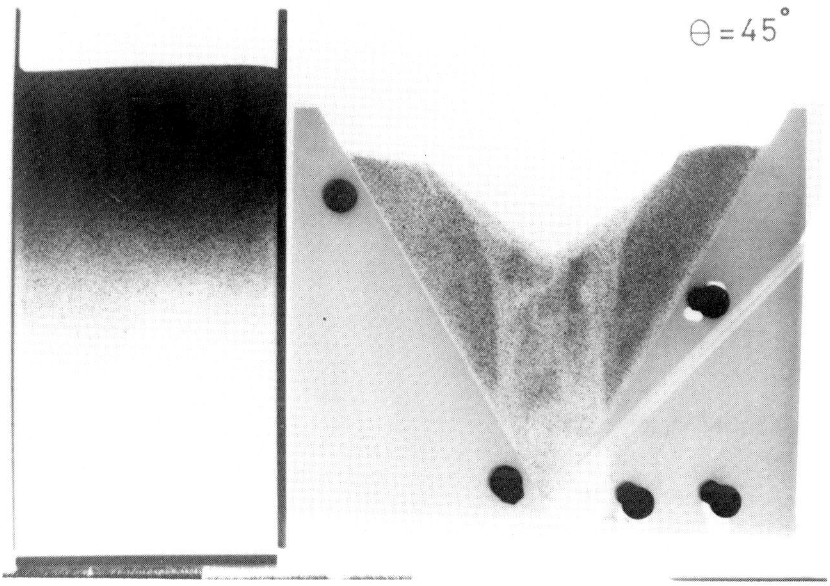

Fig.6 A typical exaple of bulk density distribution near the valve after closing it (The valve angle is 45°)

3 THE CHANGE OF BULK DENSITY NEAR SHEARING PLANE BEFORE AND AFTER AHEARING TEST

Fig.7 shows a typical bulk density distribution pictured by X-ray near shearing plane after shear test. Applying the same analyzing method above to the result,the bulk density distribution near shear plane can be quantitatively estimated.And if we estimate the powder yield locus of the powder near shearing plane,the constitution co-efficient defined by the ratio of shear stress to shearing rate , will be able to be discussed experimentally from sfear test with X-ray radiograph system.

Fig.7 A typical example of bulk density distribution near shearing plane after shearing powder

FINALS

In this paper,it was experimentally shown that the bulk density distribution can be qualitatively and quantitatively measured by using X-ray radiograph system and the analyzing method by us(1). From these results,we obtained a new experimental knowlege of the increase of bulk density after closing the valve and also pointed out the possibility to measure the constitution relation of granular materials during shear test indirectly by use of X-ray radiograph system.

REFERENCES

1) Higashiyama,A.,A.Tsukada,K.Kuramitsu and K.Makino ; J. of the Soc. of Powder Technolody Japan,24,No.5,291-295 (1987)

Micromechanics of Granular Materials, edited by M. Satake and J.T. Jenkins
Elsevier Science Publishers B.V., Amsterdam, 1988 — Printed in The Netherlands

CONSIDERATION ON THE STRESS-DILATANCY EQUATION THROUGH THE WORK INCREMENT TENSOR

M. SATAKE

Dept. of Civil Engineering, Tohoku University, Sendai (Japan)

SUMMARY

In this paper, the isotropic and anisotropic decompositions of a symmetric tensor are first explained. In 3D, it is recognized that the employment of complex number is rational and useful. Next, the stress-dilatancy equation is considered. To rewrite the Rowe's stress-dilatancy equation, which is fundamental in the mechanics of granular materials, in a tensorial expression, the work increment tensor is newly introduced. Using the work increment tensor, the common expression of the stress-dilatancy equation applicable for 2D and 3D is given and the form of the dissipative energy in granular materials is proposed. It is explained that the coefficients of the stress-dilatancy equation are determined from the stresses, where the volumetric strain takes minimum value in the given stress path. Lastly, using the modified associated flow rule, in which the stress-dilatancy equation is regarded as the constraint condition, and applying the analysis of the anisotropic decomposition, the constraining stress, which gives reasonable strain increments, is obtained.

ISOTROPIC AND ANISOTROPIC DECOMPOSITIONS OF A TENSOR

Taking the stress tensor as an example of symmetric tensor, we consider the isotropic and anisotropic decompositions of a symmetric tensor of second order.

The *isotropic decomposition* of stress tensor σ_{ij} is expressed, using the isotropic tensor $I_{ij} = \delta_{ij}$ (Kronecker's delta), as

$$\sigma_{ij} = pI_{ij} + {}_D\sigma_{ij} \tag{1}$$

where

$$p = \frac{1}{n}\sigma_{ij}I_{ij} \quad : \text{the mean stress} \tag{2}$$

(n being the number of dimensions)

and ${}_D\sigma_{ij}$ is the deviatoric stress tensor.

For the case of 2D, we can write, using the *unit deviatoric tensor* J_{ij} ($|J_{ij}| = -1$), as

$${}_D\sigma_{ij} = q\,J_{ij} \tag{3}$$

where

$$q = \sqrt{\frac{\sigma_{Dij}\,\sigma_{Dij}}{2}} = \frac{1}{2}\sigma_{ij}J_{ij} \qquad : \text{the deviatoric stress} \tag{4}$$

In the principal directions, using the principal stresses σ_1 and $\sigma_2(\sigma_1 \geq \sigma_2)$, the expressions become

$$p = \frac{1}{2}(\sigma_1 + \sigma_2), \qquad q = \frac{1}{2}(\sigma_1 - \sigma_2) \tag{5}$$

and

$$\underset{\sim}{I} = \begin{pmatrix} 1 & \\ & 1 \end{pmatrix}, \qquad \underset{\sim}{J} = \begin{pmatrix} 1 & \\ & -1 \end{pmatrix} \tag{6}$$

where the following orthogonality conditions are to be noticed:

$$\frac{1}{2}I_{ij}I_{ij} = \frac{1}{2}J_{ij}J_{ij} = 1 \tag{7}$$

$$I_{ij}J_{ij} = 0 \tag{8}$$

For the case of 3D, the similar expressions are obtained as is following (ref. 1):

$$\sigma_{Dij} = q\bar{J}_{ij} + \bar{q}J_{ij} \tag{9}$$

$$\left.\begin{aligned} p &= \frac{1}{3}\sigma_{ij}I_{ij} = \frac{1}{3}(\sigma_1 + \sigma_2 + \sigma_3) \\ q &= \frac{1}{3}\sigma_{ij}J_{ij} = \frac{1}{3}(\sigma_1 + \omega\sigma_2 + \omega^2\sigma_3) \end{aligned}\right\} \tag{10}$$

and

$$\underset{\sim}{I} = \begin{pmatrix} 1 & & \\ & 1 & \\ & & 1 \end{pmatrix}, \qquad \underset{\sim}{J} = \begin{pmatrix} 1 & & \\ & \omega & \\ & & \omega^2 \end{pmatrix} \tag{11}$$

where $\omega = \frac{1}{2}(-1+\sqrt{3}i)$, which satisfies $\omega^3 = 1$, and $-$ indicates the conjugate complex number or a tensor with the conjugate complex elements. q is termed the *(complex)deviatric stress* in 3D. The following relations are also to be noticed.

$$\frac{1}{3}I_{ij}I_{ij} = \frac{1}{3}J_{ij}\bar{J}_{ij} = 1 \tag{12}$$

and

$$I_{ij}\,J_{ij} = I_{ij}\,\bar{J}_{ij} = J_{ij}\,J_{ij} = \bar{J}_{ij}\,\bar{J}_{ij} = 0 \tag{13}$$

As q is a complex number, we can write it as

$$q = |q|\,e^{i\alpha} \tag{14}$$

where $|q|$ and α are the absolute value and the argument of q respectively. Letting S_2 and S_3 be two invariants of the deviatoric stress tensor ${}_D\sigma_{ij}$, we have

$$|q| = \sqrt{\frac{S_2}{3}} = \frac{1}{3\sqrt{2}}\sqrt{(\sigma_1-\sigma_2)^2+(\sigma_2-\sigma_3)^2+(\sigma_1-\sigma_3)^2} \tag{15}$$

and

$$\cos 3\alpha = \frac{3\sqrt{3}}{2}\;\frac{S_3}{S_2\sqrt{S_2}} \tag{16}$$

Next, we introduce an *anisotropic tensor* L_{ij} and the *accompanied tensor* M_{ij} instead of I_{ij} and J_{ij} in the case of 2D, which are expressed in the principal directions as

$$\underset{\sim}{L} = \begin{pmatrix} \lambda & \\ & \lambda^{-1} \end{pmatrix}, \qquad \underset{\sim}{M} = \begin{pmatrix} \lambda^{-1} & \\ & -\lambda \end{pmatrix} \tag{17}$$

where $0 < \lambda \le 1$ and λ is termed the *anisotropic parameter*.
We have the following orthogonality conditions:

$$\frac{1}{\ell^2}L_{ij}L_{ij} = \frac{1}{\ell^2}M_{ij}M_{ij} = 1 \tag{18}$$

$$L_{ij}\,M_{ij} = 0 \tag{19}$$

where

$$\ell = \sqrt{\lambda^2 + \lambda^{-2}} \tag{20}$$

It is easy to see that when $\lambda = 1$, eqns.(17), (18) and (19) are reduced to eqns. (6), (7) and (8) respectively. Using these relations, the following decomposition is introduced:

$$\sigma_{ij} = P'L_{ij} + q'M_{ij} \tag{21}$$

where

$$
\left.\begin{array}{ll}
p' = \frac{1}{\ell^2}\sigma_{ij} L_{ij} & : \text{the } \textit{anisotropic mean stress} \\
 & \text{ or the } \textit{weighted mean stress} \\
q' = \frac{1}{\ell^2}\sigma_{ij} M_{ij} & : \text{the } \textit{anisotropic deviatoric stress}
\end{array}\right\} \tag{22}
$$

The decomposition expressed in eqn.(21) is called the *anisotropic decomposition* of σ_{ij} with respect to the anisotropic parameter λ. The anisotropic decomposition in 3D is also introduced in a similar manner (ref. 1).

WORK INCREMENT TENSOR

For the stress tensor σ_{ij} and the strain increment tensor $d\varepsilon_{ij}$, the work increment (scalar) defined by

$$
dW = \sigma_{ij}\, d\varepsilon_{ij} \tag{23}
$$

is commonly used. For making more detailed analysis, here we introduce the *work increment tensor* dW_{ij} defined by

$$
dW_{ij} = \frac{1}{2}\left(\sigma_{ik}\, d\varepsilon_{kj} + \sigma_{jk}\, d\varepsilon_{ki}\right) \tag{24}
$$

In the case where σ_{ij} and $d\varepsilon_{ij}$ are coaxial, eqn.(24) is simplified to

$$
dW_{ij} = \sigma_{ik}\, d\varepsilon_{kj} \tag{25}
$$

In 2D, using I_{ij} and J_{ij} defined in eqn.(6), the isotropic decomposition of dW_{ij} becomes

$$
dW_{ij} = \frac{1}{2}\left(dW\, I_{ij} + dU\, J_{ij}\right) \tag{26}
$$

where

$$
\left.\begin{array}{l}
dW = dW_{ij} I_{ij} = dW_1 + dW_2 \\
dU = dW_{ij} J_{ij} = dW_1 - dW_2
\end{array}\right\} \tag{27}
$$

dW_1 and dW_2 are two *principal work increments* and dU is called the *deviatoric work increment*.

In the case of 3D, the isotropic decomposition of dW_{ij} becomes

$$dW_{ij} = \frac{1}{3}(dW\,I_{ij} + dU\,\bar{J}_{ij} + \overline{dU}\,J_{ij}) \tag{28}$$

where

$$\left.\begin{aligned} dW &= dW_{ij}\,I_{ij} = dW_1 + dW_2 + dW_3 \\ dU &= dW_{ij}\,J_{ij} = dW_1 + \omega\,dW_2 + \omega^2\,dW_3 \end{aligned}\right\} \tag{29}$$

and dW_i are the three principal work increments.

In both cases of 2D and 3D, in which σ_{ij} and $d\varepsilon_{ij}$ are coaxial, we have

$$dW_i = \sigma_i\,d\varepsilon_i \quad \text{(not summed)} \tag{30}$$

where σ_i and $d\varepsilon_i$ are the principal stresses and principal strain increments respectively.

On the other hand, assuming the coaxiality, we have the isotropic decompositions of σ_{ij} and $d\varepsilon_{ij}$ in 2D expressed as

$$\sigma_{ij} = p\,I_{ij} + q\,J_{ij} \tag{31}$$

$$d\varepsilon_{ij} = \frac{1}{2}(d\varepsilon\,I_{ij} + d\gamma\,J_{ij}) \tag{32}$$

where $d\varepsilon$ and $d\gamma$ are the volumetric and shearing strain increments respectively, which are defined as

$$\left.\begin{aligned} d\varepsilon &= d\varepsilon_{ij}\,I_{ij} = d\varepsilon_1 + d\varepsilon_2 \\ d\gamma &= d\varepsilon_{ij}\,J_{ij} = d\varepsilon_1 - d\varepsilon_2 \end{aligned}\right\} \tag{33}$$

where $d\varepsilon_1$ and $d\varepsilon_2$ are the major and minor strain increments respectively. Substituting eqns.(31) and (32) into eqn.(25) and comparing it with eqn.(26), we have

$$\left.\begin{aligned} dW &= p\,d\varepsilon + q\,d\gamma \\ dU &= p\,d\gamma + q\,d\varepsilon \end{aligned}\right\} \tag{34}$$

In the similar manner, as we have the isotropic decompositions of σ_{ij} and $d\varepsilon_{ij}$ in 3D expressed as

$$\sigma_{ij} = pI_{ij} + q\bar{J}_{ij} + \bar{q}J_{ij} \tag{35}$$

and

$$d\varepsilon_{ij} = \frac{1}{3}(d\varepsilon I_{ij} + d\gamma \bar{J}_{ij} + \overline{d\gamma} J_{ij}) \tag{36}$$

we obtain

$$\left.\begin{aligned} dW &= p\,d\varepsilon + q\,\overline{d\gamma} + \bar{q}\,d\gamma \\ dU &= p\,d\gamma + q\,d\varepsilon + \bar{q}\,\overline{d\gamma} \end{aligned}\right\} \tag{37}$$

It is of interest to see that, as is seen in eqns.(34) and (37), the work of dW is done by the usual combinations of (p, $d\varepsilon$) and (q, $d\gamma$) and on the other hand that of dU is mainly done by the unusual combinations of (p, $d\gamma$) and (q, $d\varepsilon$). eqn.(37) is further transformed as

$$\begin{aligned} dW &= p\,d\varepsilon + 2\mathrm{Re}(q\,\overline{d\gamma}) \\ &= p\,d\varepsilon + 2\,|q||d\gamma|\cos(\alpha - \beta) \\ &= p\,d\varepsilon + 2\,|q||d\gamma|\frac{3+\mu\nu}{\sqrt{(3+\mu^2)(3+\nu^2)}} \end{aligned} \tag{38}$$

where α and β are the arguments of q and $d\gamma$, and μ and ν are the Lode's parameters of σ_{ij} and $d\varepsilon_{ij}$ respectively.

ROWE'S STRESS-DILATANCY EQUATION IN 3D

Rowe's stress-dilatancy equation in 2D (ref. 2) is written as

$$\frac{\sigma_1 d\varepsilon_1}{\sigma_2 d\varepsilon_2} = \frac{dW_1}{dW_2} = -\tan^2\left(\frac{\pi}{4} + \frac{\phi_0}{2}\right) \tag{39}$$

where ϕ_0 is the value of $\phi = \sin^{-1}\frac{q}{p}$ at the minimum volume. It is noted here that eqn.(39) is not an expression in tensorial invariants and is considered to be incovenient for the extension to the case of 3D.

However, using eqn.(27), eqn.(39) is easily rewritten as

$$dW = dU \sin\phi_0 \tag{40}$$

or using an anisotropic tensor L_{ij}, which is taken to be coaxial to dW_{ij}, with the anisotropic parameter given by

$$\lambda = \tan\left(\frac{\pi}{4} - \frac{\phi_0}{2}\right) = \sqrt{\frac{1-\sin\phi_0}{1+\sin\phi_0}} \tag{41}$$

we can write as

$$dW' = L_{ij}\, dW_{ij} = \lambda\, dW_1 + \lambda^{-1}\, dW_2 = 0 \tag{42}$$

It is recognized that eqns.(40) and (42) are more appropriate expressions than eqn.(39), because they seem more convenient to consider the extension of Rowe's stress-dilatancy equation to the case of 3D. dW' is termed the *anisotropic (or weighted) work increment.*

The similar expression to eqn.(42) in 3D is given as

$$dW' = L_{ij}\, dW_{ij} = \sum_{i=1}^{3} \lambda_i\, dW_i = \sum_{i=1}^{3} \lambda_i\, \sigma_i\, d\varepsilon_i = 0 \tag{43}$$

where $\lambda_i (i = 1,2,3)$ are three anisotropic parameters in 3D, which satisfy $\lambda_1\lambda_2\lambda_3 = 1$ and $0<\lambda_1\leq\lambda_2\leq\lambda_3$.

Using two parameters expressed as

$$\left.\begin{aligned} \lambda &= \frac{1}{3}(\lambda_1 + \lambda_2 + \lambda_3) \\ \eta &= \frac{1}{3}(\lambda_1 + \omega\lambda_2 + \omega^2\lambda_3) \end{aligned}\right\} \tag{44}$$

and letting the Lode's parameters of L_{ij} and dW_{ij} be ρ and ξ respectively, we can derive the following expression of dW' similarly as eqn.(38).

$$dW' = \lambda\, dW - 2|\eta|\, |dU| \frac{3+\rho\xi}{\sqrt{(3+\rho^2)(3+\xi^2)}} = 0 \tag{45}$$

or

$$dW = 2\frac{3+\rho\xi}{\sqrt{(3+\rho^2)(3+\xi^2)}} \frac{|\eta|}{\lambda} |dU| \tag{46}$$

It is easily seen that eqn.(46) is extension of eqn.(40) to the case of 3D. To determine the parameters λ_i, we can give the following consideration.

In granular materials, the work increment dW is decomposed as

$$dW = dW^E + dW^D \tag{47}$$

where dW^E denotes the increment of elastic energy which is stored in the material and dW^D denotes the dissipative energy increment. As dW^E is considered to be negligible, we have the following equality:

$$dW^E = dW - dW^D \fallingdotseq 0 \tag{48}$$

Assuming that eqn.(48) is written in the form of eqn.(43), we may be able to consider that the coefficients λ_i in eqn.(43) are constant for the given monotonous stress path* from the principle of minimum energy ratio introduced by Rowe (ref. 2). On the other hand, we can assume that

$$\sum_{i=1}^{3} d\varepsilon_i = 0 \qquad (49)$$

in the case where the volumetric strein $\varepsilon=\sum\varepsilon_i$ takes the minimum value in the stress path. From the above consideration and using eqn.(43), we obtain

$$\lambda_i = \frac{\sqrt[3]{\underset{0}{\sigma_1}\,\underset{0}{\sigma_2}\,\underset{0}{\sigma_3}}}{\underset{0}{\sigma_i}} \qquad (i = 1, 2, 3) \qquad (50)$$

where $\underset{0}{\sigma_i}$ denotes the principal stress of $\underset{0}{\sigma_{ij}}$ in which the volumetric strain ε takes the minimum value.

Using the relation

$$\frac{|\eta|}{\lambda} = \sqrt{1 - 3\,\frac{\underset{0}{I_1}\,\underset{0}{I_3}}{\underset{0}{I_2^2}}} \qquad (51)$$

where $\underset{0}{I_1}$, $\underset{0}{I_2}$, and $\underset{0}{I_3}$ are the three invariants of $\underset{0}{\sigma_{ij}}$, we can assume from eqns. (46) and (48) that

$$dW^D = 2\,\frac{3+\rho\xi}{\sqrt{(3+\rho^2)(3+\xi^2)}}\,\sqrt{1 - 3\,\frac{\underset{0}{I_1}\,\underset{0}{I_3}}{\underset{0}{I_2^2}}}\,|dU| \qquad (52)$$

where ρ is identical with the Lode's parameter of $\underset{0}{\sigma^{-1}_{ij}}$ and is written by using $\underset{0}{\mu}$, the Lode's parameter of $\underset{0}{\sigma_{ij}}$, as

$$\rho = \frac{(1+\underset{0}{\mu})\underset{0}{\sigma_1}-(1-\underset{0}{\mu})\underset{0}{\sigma_3}}{(1+\underset{0}{\mu})\underset{0}{\sigma_1}+(1-\underset{0}{\mu})\underset{0}{\sigma_3}} \qquad (53)$$

---

* The monotonous stress path is considered as the case where a linear relation is established between p and $|q|$ (or q in 2D) expressed as

$$p = a + b\,|q| \qquad \text{(a, b : constants)}$$

and the argument α and the Lode's parameter μ are kept constant.

ASSOCIATED FLOW RULE FOR GRANULAR MATERIALS

A simple form of the associated flow rule is given by

$$d\varepsilon_{ij} = \frac{\partial \Phi}{\partial \sigma_{ij}} d\Lambda \tag{54}$$

where Φ is the yield function and Λ is a scalar parameter. Eqn.(54) is originated from Drucker's postulate. It has been known, however, that the simple associated flow rule gives an unrealistic dilatancy in granular materials. Replying this problem, Kanatani (ref. 3) pointed out that if there is any constraint condition for the strain increment, the differentiation of $\frac{\partial}{\partial \sigma_{ij}}$ must be done in keeping the constraining stress constant. The constraining stress is defined as a stress which does a work against the strain which vanishes by the given constraint condition. Thus eqn.(54) is modified as

$$d\varepsilon_{ij} = \left(\frac{\partial}{\partial \sigma_{ij}}\right)^{*} \Phi \, d\Lambda \tag{55}$$

where $\left(\frac{\partial}{\partial \sigma_{ij}}\right)^{*}$ indicates the differentiation in keeping the constraining stress constant.

For the case of 2D, it is easy to have the anisotropy parameter λ^{ε} for the strain increment tensor $d\varepsilon_{ij}$ from the stress-dilatancy equation (39), such that

$$\lambda^{\varepsilon} = \sqrt{-\frac{d\varepsilon_2}{d\varepsilon_1}} = \lambda \sqrt{\frac{\sigma_1}{\sigma_2}} \tag{56}$$

where λ is given in eqn.(41). This anisotropy of the strain increment is considered as a constraint condition. In this case, the anisotropic decomposition of the stress tensor becomes the form expressed in eqn.(21) and the constraining stress becomes $p' L^{\varepsilon}_{ij}$, where

$$p' = \frac{1}{\ell^2} \sigma_{ij} L^{\varepsilon}_{ij} = \sqrt{\sigma_1 \sigma_2} \; \frac{\lambda \sigma_1{}^2 + \lambda^{-1} \sigma_2{}^2}{\lambda^2 \sigma_1^2 + \lambda^{-2} \sigma_2{}^2} \tag{57}$$

$$\underset{\sim}{L}^{\varepsilon} = \begin{pmatrix} \lambda \sqrt{\frac{\sigma_1}{\sigma_2}} & \\ & \lambda^{-1} \sqrt{\frac{\sigma_2}{\sigma_1}} \end{pmatrix} \tag{58}$$

Quite similarly, the constraint condition in 3D is given by eqn.(43) and the constraining stress in 3D becomes also $p' L^{\varepsilon}_{ij}$, where

$$p' = \frac{1}{\ell^2} \sigma_{ij} L^{\varepsilon}_{ij} = \sqrt[3]{\sigma_1 \sigma_2 \sigma_3} \frac{\lambda_1 \sigma_1^2 + \lambda_2 \sigma_2^2 + \lambda_3 \sigma_3^2}{\lambda_1^2 \sigma_1^2 + \lambda_2^2 \sigma_2^2 + \lambda_3^2 \sigma_3^2} \tag{59}$$

$$\underset{\sim}{L}^{\varepsilon} = \frac{1}{\sqrt[3]{\sigma_1 \sigma_2 \sigma_3}} \begin{pmatrix} \lambda_1 \sigma_1 & & \\ & \lambda_2 \sigma_2 & \\ & & \lambda_3 \sigma_3 \end{pmatrix} \tag{60}$$

CONCLUDING REMARKS

This paper first discusses the isotropic and anisotropic decompositions of tensor in the cases of 2D and 3D. It is seen that in 3D, the introduction of a complex number ω is useful and effective. Using the newly introduced work increment tensor, Rowe's stress-dilatancy equation is expressed in terms of tensorial invariants, which makes easy the extension of the stress-dilatancy equation to the case of 3D. It is recognized that in such an analysis, the introduction of work increment tensor is indispensable. Lastly, the modified associated flow rule is applied in considering the Rowe's stress-dilatancy equation as a constraint condition, where the anisotropic decompositions of the stress and strain increment tensors are used as the basis of the analysis.

REFERENCES

1 M. Satake, Expression of Three Dimensional Deviatoric Tensor through Complex Number ω, to appear in Theoretical and Applied Mechanics, 37, Univ. of Tokyo Press.

2 P. W. Rowe, The Stress-Dilatancy Relation for Static Equilibrium of an Assembly of Particles in Contact, Proc. Roy. Soc. A 269, 1962, pp. 500-527.

3 K. Kanatani, Mechanical Foundation of the Plastic Deformation of Granular Materials, in Deformation and Failure of Granular Materials, P. A. Vermeer and H. J. Luger(Eds.), A. A. Balkema, Rotterdam, 1982, pp. 119-127.

Micromechanics of Granular Materials, edited by M. Satake and J.T. Jenkins
Elsevier Science Publishers B.V., Amsterdam, 1988 — Printed in The Netherlands

A SIMPLE THEORY OF INSTANTANEOUSLY INDUCED ANISOTROPY

S. C. COWIN

Department of Biomedical Engineering, Tulane University, New Orleans, LA, 70118 (USA)

ABSTRACT

This paper presents stress-strain relations for an ideal material which, in the plastic domain, instantaneously adapts its material anisotropy to the strain state it is experiencing. The objective of this work is to obtain a simple model for predicting strain-induced anisotropy.

INTRODUCTION

There are many examples of the elastic symmetry of a material being changed by plastic deformation. The concept of stress-induced anisotropy in granular materials is well known (ref. 1,2). The rolling or extrusion processes used in making structural steel shapes induce anisotropy in the steel. Before extrusion or rolling the polycrystalline steel is likely to be isotropic, that is to say the distribution of angular orientation of the crystal grains in the steel is uniform. The plastic strains of extrusion or rolling give a preferred angular orientation to the crystal grains and induce transversely isotropic or orthotropic material symmetry.

The elastic symmetries of materials are divided here into two general classes, crystallographic and textural. <u>Crystallographic</u> symmetries are characteristic of single crystals and are generally determined by the chemical constituents of the materials. The force systems that form the crystallographic symmetries are internal, molecular-level force systems. The crystallographic symmetries are associated with pure crystals and are not of interest in the present work. <u>Textural symmetries</u> are the symmetries of materials determined by their mesostructural organization. These symmetries are induced by the method of fabrication of the material or by the stress history of the material. The force systems that form the textural symmetries are external force systems applied to the material by the agent of fabrication or by the stress history. The textural symmetries include orthotropy, transverse isotropy and isotropy. The examples of textural symmetries of interest here are the stress-induced symmetries of granular materials and structural steels mentioned above.

Quantitative stereological methods may be used to measure the textural anisotropy of materials (ref. 3,4). The stress-induced anisotropy in a granular material was quantified by Oda (ref. 5). Oda introduced the concept of a fabric ellipsoid, an ellipsoid determined by the three-dimensional distribution of normals to the tangential contact planes between particles in a tested specimen. The deformation of this ellipsoid with different stress histories has been demonstrated (ref. 1,2,5). Similar fabric ellipsoids have been defined or measured for other materials (ref. 6). It has been shown (ref. 7) that the fabric ellipsoid is associated with orthotropic symmetry if all three axes of the ellipsoid are unequal. If two of the axes are equal, then the elastic symmetry of the material is transversely isotropic. If the fabric ellipsoid degenerates to a sphere, the elastic symmetry of the material is isotropic. The fabric ellipsoid of a material can be represented by a positive definite symmetric tensor **A** called the fabric tensor (ref. 8).

In the following two sections the dependence of the linear anisotropic form of Hooke's law upon the fabric tensor and the dependence of a tensor polynomial yield condition upon the fabric tensor, respectively, are summarized. The constitutive relations for strain-induced instantaneous anisotropy are then described and developed in the subsequent sections.

FABRIC DEPENDENT ELASTIC STRESS-STRAIN RELATIONS

The linear anisotropic form of Hooke's law is written in the form of a six-by-six matrix of elastic coefficients **C**[ ] multiplying a column vector **E** of strain components E_{11}, E_{22}, E_{33}, $2E_{23}$, $2E_{13}$ and $2E_{12}$ to obtain a column vector **T** of stress components T_{11}, T_{22}, T_{33}, T_{23}, T_{13} and T_{12}, thus

$$\mathbf{T} = \mathbf{C}[\mathbf{E}]\,. \tag{1}$$

The components of the six-by-six matrix are denoted by c_{ij}, i,j = 1,2,...,6 and since the matrix is symmetric, there are, at most, only 21 distinct elastic constants c_{ij}. If it is assumed that all the material anisotropy is due to the fabric, then Hooke's law (1) can be rewritten with an explicit dependence on the fabric tensor in the form (ref. 7)

$$\mathbf{T} = \mathbf{C}[\mathbf{E}] = a_1\mathbf{1}\,\mathrm{tr}\mathbf{E} + a_2((\mathrm{tr}\mathbf{E})\mathbf{A} + \mathrm{tr}(\mathbf{AE})\mathbf{1}) + a_3((\mathrm{tr}\mathbf{A}^2\mathbf{E})\mathbf{1} + (\mathrm{tr}\mathbf{E})\mathbf{A}^2) + b_1(\mathrm{tr}\mathbf{AE})\mathbf{A} +$$
$$+ b_2((\mathrm{tr}\mathbf{A}^2\mathbf{E})\mathbf{A} + (\mathrm{tr}\mathbf{AE})\mathbf{A}^2) + b_3(\mathrm{tr}\mathbf{A}^2\mathbf{E})\mathbf{A}^2 + 2c_1\mathbf{E} + 2c_2(\mathbf{AE} + \mathbf{EA}) + 2c_3(\mathbf{A}^2\mathbf{E} + \mathbf{EA}^2), \tag{2}$$

where a_1, a_2, a_3, b_1, b_2, b_3, c_1, c_2, c_3 are functions of $\mathrm{tr}\mathbf{A}$, $\mathrm{tr}\mathbf{A}^2$, $\mathrm{tr}\mathbf{A}^3$ and the volume fraction or phase concentration denoted by ν.

FABRIC DEPENDENT TENSOR POLYNOMIAL YIELD CRITERION

A yield criterion may be formulated in terms of the fabric tensor using the tensor polynomial yield strength formalism developed by Malmeister (ref. 9) and Tsai and Wu (ref. 10). For materials of interest here the phenomenological strength criterion is assumed to depend upon the fabric tensor $\mathbf{A}$, the volume fraction ν and the strain tensor $\mathbf{E}$. The strain state $\mathbf{E}$ is then determined by the value of $\mathbf{E}$ for which the functional relationship

$$f(\mathbf{A},\mathbf{E},\nu) = 1 , \tag{3}$$

is satisfied. Values of $f(\mathbf{A},\mathbf{E},\nu)$ less than one correspond to unyielded strain states, and values greater than or equal to one correspond to yielded states. Assuming that all the anisotropy of the material is determined by the fabric, then expanding the function (3) in a polynomial that retains only second order terms in strain, the following representation (ref. 6) is obtained for the function $f(\mathbf{A},\mathbf{E},\nu)$,

$$\begin{aligned} f(\mathbf{A},\mathbf{E},\nu) = {} & g_1\mathrm{tr}\mathbf{E} + g_2\mathrm{tr}(\mathbf{A}\mathbf{E}) + g_3\mathrm{tr}(\mathbf{A}^2\mathbf{E}) + f_1\mathrm{tr}\mathbf{E}^2 + f_2(\mathrm{tr}\mathbf{E})(\mathrm{tr}\mathbf{A}\mathbf{E}) + f_3(\mathrm{tr}\mathbf{E})\mathrm{tr}(\mathbf{A}^2\mathbf{E}) \\ & + f_4\,(\mathrm{tr}\mathbf{A}\mathbf{E})^2 + f_5\mathrm{tr}(\mathbf{A}\mathbf{E})\mathrm{tr}(\mathbf{A}^2\mathbf{E}) + f_6(\mathrm{tr}\mathbf{A}^2\mathbf{E})^2 + f_7\mathrm{tr}\mathbf{E}^2 + f_8\mathrm{tr}(\mathbf{A}\mathbf{E}^2) + f_9\mathrm{tr}(\mathbf{A}2\mathbf{E}^2) . \end{aligned} \tag{4}$$

where g_1, g_2, g_3 and f_1 through f_9 are functions of $\mathrm{tr}\mathbf{A}$, $\mathrm{tr}\mathbf{A}^2$, $\mathrm{tr}\mathbf{A}^3$ and ν. This yield condition could also be formulated in terms of the stress tensor $\mathbf{T}$ rather than the strain tensor $\mathbf{E}$ (ref. 6).

A THEORY OF INSTANTANEOUSLY INDUCED ANISOTROPY

In this section a theory of plastic strain-induced instantaneous anisotropy is introduced. The constitutive behavior for three regions of stress-strain behavior will be considered. These regions are (A) the elastic domain before yielding, (B) the plastic domain and (C) the elastic domain experienced after unloading from the plastic domain. In the region (A) the constitutive relation is the anisotropic form of Hooke's law (2) with the fabric given by a constant tensor $\mathbf{A}^o$. The strain states of this elastic region are those that satisfy the relation $f(\mathbf{A}^o, \mathbf{E},\nu) < 1$. In the plastic region, region (B), it is assumed that the fabric tensor instantaneously adapts to the strains it is experiencing. To express this idea in a mathematically simple fashion it is assumed that the principal axes of $\mathbf{A}$ and $\mathbf{E}$ are coincident ($\mathbf{A}\mathbf{E} = \mathbf{E}\mathbf{A}$) and that the magnitude of each eigenvalue of **A** is equal to the absolute value of the corresponding eigenvalue of **E** divided by the sum of the three eigenvalues of **E**. The tensor **S** is defined so that each diagonal element has the sign of the corresponding strain component and an absolute magnitude of one. The tensor S is then a symmetric orthogonal tensor whose square is the unit tensor:

$$\mathbf{S}\mathbf{S}^T = \mathbf{S}^T\mathbf{S} = \mathbf{1}, \ \mathbf{S} = \mathbf{S}^T, \ \mathbf{S}^2 = \mathbf{1}. \tag{5}$$

If all the eigenvalues of **E** are non-negative, $\mathbf{S} = \mathbf{1}$ and if all the eigenvalues of **E** are non-positive, $\mathbf{S} = -\mathbf{1}$, and if they are of mixed sign, the signs of **S** are constructed so that the components of $\mathbf{SE} = \mathbf{ES}$ are non-negative. The instantaneous fabric tensor **A** in the plastic domain (B) is then defined by

$$\mathbf{A} \equiv (\mathbf{SE}/\mathrm{tr}\mathbf{SE}), \ \mathrm{tr}\mathbf{A} = 1, \ \text{for } f(\mathbf{A},\mathbf{E},\nu) \geq 1. \tag{6}$$

The stress-strain relation in the plastic domain (B) is given by (2) with **A** given by (6), thus

$$\begin{aligned}\mathbf{T} = \mathbf{C}[\mathbf{E}] = {} & (1/2)\mathrm{tr}\mathbf{E}((2a_1+a_3(I_2-1))\mathbf{1} + (2a_2+b_2(I_2-1))\mathbf{A} + (2a_3+b_3(I_2-1))\mathbf{A}^2 + \\ & + (1/6)(\mathrm{tr}\mathbf{SE})(\mathrm{tr}\mathbf{S})(2I_3-3I_2+1)(a_3\mathbf{1} + b_2\mathbf{A}+b_3\mathbf{A}^2)+2c_3\mathbf{S}(\mathrm{tr}\mathbf{SE})(2I_3-2I_2+1)+ \\ & +(\mathrm{tr}\mathbf{AE})(a_2+a_3)\mathbf{1} + (b_1+b_2)\mathbf{A} + (b_2+b_3)\mathbf{A}^2 + (2c_1 + c_3(I_2-1))\mathbf{E} + 2(c_2+c_3)(\mathbf{AE} + \mathbf{EA}),\end{aligned} \tag{7}$$

where a_1, a_2, a_3, b_1, b_2, b_3 and c_1, c_2, c_3 are functions of ν, I_2 and I_3 where

$$I_2 = \mathrm{tr}\mathbf{A}^2, \ I_3 = \mathrm{tr}\mathbf{A}^3, \tag{8}$$

and where the Cayley Hamilton theorem has been employed.

The stress-strain behavior in region (C), the elastic domain experienced after unloading from the plastic domain, is again a linear anisotropic elastic behavior, but with a new fabric tensor. As unloading from the plastic domain occurs, let $\mathbf{A}^*$ represent the fabric tensor for which $f(\mathbf{A}^*,\mathbf{E},\nu)=1$. This fabric tensor is then taken as the fabric tensor of the post-plastic elastic domain and the applicable stress-strain relation is given by (2) with **A** set equal to $\mathbf{A}^*$. The material then has a constant fabric represented by $\mathbf{A}^*$ until it is again yielded.

To summarize these constitutive assumptions it is noted that, in the theory described, the material in the elastic domain, $f(\mathbf{A}^o,\mathbf{E},\nu)<1$, behaves as an anisotropic elastic material with a stress-strain law given by (2) with $\mathbf{A} = \mathbf{A}^o$ until the condition

$f(\mathbf{A}^o,\mathbf{E},\nu) \geq 1$ is satisfied. At that instant the constitutive relation (7) will then apply as long as the condition $f(\mathbf{A},\mathbf{E},\nu)\geq 1$, with $\mathbf{A}$ given by (6) is satisfied. In the post-plastic elastic domain the constitutive equation (2) with $\mathbf{A} = \mathbf{A}^*$ is applicable. The constant fabric tensor $\mathbf{A}^*$ is determined from the condition $f(\mathbf{A}^*,\mathbf{E},\nu)=1$ in the unloading process.

The theory described above represents a considerable simplification of the physical processes involved. There is probably no real material in which the elastic symmetry instantaneously adapts to the strain state. In real materials physical experience suggests that the elastic symmetry adjusts to the strain state only with continued plastic straining. The constitutive model assumes that this adaptation occurs instantaneously in order to obtain a simpler model.

THE STRESS-STRAIN RELATION FOR INSTANTANEOUSLY INDUCED ANISOTROPY IN THE CASE OF NON-NEGATIVE OR NON-POSITIVE STRAINS

The constitutive relation (7) for the plastic domain simplifies considerably if it is assumed that all the normal strains are either non-negative ($\mathbf{S} = \mathbf{1}$) or non-positive ($\mathbf{S} = -\mathbf{1}$). In this case the constitutive relation (7) simplifies to

$$\mathbf{T} = \lambda(\nu,I_2,I_3)\theta\mathbf{1} + 2\mu(\nu,I_2,I_3)\mathbf{E} + 2\beta(\nu,I_2,I_3)(1/\theta)\mathbf{E}^2 , \tag{9}$$

where

$$\theta = \mathrm{tr}\mathbf{E}, \quad I_2 = (\mathrm{tr}\mathbf{E}^2/\theta) , \quad I_3 = (\mathrm{tr}\mathbf{E}^3/\theta). \tag{10}$$

Because the assumptions of all non-negative or non-positive normal strains are equivalent, we arbitrarily take the normal strains to be all non-negative, thus $E_{11}\geq 0$, $E_{22}\geq 0$ and $E_{33}\geq 0$. The constitutive relation (9) is, in some ways, similar to the constitutive equation for non-linear isotropic elastic materials. Thus some solutions can be obtained from those known in non-linear elasticity. There are many important aspects of the constitutive relation (9) that are different from the constitutive equation for non-linear isotropic elastic materials. These differences include the fact that (9) applies only: (a) for small strains, (b) for situations in which all the normal strains are non-negative or non-positive, and (c) has a special and explicit dependence upon $\theta=\mathrm{tr}\mathbf{E}$.

If the constitutive relation (9) were viewed as a linear anisotropic constitutive relation of the form (1) with variable elastic coefficients c_{ij}, $i,j=1,2,...6$, depending on the principal strains E_1,E_2,E_3, the variable elastic coefficients are given by (ref. 7)

$$c_{11} = \lambda+2\mu - 2\omega(1+I_2) + 2(\beta+2\omega)\ (E_1/\theta)\ ,$$

$$c_{22} = \lambda+2\mu - 2\omega(1+I_2) + 2(\beta+2\omega)\ (E_2/\theta)\ ,$$

$$c_{33} = \lambda+2\mu - 2\omega(1+I_2) + 2(\beta+2\omega)\ (E_3/\theta)\ , \tag{11}$$

$$c_{44} = \mu - \omega + (\beta/2\theta)\ (E_2+E_3)\ , \quad c_{23} = \lambda - 2\omega I_2 + (2\omega/\theta)\ (E_2+E_3)\ ,$$

$$c_{55} = \mu - \omega + (\beta/2\theta)\ (E_1+E_3)\ , \quad c_{13} = \lambda - 2\omega I_2 + (2\omega/\theta)\ (E_1+E_3)\ ,$$

$$c_{66} = \mu - \omega + (\beta/2\theta)\ (E_1+E_2)\ , \quad c_{12} = \lambda - 2\omega I_2 + (2\omega/\theta)\ (E_1+E_2)\ ,$$

where all other c_{ij} are zero, and where ω is a function of ν, I_2 and I_3 as are λ, μ and β. It is interesting to note that the material function $\omega(\nu,I_2,I_3)$ does not appear in the constitutive relation (9), but that it does appear in the relations (11) relating the elastic constants c_{ij} to the principal strains. In order that the strain energy of a material characterized by c_{ij} be positive definite it is necessary and sufficient that all the minor determinants formed from c_{ij}, and which contain along their diagonals the diagonal elements of c_{ij}, be positive. If the post-plastic state of the material is elastic, then these relations must be satisfied by the strain state that determines the post-plastic fabric $\mathbf{A}^*$.

The constitutive relation (9) can be written in a more convenient form if I_2 and I_3 are expressed in terms of the deviator strain tensor $\mathbf{E}'$,

$$\mathbf{E}' = \mathbf{E} - (1/3)\ (\mathrm{tr}\mathbf{E})\mathbf{1}. \tag{12}$$

It is easy to show that

$$I_2 = (1/3) + (\Omega/\theta^2)\ , \quad I_3 = (3\Omega/2\theta^2) + (\Delta/\theta^3)\ , \tag{13}$$

where

$$\Delta = 3\det\mathbf{E} = 3E_1E_2E_3$$

$$\Omega = \mathrm{tr}(\mathbf{E}')^2 = (1/3)\ [(E_1-E_2)^2 + (E_1-E_3)^2 + (E_2-E_3)^2]. \tag{14}$$

With the new notations Ω and Δ, the stress-strain relation (9) is rewritten as

$$\mathbf{T} = \lambda(\nu, (\Omega/\theta^2), (\Delta/\theta^3))\ \theta\,\mathbf{1} + 2\mu\,(\nu, (\Omega/\theta^2), (\Delta/\theta^2))\mathbf{E} + 2\beta\,(\nu, (\Omega/\theta^2), (\Delta/\theta^3))\mathbf{E}^2\ . \tag{15}$$

The functions λ, μ and β must satisfy conditions which ensure that the work done by the stresses acting through the strains is positive. The condition that $\mathrm{tr}\mathbf{T}\,\mathbf{E}$ be positive is

$$\mathrm{tr}\mathbf{T}\mathbf{E} = (\lambda + (2\mu/3))\theta^2 + (2\mu + 3\beta)\,\Omega + (2\beta\Delta/\theta)\,\Delta \geq 0. \tag{16}$$

Since θ, Ω and Δ must all be positive, sufficient conditions for $\mathrm{tr}\mathbf{T}\mathbf{E}$ to be positive are that

$$3\lambda + 2\mu \geq 0 \quad , \quad 2\mu + 3\beta \geq 0 \quad , \quad \beta \geq 0 \tag{17}$$

where λ, μ and β depend upon ν, θ, Ω and Δ.

FABRIC INDUCED BY A HYDROSTATIC STRESS STATE IN THE PLASTIC DOMAIN

In order to determine the fabric induced by a hydrostatic state of stress, the constitutive relations (9) must be solved for E_1, E_2 and E_3 with $\mathbf{T} = p\mathbf{1}$. Using the inequalities (17) and the fact that all the strains are nonnegative, it is easy to show that all the strains are equal. Thus $\Omega = 0$, $\Delta = (1/9)$, $\mathbf{E} = (\theta/3)\ \mathbf{1}$, and θ is given by

$$\theta = 9p/[9\lambda(\nu,0,(1/9)) + 6\mu(\nu,0,(1/9)) + 2\beta(\nu,0,(1/9))]. \tag{18}$$

This result shows that a hydrostatic stress state will induce a fabric tensor A of the form $(1/3)\mathbf{1}$; thus, from (2), a hydrostatic stress state will induce isotropic symmetry.

THE FABRIC INDUCED BY A STRESS STATE WITH THREE NON-POSITIVE OR NON-NEGATIVE PRINCIPAL STRESSES, TWO OF WHICH ARE EQUAL

The problem addressed in this section is summarized in the section title. It is assumed that the magnitudes of the stresses are such that a non-positive (non-negative) stress state induces only a non-positive (non-negative) strain state. The constitutive equation (15) written in the principal coordinate system and with $T_{22} = T_{33}$ takes the form

$$\theta T_{11} = \lambda(\nu, (\Omega/\theta^2), (\Delta/\theta^3))\, \theta^2 + 2\mu\, (\nu, (\Omega/\theta^2), (\Delta/\theta^3))\, \theta E_{11} + 2\beta(\nu, (\Omega/\theta^2), (\Delta/\theta^3))\, E_{11}^2,$$

$$\theta T_{22} = \lambda(\nu, (\Omega/\theta^2), (\Delta/\theta^3))\, \theta^2 + 2\mu\, (\nu,(\Omega/\theta^2), (\Delta/\theta^3))\, \theta E_{22} + 2\beta\, (\nu, (\Omega/\theta^2), (\Delta/\theta^3))\, E_{22}^2, \tag{19}$$

$$\theta T_{22} = \lambda(\nu, (\Omega/\theta^2), (\Delta/\theta^3))\, \theta^2 + 2\mu\, (\nu, (\Omega/\theta^2), (\Delta/\theta^3))\, \theta E_{33} + 2\beta\, (\nu, (\Omega/\theta^2), (\Delta/\theta^3))\, E_{33}^2 .$$

Subtracting the last of these equations from the middle one, it follows that

$$(E_{22}-E_{33})\,(2\mu(\nu,(\Omega/\theta^2),(\Delta/\theta^3)) + (E_{22}+E_{33})\,2\beta\,(\nu,(\Omega/\theta^2),(\Delta/\theta^3)) = 0, \tag{20}$$

from which it is concluded that

$$E_{22} = E_{33}, \tag{21}$$

since the expression in the second set of parentheses cannot be zero due to the inequalities (17) and the requirement that all the strains be non-negative or non-positive. Taking advantage of the result (21) the dimensionless parameters ξ and τ are introduced,

$$\xi = (\theta/E_{11}) = 1 + (2E_{22}/E_{11}), \quad \tau = (T_{22}/T_{11}). \tag{22}$$

The parameter τ is a constant fixed by the stress boundary conditions. The parameter ξ is a variable whose value is sought as part of the solution to the problem. The variable parameter ξ has the range from 1 (when $E_{22} = 0$) to 3 (when $E_{22}=E_{11}$). The functions (Ω/θ^2) and (Δ/θ^3) can be expressed in terms of ξ as

$$(\Omega/\theta^2) = (1/6\xi^2)(3-\xi)^2, \quad (\Delta/\theta^3) = (3/4\xi^3)(\xi-1)^2, \tag{23}$$

and henceforth the shorthand notation $\lambda(\nu,\xi)$, $\mu(\nu,\xi)$ and $\beta(\nu,\xi)$ will be employed for the three material functions appearing in (19). Employing these notations, equations (19) are rewritten as

$$\begin{aligned}(\xi T_{11}/E_{11}) &= \lambda(\nu,\xi)\xi^2 + 2\mu(\nu,\xi)\xi + \beta(\nu,\xi),\\ (\tau\xi T_{11}/E_{11}) &= \lambda(\nu,\xi)\xi^2 + 2\mu(\nu,\xi)\,\xi(\xi/2)(\xi-1) + 2\beta(\nu,\xi)(1/4)(\xi-1)^2.\end{aligned} \tag{24}$$

Multiplying the first of these equations by τ and then subtracting it from the second, one obtains the condition

$$F(\xi)=0, \tag{25}$$

where

$$F(\xi) = (1-\tau)\,\lambda(\nu,\xi)\,\xi^2 + 2\mu(\nu,\xi)((1/2)(\xi-1)-\tau)\xi + 2\beta(\nu,\xi)((1/4)(\xi-1)^2-\tau). \tag{26}$$

The values of $F(\xi)$ at $\xi = 1$ and $\xi = 3$ are given by

$$F(1) = \lambda(\nu,1) - \tau\,(\lambda(\nu,1) + 2\mu\,(\nu,1) + 2\beta(\nu,1)),$$

and (27)

$$F(3) = (1-\tau)\,(9\lambda(\nu,3) + 6\mu(\nu,3) + 2\beta(\nu,3)),$$

respectively. Using the inequalities (17) it is easy to see that F(1) is negative and F(3) is positive under only a minor restriction. Specifically, if $\lambda(\nu,1) \leq 0$ and $1 \geq \tau \geq 0$ or if $\lambda(\nu,1) > 0$ and

$$1 \geq \tau \geq \lambda(\nu,1)/\,(\lambda(\nu,1)+2\mu(\nu,1)+2\beta(\nu,1)) \tag{28}$$

then $F(1) < 0$ and $F(3) > 0$. It follows that the condition (25) is satisfied at least once in the domain $3 \geq \xi \geq 1$ and that it could be satisfied an odd number of times greater than one depending on the functions $\lambda(\nu,\xi)$, $\mu(\nu,\xi)$ and $\beta(\nu,\xi)$ appearing in (26). Assume that the material functions $\lambda(\nu,\xi)$, $\mu(\nu,\xi)$ and $\beta(\nu,\xi)$ are known and that all the solutions to (25) have been determined. Let ξ_i denote the ith solution. The strains corresponding to this solution are given by

$$(E_{11})_i = \xi_i\, T_{11}/(\lambda(\nu,\xi_i)\xi_i^2 + 2\mu(\nu,\xi_i)\xi_i + 2\beta(\nu,\xi_i))$$

$$(E_{22})_i = (1/2)(\xi_i-1)E_{11}\,, \tag{29}$$

and the fabric tensor $\mathbf{A}_{(i)}$ by

$$\mathbf{A}_{(i)} = \begin{bmatrix} (1/\xi_i) & 0 & 0 \\ 0 & (1/2)(1-(1/\xi_i)) & 0 \\ 0 & 0 & (1/2)(1-(1/\xi_i)) \end{bmatrix} \tag{30}$$

These results show that at least one transversely isotropic fabric will be induced by a stress state with three non-positive or non-negative principal stresses, two of which are equal. These results also show the possibility of multiple solutions, an odd number of multiple solutions. The number of the multiple solutions depends upon the functional forms of $\lambda(\nu,\xi)$, $\mu(\nu,\xi)$ and $\beta(\nu,\xi)$.

DISCUSSION

The theory presented appears to be the first continuum theory for the determination of strain induced anisotropy. The features of the theory include instantaneously induced anisotropy in the plastic domain and coupled shear-volumetric deformation (dilatancy) in the elastic, plastic and post-plastic elastic domain. The roles for the coupled shear-volumetric deformation are different in each of the three domains and change abruptly at the elastic-plastic and post plastic-elastic transitions. The stress strain relations in the plastic domain are of three different types depending on the signs of principal strains. The three different types of stress strain relations correspond to the case when all three strains are of the same sign, the case when two are negative and one is positive and the case when two are positive and one is negative. It is shown here that a hydrostatic state of stress induces isotropy and that a stress state with all non-positive or non-negative stresses, two of which are equal, induces transversely isotropic symmetry. The possibility of multiple solutions was demonstrated in the latter case. It is conjectured that a stress state with all non-equal, non-positive or non-negative stresses will induce orthotropic material symmetry and the possibility of multiple solutions will occur again in this case.

ACKNOWLEDGMENT

This work was done when I was supported on a fellowship from the Japan Society for the Promotion of Science (JSPS) and a guest in the laboratory of Professor M. Satake at Tohoku University. I thank JSPS for their support and Professor Satake and his laboratory for their great hospitality.

REFERENCES

1. M. Oda, in S. C. Cowin and M. Satake (Eds.), Continuum Mechanical and Statistical Approaches in the Mechanics of Granular Materials, Gakujutsu Bunken Fukyu-Kai, Tokyo, 1978, pp. 7-26.
2. J. Konishi, in S. C. Cowin and M. Satake (Eds.), Continuum Mechanical and Statistical Approaches in the Mechanics of Granular Materials, Gakujutsu Bunken Fukyu-Kai, Tokyo, 1978, pp. 27-46.
3. J. E. Hilliard, in Stereology - Proc. 2nd Int. Congress for Stereology, Springer, Berlin, 1967, pp. 219-225.
4. H. J. Bunge, Texture Analysis in Materials Science, Butterworths, London, 1982.
5. M. Oda, Fabrics and their effects on the deformation behaviors of sand, Dept. of Foundation Engr., Saitama University, 1976.
6. S. C. Cowin, Mech. of Materials, 5 (1986) 251-260.
7. S. C. Cowin, Mech. of Materials, 4 (1985) 137-147.
8. S. C. Cowin, in S. C. Cowin and M. Satake (Eds.), Continuum Mechanical and Statistical Approaches in the Mechanics of Granular Materials, Gakujutsu Bunken Fukyu-Kai, Tokyo, 1978, pp. 162-170.
9. A. K. Malmeister, Polymer Mechanics, 1 (1968) 324-331.
10. S. W. Tsai and E. M. Wu, J. Composite Mat., 5(1971) 58-72.

Micromechanics of Granular Materials, edited by M. Satake and J.T. Jenkins
Elsevier Science Publishers B.V., Amsterdam, 1988 — Printed in The Netherlands

INTRODUCTION OF INHERENT ANISOTROPY OF SOILS IN THE YIELD FUNCTION

M. Oda and H. Nakayama
Department of Foundation Engineering, Saitama University, Saitama 338 (Japan)

SUMMARY

Fabric tensor F_{ij} for three-dimensional assemblies of granular soils is introduced as an index showing the anisotropy due to the preferred orientation of constituent particles. Using the fabric tensor, Drucker-Prager type of yield function is extended so as to take into account the anisotropic yielding behavior of granular soils. Plane strain tests on Toyoura sand are analyzed with a result that the anisotropic shear strength is well fitted by the extended Drucker-Prager's yield function. Based on this, it is concluded that this study provides a first step to link the material science approach of soils, in which the spatial arrangement of particles and associated voids plays an important role, to the continuum theory of plasticity.

INTRODUCTION

There seems to be two major trends in the recent development of soil science. One trend has been based on the material science in which soils are treated as assemblies of discrete particles. The spatial arrangement of particles and associated voids (called "fabric") plays an important role in the development (e.g., refs. 19, 2). Another trend has been motivated by the recent development of soil plasticity in which soils are treated like continuums (e.g., refs. 3-4). In spite of the fact that the knowledge concerning soil behavior is still accumulating, there seems to be something missing. This is partially because both trends have been developing in completely separate manner.

In this study, a tensor (called the fabric tensor) is introduced to express concretely the geometry made by discrete particles, and a continuum model is formulated in such that the tensor is explicitely used in the framework of soil plasticity. Our present interest is not to show a complete theory covering the whole range of soil behavior, but rather to exemplify a general method for combining the discrete approach with the continuum model.

INHERENT ANISOTROPY AND FABRIC TENSOR

Inherent anisotropy

Anisotropy is one of the fundamental aspects to be taken into account for better understanding of soil properties. In relation to the soil anisotropy, Casagrande and Carillo (ref. 5) first discriminated between the inherent anisotropy and the induced anisotropy. The former is produced through the sedimentation of particles, while the latter is induced afterward during the process of non-elastic deformation.

Soils consist of two phases; solid particles and associated voids. In order to discuss the anisotropy of soils in general, therefore, we may have to take into account the following three sources of anisotropy at least; (1) anisotropic distribution of contact normals which is indicative of the mutual relation among particles, (2) preferred orientation of non-spherical voids and (3) preferred orientation of non-spherical particles. In biaxial compression tests on two-dimensional assemblies of rods, Oda, Nemat-Nasser and Konish (ref. 6) have observed that the inherent anisotropy by (1) and (2) tends to be completely altered during the relatively early stage of non-elastic deformation. The inherent anisotropy by (3), on the other hand, still remains at the later stage of deformation. This is the reason why the strength is seriously affected by the direction of compression with respect to the axis of sedimentation (strength anisotropy).

Here, the inherent anisotropy by (3) is only discussed with restricting our interest to the relation between the inherent anisotropy and the anisotropic yield function. (The experimental evidence by Oda et al.(ref. 6) has also proved that the induced anisotropy is continuously produced by changing the distribution of contact normals in such that a greater concentration of contact normals is produced along the orientation parallel to the direction of the maximum principal axis. It can be believed that the induced anisotropy becomes very important when dealing with non-elastic behavior of soils subjected to cyclic stresses.)

Definition of fabric tensor

Particles even in granular soils are seldom spherical in shape. Accordingly, anisotropy is produced by the preferred orientation of non-spherical particles. A tensor quantity is introduced here to express the anisotropy in general. A typical direction for each particle is labelled by two unit vectors $\underset{\sim}{n}^{(+)}$ and $\underset{\sim}{n}^{(-)}$ (Fig.1). (For simplicity, $\underset{\sim}{n}$ is used to denote the both.) If a particle like an ellipsoid in shape is concerned, $\underset{\sim}{n}$ is taken as being parallel to the longest axis. (If a platy particle like a clay mineral is concerned, $\underset{\sim}{n}$ is taken perpendicular to the major plane.) Then, a tensor F_{ij} is introduced as

$$F_{ij} = \int_{\Omega} n_i n_j E(\underset{\sim}{n}) d\Omega \qquad (i,j=1,2,3) \tag{1}$$

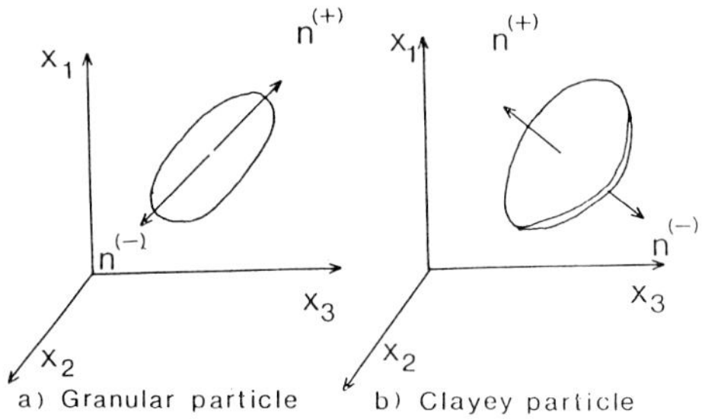

Fig.1 Unit vectors $\underset{\sim}{n}$ to show the typical orientations in granular and clayey particles.

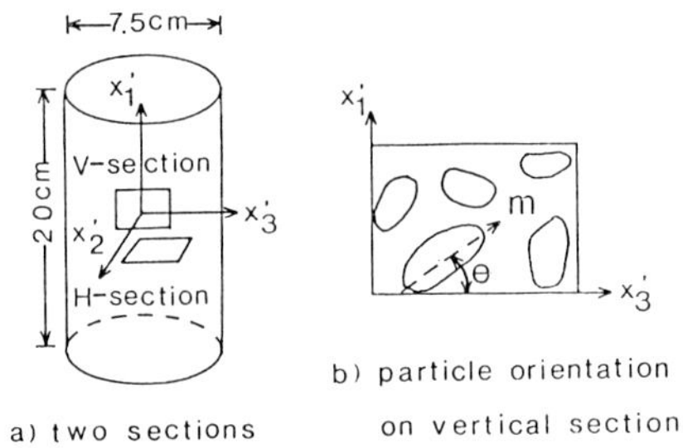

Fig.2 Two sections cut through a particle assembly fixed by a glue and unit vector $\underset{\sim}{m}$ to show a particle orientation on a vertical section.

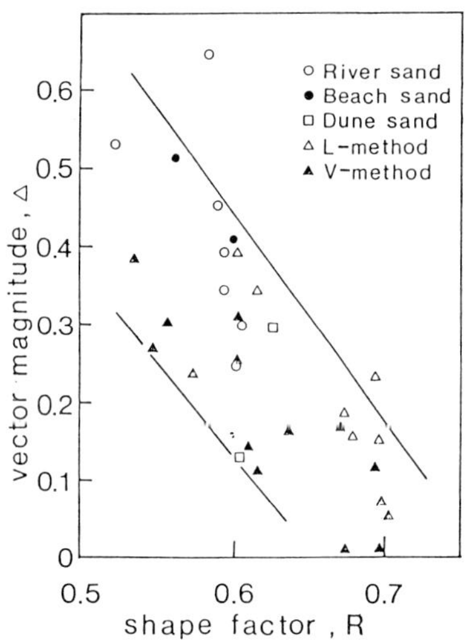

Fig.3 Effect of the particle shape R on the vector magnitude Δ showing the intensity of particle orientation.

where n_i (i=1,2,3) are components of a unit vector $\underset{\sim}{n}$ projected on the orthogonal reference axes x_i (i=1,2,3) and Ω is a solid angle corresponding to the entire surface of a unit sphere ($\Omega=4\pi$). (Because of the two unit vectors in each particle, $\underset{\sim}{n}$ is distributed over the whole solid angle 4π.) To describe the spatial distribution of $\underset{\sim}{n}$, a density function $E(\underset{\sim}{n})$ is introduced in such that $E(\underset{\sim}{n})d\Omega$ corresponds to the rate of unit vectors oriented within a small solid angle $d\Omega$. By definition, $E(\underset{\sim}{n})$ must satisfy

$$\int_{\Omega} E(\underset{\sim}{n})d\Omega = 1 \qquad (2)$$

To characterize the granular fabric, some tensors similar to F_{ij} have already been introduced (refs. 7-9). Among them, the mathematical study by Kanatani (ref. 9) is important. He has pointed out that $E(\underset{\sim}{n})$ can be rewritten in terms of the tensor F_{ij} as a first approximation, as follows:

$$E(\underset{\sim}{n}) = (15/8\pi)(F_{ij}-1/5\delta_{ij})n_i n_j \qquad (3)$$

where δ_{ij} is Kronecker's delta. (The summation convention is applied if any subscript appears twice. This is not the case if superscripts appear in parentheses.) It can be concluded from eqn.(3) that the tensor F_{ij} is an index measure showing the anisotropy due to the particle orientation. It is worthy of note that F_{ij} is symmetric in the sense of $F_{ij}=F_{ji}$, with the trace F_{ii} being equal to unity.

Two-dimensional equivalent

One difficulty may arise when eqn.(1) is applied to natural soils. In fact, it is almost impossible to identify $\underset{\sim}{n}$ in each particle. An alternative method is given here without using $E(\underset{\sim}{n})$ in the definition.

To this end, an assembly of particles is fixed in place by a glue, and is cut along a vertical and a horizontal sections by a saw (Fig. 2a). Let $\underset{\sim}{m}$ be a unit vector pointing to a longest axis of a particle visible on one of the sections (Fig.2b). A tensor $\bar{F}_{ij}$, which is a two-dimensional equivalent of eqn.(1), is defined as follows:

$$\bar{F}_{ij} = \int_{\Omega} m_i m_j E(\underset{\sim}{m})d\Omega \qquad (i,j=1,2) \qquad (4)$$

where Ω equals 2π since all $\underset{\sim}{m}$ are placed on a plane.

Let us consider, for example, a vertical section including a horizontal (x_3') and a vertical (x_1') axes (Fig.2b). Let $\theta^{(k)}$ be an inclination angle of a unit vector $\underset{\sim}{m}^{(k)}$ (corresponding to a kth particle) to the axis x_3'. Since $m_1^{(k)}$ and $m_3^{(k)}$ are $\sin\theta^{(k)}$ and $\cos\theta^{(k)}$ respectively, the components of $\bar{F}_{ij}$ are calculated by additive equivalents of eqn.(4), as follows:

$$\bar{F}_{11} = (1/M)\sum_{k=1}^{M} \sin^2\theta^{(k)}$$

$$\bar{F}_{13} = (1/M)\sum_{k=1}^{M} \cos\theta^{(k)}\sin\theta^{(k)} \tag{5}$$

$$\bar{F}_{33} = (1/M)\sum_{k=1}^{M} \cos^2\theta^{(k)}$$

Since $\bar{F}_{ij}$ is a symmetric, second-rank tensor, the components can always be converted to two principal values $\bar{F}_1$ and $\bar{F}_3$ in the corresponding principal directions θ_1 and θ_3, as follows:

$$\left\{ \begin{matrix} \bar{F}_1 \\ \bar{F}_3 \end{matrix} \right\} = (\bar{F}_{11} + \bar{F}_{33})/2 \pm \left[(\bar{F}_{11} - \bar{F}_{33})^2/4 + \bar{F}_{13}^{\,2} \right]^{1/2} = (1 \pm \Delta)/2 \tag{6}$$

$$\left\{ \begin{matrix} \theta_1 \\ \theta_3 \end{matrix} \right\} = (1/2)\arctan 2\bar{F}_{13}/(\bar{F}_{33} - \bar{F}_{11}) \tag{7}$$

where $$\Delta = (1/M)\left[\sum_{k=1}^{M} (\cos 2\theta^{(k)})^2 + \sum_{k=1}^{M} (\sin 2\theta^{(k)})^2 \right]^{1/2} \tag{8}$$

is an index measure to show the intensity of the preferred orientation of particles, which has first been introduced by Curray (rer. 10) with the name of "the vector magnitude".

Examples

Before discussing the fabric tensors F_{ij} of natural soils in three-dimension, a few comments on the general characteristics of the vector magnitude are helpful.

Oda et al. (ref. 11) observed the particle orientation using vertical and horizontal sections of ten undisturbed sands which were sampled from four rivers, two beaches and two dunes, and of fifteen disturbed sands which were artificially deposited in a laboratory by either L- or V-method. (L-method; sand particles were placed with a spoon to make a loose assembly in a mold. V-method; the loose assembly by L-method was compacted further by vibrating it by hand to make a dense assembly.) The results of their observation are briefly summarized, as follows:

1) In sands artificially deposited in a laboratory, the fabric is all characterized by axial symmetry, with a rotational symmetry axis parallel to the vertical which accords exactly with the symmetry of the gravitational force. No preferred orientation is observed on the horizontal section so that the corresponding vector magnitude is almost zero.

2) The vector magnitudes of vertical sections depend not only on the packing methods of particles but also on the shape characteristic of particles. If particles are like a flat ellipsoid in shape and are deposited freely under the

action of the gravitational force, they tend to be oriented in parallel to yield a high vector magnitude. In Fig.3, the vector magnitudes of the vertical sections are shown as a function of a shape factor R. Here, R=1 means that the sand is composed of spheres. If R becomes smaller toward zero, the particle looks more like a flat ellipsoid. In spite of the large fluctuation in the relation between Δ and R, it can be said that the vector magnitude is inversely related to the shape factor.

3) The similar characteristics are observed in the undisturbed sands. The axial symmetry characteristic, which is common in artificially deposited sands, disappears in some undisturbed sands. This is probably because water (or air) flow at the time of sedimentation may give an effect on the particle orientation.

Three-dimensional fabric tensor

In order to formulate the fabric tensor in three-dimension, it is firstly assumed that the inherent anisotropy exhibits axial symmetry with a symmetry axis parallel to the vertical direction. In other words, the axes x_1', x_2' and x_3' in Fig.2 are the principal axes, and the principal values satisfy $F_2=F_3$. Then, we have

$$F_1 + F_2 + F_3 = F_1 + 2F_3 = 1 \tag{9}$$

One more assumption is that

$$F_1/F_3 = \bar{F}_1/\bar{F}_3 \tag{10}$$

which seems reasonable since the principal values $\bar{F}_1$ and $\bar{F}_3$ in two-dimension should be proportional to the principal values F_1 and F_3. Eqns.(9) and (10), together with eqn.(6), lead to the fabric tensor F_{ij}, as follows:

$$F_{ij} = \begin{pmatrix} \frac{1-\Delta}{3+\Delta} & 0 & 0 \\ 0 & \frac{1+\Delta}{3+\Delta} & 0 \\ 0 & 0 & \frac{1+\Delta}{3+\Delta} \end{pmatrix} \tag{11}$$

Eqn.(11) is useful because the fabric tensor is rewritten in terms of the actually measureable quantity Δ.

YIELD FUNCTION FOR ANISOTROPIC SOILS

In the current plasticity theory for an anisotropic metal, a fourth-rank tensor A_{ijkl} is introduced in such that the stress invariants, which appear in a yield function for an isotropic metal, are multiplied by the tensor to make some correction for anisotropy (refs. 12-14). The similar approach has been applied to anisotropic rocks and soils (refs. 15-17). It should be noted, however, that the actual determination of the anisotropic tensor A_{ijkl} is a crucial point to stimulate the practical use of anisotropic yield function in the soil plasticity.

Here, a yield function for anisotropic soils is given in a practically tractable form on the assumption that A_{ijkl} is a unique function of F_{ij}.

Yield function

Let σ_{ij} and S_{ij} be respectively a stress tensor and the corresponding deviatoric tensor defined by $S_{ij} = \sigma_{ij} - 1/3\sigma_{kk}\delta_{ij}$. As far as our interest is limited to isotropic soils, the yield function is generally written using three invariants I_1, J_2 and J_3, as follows:

$$f = f(I_1, J_2, J_3) \tag{12}$$

$$\text{where } I_1 = \sigma_{ii}$$
$$J_2 = (1/2)S_{ij}S_{ji} \tag{13}$$
$$\text{and } J_3 = (1/3)S_{ij}S_{jk}S_{ki}$$

In order to deal with anisotropic soils in similar manner, other invariants must be taken into account under the condition that functional form remains unchanged even if any orthogonal axes are referred. For example, eqn.(12) is generalized by introducing the following additional invariants:

$$f = f(I_1, J_2, J_3, \bar{I}_1, \bar{J}_2, \bar{J}_3, ..) \tag{14}$$

$$\text{where } \bar{I}_1 = C_{ij}\sigma_{ji}$$
$$\bar{J}_2 = C_{ijkl}S_{ij}S_{kl} \tag{15}$$
$$\text{and } \bar{J}_3 = C_{ijklmn}S_{ij}S_{kl}S_{mn}$$

Here, C_{ij}, C_{ijkl}, C_{ijklmn} are second-, fourth- and sixth-rank tensors respectively.

A simple, but still general, assumption is accepted to give a special form of eqn.(14). That is, any yield functions which have been originally proposed for isotropic materials are generalized in such that I_1, J_2 and J_3 are replaced by $\bar{I}_1$, $\bar{J}_2$ and $\bar{J}_3$ respectively. Drucker-Prager's yield function, for example, is extended as follows (ref. 18);

$$f = \bar{J}_2^{1/2} - k\bar{I}_1 - c \tag{10}$$

where k and c are considered as parameters independent of the anisotropic state of soils (called the extended Drucker-Prager's yield function). Based on the assumption that C_{ij} and C_{ijkl} are unique functions of F_{ij}, then we have

$$C_{ij} = a_1\delta_{ij} + a_2F_{ij} + a_3F_{ik}F_{kj} \tag{17}$$
$$C_{ijkl} = b_1\delta_{ij}\delta_{kl} + b_2(F_{ij}\delta_{kl} + F_{kl}\delta_{ij}) + b_3(\delta_{ij}F_{km}F_{ml} + \delta_{kl}F_{im}F_{mj})$$

$$+ b_4(F_{ij}F_{km}F_{ml} + F_{im}F_{mj}F_{kl}) + b_5F_{im}F_{mj}F_{kn}F_{nl} + b_6(\delta_{ki}\delta_{lj}+\delta_{li}\delta_{kj}) \quad (18)$$
$$+ b_7(F_{ik}\delta_{lj} + F_{kj}\delta_{li} + F_{il}\delta_{kj} + F_{lj}\delta_{ki}) + b_8F_{ij}F_{kl}$$
$$+ b_9(F_{im}F_{mk}\delta_{lj} + F_{km}F_{mj}\delta_{li} + F_{im}F_{ml}\delta_{kj} + F_{lm}F_{mj}\delta_{ik})$$

where a_i (i=1,2,3) and b_i (i=1,2,...,9) are parameters depending only on the three invariants of F_{ij} (e.g., ref. 1).

It is inconvenient for all terms in eqns.(17) and (18) to be taken into account. For the sake of simplicity, therefore, only the first terms with respect to F_{ij} are considered. Then, we have

$$\bar{I}_1 = (a_1\delta_{ij} + a_2F_{ij})\sigma_{ij} \quad (19)$$
$$\bar{J}_2 = (2b_6\delta_{ik}\delta_{jl} + 4b_7F_{ik}\delta_{lj})S_{ij}S_{kl} \quad (20)$$

If an isotropic soil with an isotropic tensor of $F_{ij} = (1/3)F_{kk}\delta_{ij}$ is only concerned, eqn.(16) is reduced to

$$f = \left[(2b_6 + 4b_7/3)S_{ij}S_{ij}\right]^{1/2} - (a_1 + a_3/3)\sigma_{ii} - c \quad (21)$$

which should exactly accord with the original form of Drucker-Prager's yield function. Accordingly, the following additional restrictions among the parameters are imposed:

$$2b_6 + 4b_7/3 = 1/2$$
$$a_1 + a_2/3 = 1 \quad (22)$$

In addition to eqn.(22), it is known that a_1, a_2, b_6 and b_7 are functions of the following invariants:

$$F_o = F_{ii} = 1$$
$$D_2 = (1/2)F'_{ij}F'_{ji} \quad (23)$$
$$D_3 = (1/3)F'_{ij}F'_{jk}F'_{ki}$$

where F'_{ij} is the deviatoric fabric tensor defined by $F'_{ij}=F_{ij}-(1/3)F_{kk}\delta_{ij}$. Using eqn.(11), the second invariant D_2 is rewritten as

$$D_2 = \frac{4\Delta^2}{3(3 + \Delta)^2} \quad (24)$$

Eqn.(24) means that the vector magnitude can be interchanged with the second invariant D_2. Noting F_o =1 and neglecting the effect of the third invariant D_3 onthe final result, we have

$$a_i = a_i(F_o, D_2)$$
$$b_i = b_i(F_o, D_2) \quad (25)$$

It should be noted that eqn.(21) is reduced to Drucker-Prager's equation if a_1, a_2, b_6 and b_7 are set to 1, 0, 1/4 and 0 respectively. Now, let us assume that a_1, a_2, b_6 and b_7 are changed in proportion to $(2D_2)^{1/2}$, under the boundary condition that $a_1=1$, $a_2=0$, $b_6=1/4$ and $b_7=0$ is satisfied when $(2D_2)^{1/2}$ equals zero. This assumption, together with eqn.(22), is accepted to give

$$\begin{aligned} a_1 &= 1 - \lambda(2D_2)^{1/2} \\ a_2 &= 3\lambda(2D_2)^{1/2} \\ b_6 &= 1/4 - \mu(2D_2)^{1/2}/2 \\ b_7 &= (3/4)\mu(2D_2)^{1/2} \end{aligned} \qquad (26)$$

where λ and μ are constants to be experimentally determined. It is worthy of note that the anisotropic yield function can be completely expressed in terms of the anisotropic index Δ once the constants λ and μ are determined. (In this paper, we are only concerned with perfectly plastic behavior of anisotropic soils with a potential function similar form to the corresponding yield function.)

CONCLUDING REMARKS

Fabric tensor F_{ij} for a three-dimensional assembly of soil particles is actually determined by measuring the long axes of particles visible on a vertical section. In order to demonstrate that the tensor is useful to link the discrete approach of soils to the continuum approach, a yield and a potential functions are formulated in terms of the tensor. The extended Drucker-Prager's function is not complete to describe all aspects of the inelastic behavior of anisotropic soils. However, it is shown through this study that this approach can be a first step to discuss the soil plasticity on the basis of material science.

ACKNOWLEDGEMENTS

M.M. Mehrabadi, who was working together with the senior author in Northwestern University in 1979, has first suggested that the vector magnitude by Curray (1956) can be related to a tensor quantity. We would like to express sincere thanks to him because Eq.(6) is basically from his suggestion. Thanks are due to Prof. Chuji Tanaka of Meiji University and Prof. Fumio Tatsuoka of Institution of Industrial Science, the University of Tokyo, for their valuable discussion and suggestion.

REFERENCES

1. Cowin, S.C. (1985). "The relationship between the elasticity tensor and the fabric tensor." Mechanics of Materials, 4, 1-11.
2. Jenkins, J.T. and Satake, M. (1983). "Mechanics of Granular Materials: New Models and Constitutive Relations." Studies in Applied Mechanics 7. Elsevier, Amsterdam.
3. Chen, W.F. and Saleeb, A.F.(1982) "Constitutive Equations for Engineering

Materials: Elasticity and Modeling." John Wiley & Sons, New York, N.Y.

4. Desai, C.S. and Siriwardane, H.J. (1984). "Constitutive Laws for Engineering Materials with Emphasis on Geologic Materials." Pentice-Hall, Inc., Englewood Cliffs, NJ07632.
5. Casagrande, A. and Carillo, N. (1944). "Shear failure of anisotropic materials." Proc. Boston Society of Civil Engrs., 31, 74-87.
6. Oda, M., Nemat-Nasser, S. and Konishi, J. (1985). "Stress-induced anisotropy in granular masses." Soils and Foundations, 25(3), 85-97.
7. Satake, M. (1982). "Fabric tensor in granular materials." In the book entitled "Deformation and Failure of Granular Materials (edited by Vermeer, P.A. and Luger, H.J.), Balkema, Rotterdam, 63-68.
8. Oda, M., Nemat-Nasser, S. and Mehrabadi,M.M.(1982). "A statistical study of fabric in a random assembly of spherical granules." Int. J. Numerical and Analytical Methods in Geomechanics, 6, 77-94.
9. Kanatani, K. (1983). "Characterization of structural anisotropy by fabric tensors and their statistical test." J. of Japanese Society of Soil Mechanics and Foundation Engineering, 23(4), 171-177.
10. Curray, J.R.(1956). "The anakysis of two-dimensional orientation data." J. Geol., 64, 117-131.
11. Oda, M., Koishikawa, I. and Higuchi, T. (1978). "Experimental study of anisotropic shear strength of sand by plane strain test." Soils and Foundations, 18(1), 25-38.
12. Baltov, A. and Sawczuk, A. (1965). "A rule of anisotropic hardening." Acta Mechanica, 1, 81-92.
13. Hill, R. (1948). "A theory of yielding and plastic flow of anisotropic metals." Proc. of Royal Society of London. Series A, 193.
14. Katchanov, L.M. (1974). "Osnovy mechaniki razrutecheniya." Nauka, Moscow (translated into Japanese by Y. Oohashi in 1977).
15. Pariseau, W.G. (1968). "Plasticity theory for anisotropic rocks and soils." Proc. of 19th Symp. on Rock Mechanics, Austin, 267-295.
16. Reinicke, K.M. and Ralston, T.D. (1977). "Plastic limit analysis with an anisotropic, parabolic yield function." Int. J. Rock Mech. Min. Sci. & Geomech. Abstr., 14, 147-154.
17. Ghaboussi, J. and Momen, H. (1984). "Plasticity model for inherently anisotropic behaviour of sand." Int. J. for Numerical Methods in Geomechanics, 8, 1-17.
18. Drucker,D.C. and Prager, W. (1952). "Soil mechanics and plastic analysis of limit design." Quart. Applied Mathematics, 10(2), 157-165.
19. Cowin, S.C. and Satake, T. (1978). "Continuum Mechanical and Statistical Approaches in the Mechanics of Granular Materials." Gakujutsu Bunken Fukyu-Kai, Tokyo, Japan.

Micromechanics of Granular Materials, edited by M. Satake and J.T. Jenkins
Elsevier Science Publishers B.V., Amsterdam, 1988 — Printed in The Netherlands

IS THE FABRIC TENSOR SUFFICIENT?

H. Horii, K. Sahasakmontri

Dept. Civil Engineering, Univ. of Tokyo, Bunkyo-ku, Japan

SUMMARY

In this paper discussion is made on the sufficiency of the fabric tensor in describing the effect of fabric on the mechanical properties of geological materials. For the case of cracked solids, it is shown that the effect of the fabric structure on the overall moduli can be characterized with the fabric tensor alone provided that the interaction effect is neglected. Including the interaction, additional parameter is found necessary and its essential characteristic is presented.

THE FABRIC TENSOR AND THE OVERALL MODULI OF CRACKED SOLIDS

It is known that the fabric tensor of a cracked body is defined by taking into account:

1) volume density of cracks,

2) size of cracks,

3) orientation of cracks,

see [1,2]. The usefulness of the fabric tensor in characterizing the anisotropy of the fabric structure (which is the crack pattern in the present discussion) has been recognized in many experiments. As an example, [2] has shown that the supersonic wave velocity propagating through cracked bodies is closely related to the characteristic of the fabric tensor. This seems to suggest the possibility of estimating the fabric tensor of in situ rock masses by field measurement of the wave velocity.

However, since the distribution of cracks in space is not included in the definition of the fabric tensor, it is possible that cracked bodies having the crack geometries being described by the same fabric tensor may have different overall moduli. One therefore may wonder whether or not the fabric tensor is sufficient in describing the effect of cracks on the mechanical properties of cracked solids.

1) Without interaction

Following the method given in [3], it can be shown that if the interaction between neighboring cracks is neglected, the constitutive equation of the two-dimensional cracked solid shown in Fig. 1 is given by

$$\varepsilon_i = C_{ij}(\underset{\sim}{F})\sigma_j \ , \quad i,j=1,2,6 \tag{1}$$

in which $\varepsilon_1 = \varepsilon_{11}$, $\varepsilon_2 = \varepsilon_{22}$, $\varepsilon_6 = 2\varepsilon_{12}$ are the overall strain components; $\sigma_1 = \sigma_{11}$, $\sigma_2 = \sigma_{22}$, $\sigma_6 = \sigma_{12}$ are the overall stress components; and $[C_{ij}(\underset{\sim}{F})]$ is the overall compliance matrix which is a function of $\underset{\sim}{F}$ only. $\underset{\sim}{F}$ is the 2nd-rank fabric tensor whose components are given by (see[1])

$$F_{ij} = \rho\langle 4c^2\rangle\langle n_i n_j\rangle \ , \tag{2}$$

$$\text{in which } \langle 4c^2\rangle = 4\int_0^\infty c^2 f(c)\,dc \ ,$$

$$\langle n_i n_j\rangle = \int_0^{2\pi} n_i n_j e(\hat{n})\,d\beta \ , \quad i,j=1,2 \ ,$$

$f(c)$ is a probability density function of c, representing the distribution of the crack of length 2c; $e(\hat{n})$ is a probability density function of $\hat{n}$, representing the distribution of the crack having the unit normal vector $\hat{n}$; $n_1 = \sin\beta$, $n_2 = \cos\beta$ are the components of $\hat{n}$; and ρ is number of cracks per unit volume. Here, we assume that $\hat{n}$ and c are mutually independent and the solid is of unit thickness.

For plane stress, the overall compliance $[C_{ij}(\underset{\sim}{F})]$ is given by

$$[C_{ij}(\underset{\sim}{F})] = \frac{1}{E}\begin{bmatrix} 1+\frac{\pi}{2}F_{11} & -\nu & \frac{\pi}{2}F_{12} \\ -\nu & 1+\frac{\pi}{2}F_{22} & \frac{\pi}{2}F_{12} \\ \frac{\pi}{2}F_{12} & \frac{\pi}{2}F_{12} & 2(1+\nu)+\frac{\pi}{2}(F_{11}+F_{22}) \end{bmatrix} , \tag{3}$$

in which E and ν are respectively the Young's modulus and the Poisson's ratio of the homogeneous isotropic matrix. It is seen that the fabric tensor is the only parameter needed to characterize the influence of the crack structure on the overall compliance of the solid if the interaction effect is ignored. In the following, however, it will be shown that if the interaction is taken into account, this is no longer true.

2) <u>Doubly periodic array of cracks</u>

Consider the solid containing a periodic array of cracks of equal length 2c shown in Fig. 2, it follows from eqn. (2) that the fabric tensor is

$$[F_{ij}] = \begin{bmatrix} 0 & 0 \\ 0 & \frac{4c^2}{HW} \end{bmatrix}, \quad i,j=1,2 . \tag{4}$$

(Note that $\frac{c^2}{HW}$ measures the crack density.) From eqns. (3) and (4) the overall moduli of the solid is found to be

$$E_y = \frac{E}{1 + 2\frac{\pi c^2}{HW}}, \quad G_{xy} = \frac{G}{1 + 2\,\frac{G}{E}\,\frac{\pi c^2}{HW}}, \tag{5}$$

where G is the shear modulus of the matrix surrounding the cracks; E_y and G_{xy} are respectively the overall tensile modulus in the direction normal to the crack faces, and the overall shear modulus of the cracked solid. However, the asymptotic solution for the overall moduli of the solid containing a doubly periodic array of cracks has been shown by [3] to be

$$E_y = \frac{E}{1 + 2\frac{\pi c^2}{HW}[1 + \frac{c^2}{HW} H_I(\Phi)]},$$

$$G_{xy} = \frac{G}{1 + 2\,\frac{G}{E}\,\frac{\pi c^2}{HW}[1 + \frac{c^2}{HW} H_{II}(\Phi)]}, \tag{6}$$

in which Φ measures the shape of the crack array and is defined in Fig. 2; $H_I(\Phi)$ and $H_{II}(\Phi)$, representing the interaction effect, are given in [3] as explicit functions of Φ and are shown in Figs. 3 and 4.

It is seen from eqn. (6) that including the interaction, the overall moduli of the cracked body becomes dependent on the shape of the crack array or, in other words, the distribution of cracks in the solid. The fabric tensor cannot, in this case, completely describe the effect of the crack geometry on the overall moduli of the solid (even though a higher rank fabric tensor is employed), and the additional parameter $[H_I(\Phi), H_{II}(\Phi)]$ is needed that takes into account the distribution of the cracks. It is important to notice that this new parameter depends on the loading condition and is obtained by considering the interaction effect based on micromechanics.

REFERENCES

[1] Oda, M.: FABRIC TENSOR FOR DISCONTINUOUS GEOLOGICAL MATERIALS, Soils and Foundations, Vol.22, No.4, pp.96-108, 1982.

[2] Oda, M., Suzuki, K., and Maeshibu, T.: ELASTIC COMPLIANCE FOR ROCK-LIKE MATERIALS WITH RANDOM CRACKS, Soils and Foundations, Vol.24, No.3, pp.27-40, 1984.

[3] Horii, H. and Nemat-Nasser, S.: OVERALL MODULI OF SOLIDS WITH MICROCRACKS: LOAD-INDUCED ANISOTROPY, J. Mech. Phys. Solids, Vol.31, No.2, pp.155-177, 1983.

[4] Sahasakmontri, K., Horii, H., Hasegawa, A., and Nishino, F.: MECHANICAL PROPERTIES OF SOLIDS CONTAINING A DOUBLY PERIODIC RECTANGULAR ARRAY OF CRACKS, Structural Eng./Earthquake Eng., Proc. of JSCE, Vol.4, No.1, pp.125s-135s, 1987.

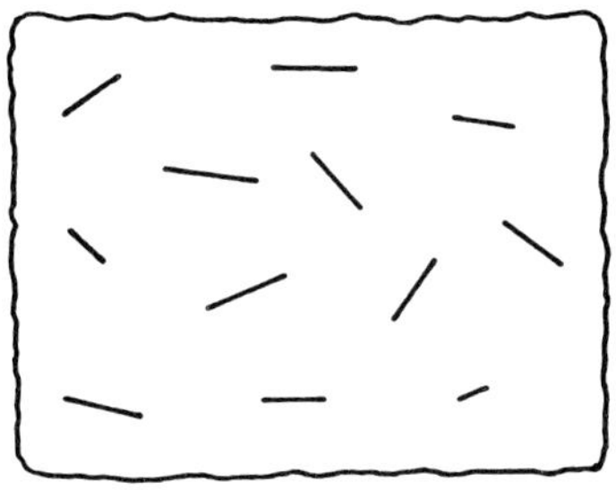

Fig. 1. A cracked solid.

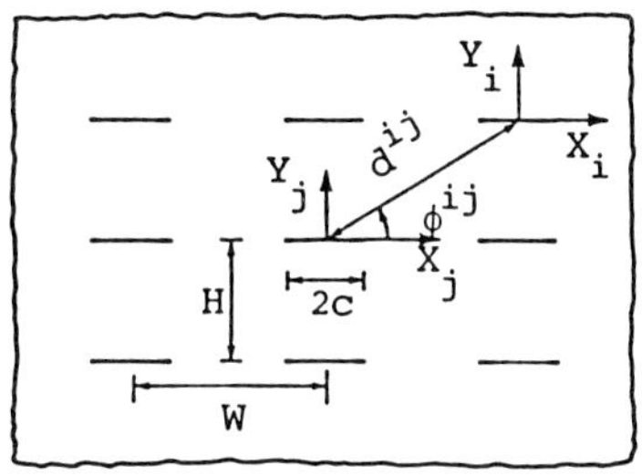

Fig. 2. A doubly periodic rectangular array of cracks.

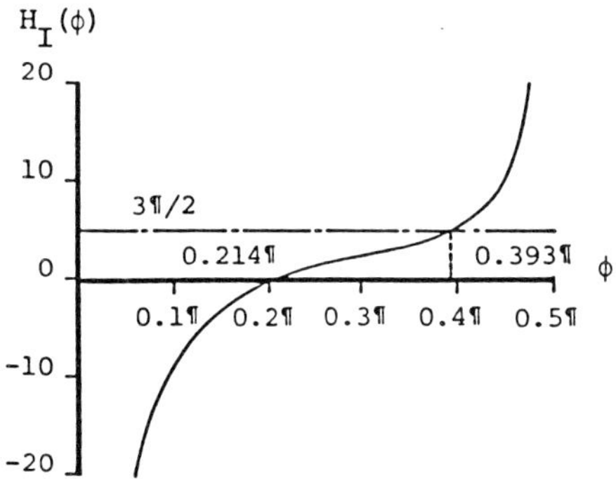

Fig. 3. Plot of $H_I(\phi)$

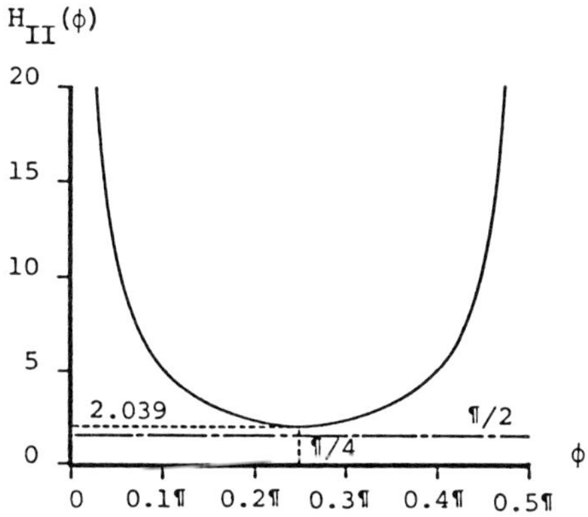

Fig. 4. Plot of $H_{II}(\phi)$

Micromechanics of Granular Materials, edited by M. Satake and J.T. Jenkins
Elsevier Science Publishers B.V., Amsterdam, 1988 — Printed in The Netherlands

DYNAMIC SHEAR MODULI AND THEIR RELATIONSHIP TO FABRIC OF GRANULAR MATERIALS

Isao Ishibashi[1] and Yao-Chung Chen[2]

[1]Department of Civil Engineering, Old Dominion University, Norfolk, Virginia 23529, U.S.A.

[2]Department of Construction Engineering and Technology, National Taiwan Institute of Technology, P.O. Box 90-130, Taipei, Taiwan, Republic of China

ABSTRACT

This paper presents experimental results from the investigation of the change of granular fabric during shear deformation and the determination of its relationship with the shear modulus measured at low strain amplitude. A torsional simple shear/resonant column device was used to perform experiments on hollow cylindrical specimens under a variety of stress conditions. The effects of the fabric on the shear moduli of different stress paths, stress reversals, rotations of principal stress axes, and preshearing were investigated.

INTRODUCTION

The mutual arrangement of particles (geometric fabric) has been found to be an important component in the anisotropic behavior of granular materials, and direct measurements of granular fabric by using the microphotographic and photoelastic techniques are extremely difficult, time consuming and often impossible for complex stress conditions. It would be very useful if the fabric of granular material could be related to some measurable quantities.

The purpose of this investigation is to find an indirect measure of fabric under a variety of stress conditions. It is anticipated that the speed of shear wave propagation through the specimen depends on the average number of particle contacts in the plane of the shear waves and on the contact forces at those contact points. Therefore, the shear moduli measured by resonant column test may provide a quantitative or qualitative measure of the fabric.

TORSIONAL SIMPLE SHEAR/RESONANT COLUMN APPARATUS

All experiments were performed in a torsional simple shear/resonant column apparatus. The apparatus has the capability of testing hollow cylindrical specimen (inner radius = 35.6 mm, outer radius = 50.8 mm, and height = 193 mm) by applying simultaneously static triaxial/torsional stresses and horizontal torsional resonant vibrations. Since the resonance system is separated from the static loading system, both static and resonant vibrations can be controlled independently.

The specimen was subjected to the axial stress σ_z, equal inner and outer lateral confining stresses σ_r, and the horizontal torsional shear stress $\tau_{z\theta}$ on the top of the specimen. Since these stresses can be controlled independently, it is possible to select the desired direction of the major principal stresses θ from the vertical direction by different combinations of σ_z, σ_r, $\tau_{z\theta}$.

A detailed discussion about a similar torsional simple shear device is given by Ishibashi et al. (Ref. 1).

MATERIAL PROPERTIES

Glass beads were used. Poisson's ratio is 0.21, shear modulus is 3.0 X 10^7 kP_a, Young's modulus is 7.2 X 10^7 kPa and the specific gravity of the glass beads is 2.472. The coefficient of friction was measured as 0.15. Specimen are composed of two sizes of glass beads: sieve Nos. 40 to 50 (0.43 mm to 0.30 mm) and Nos. 60 to 80 (0.25 mm to 0.18 mm) with weight ratio of 1 to 2.

To achieve uniform density, all specimens were prepared in five layers. Each layer was constructed by raining an equal amount of dry glass beads through the air into the molds and compacting the layer to the desired height by a tamper. The specimens thus prepared had about the same void ratio (e = 0.58) or the same relative density (D_r = 60%). The specimens were later fully saturated with water for the purpose of measuring the volume change in drained conditions during the shearing process.

TEST CONDITIONS

Each specimen was first isotropically consolidated under 138 kPa prior to being sheared along the desired stress paths. For all the tests, the mean effective normal stress, $(\sigma_1 + \sigma_2 + \sigma_3)/3$, was kept constant throughout the shearing stage. The shear modulus governing the propagation of a shear wave down the axis of the specimen was obtained by measuring the frequency of this shear wave at resonance. The vibrating amplitude at resonance was in a range of $1 \sim 5 \times 10^{-7}$.

Several series of tests were conducted in order to investigate the effects of various stress paths and fabric on the dynamic shear modulus measurement and those loading sequences are individually described in the following sections.

TEST RESULTS AND DISCUSSION

Throughout the analyses in this paper if not mentioned otherwise, the measured shear moduli were first divided by the value $[(2.17 - e)^2/(1 + e)].[(\sigma_1 + \sigma_2 + \sigma_3)/3]^{0.40}$ to eliminate the effect of the void ratio (e) and the mean normal stress on shear modulus based on Hardin's earlier investigations (Ref. 2). The above shear moduli were further divided by their values prior to shearing for easy comparisons of the data. Therefore, the initial normalized shear moduli are always unity.

<u>Series 1. Effect of Shearing Direction Relative to the Vertical Direction</u>

During this series of tests, specimens were sheared along a constant θ angle until failure. Various θ angles were selected from 0°, 45°,- 45°, and 90°. In order to investigate the effect of depositional fabric and induced fabric on dynamic shear modulus, some of the specimens were presheared prior to shearing: Series 1a - no preshearing, Series 1b - one half cycle preshearing along θ = 45° up to τ_{max} = 62 kPa.

Figs. 1 and 2 show relationships between the normalized shear modulus and the maximum shear stress obtained respectively from Series 1a and 1b experiments. In the figures, in general, the normalized shear modulus decreases with increasing maximum shear stress. The decrease of shear modulus indicates a decrease in rigidity with increasing maximum shear stress. According to Stokoe, et al. (Ref. 3), shear waves propagate with a higher velocity in a plane of isotropy than in a plane of anisotropy. Therefore, samples with isotropic fabric probably have more resistance to shear vibration and so have higher shear modulus than those with anisotropic fabric under the same stress conditions and the same density.

When a sample is sheared under anisotropic stress conditions, it deforms and the contact number generally decreases with increasing shear stress. This was pointed out by Cundall, et al. (Ref. 4) in their numerical simulations. The contact normals, however, tend to concentrate in the direction parallel to the major principal stress direction. Because of the decreasing number of contacts and the increasing anisotropy of the fabric the shear modulus decreases with

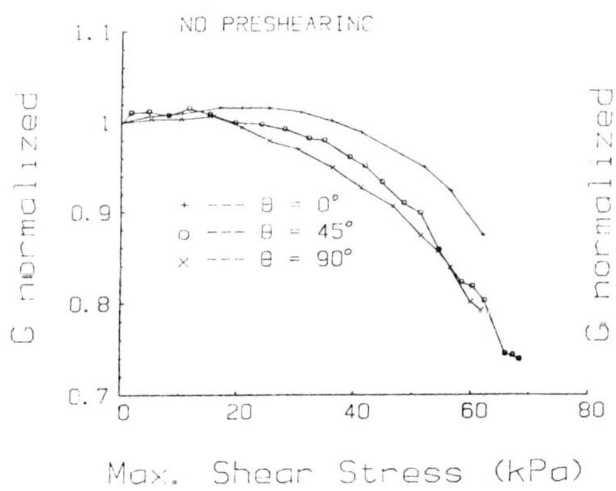

Fig. 1. Normalized shear modulus versus maximum shear stress (no preshearing)

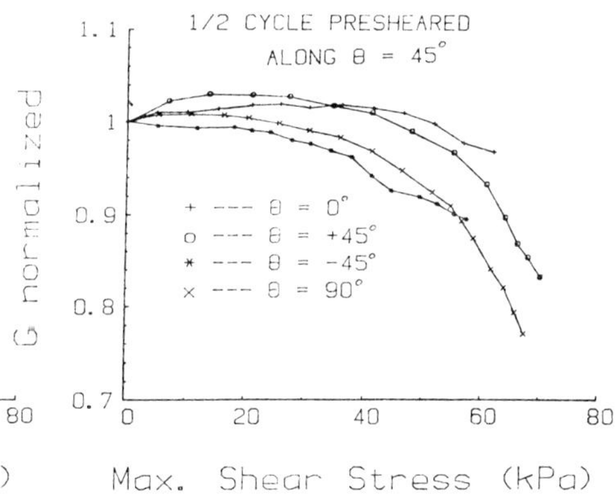

Fig. 2. Normalized shear modulus versus maximum shear stress (presheared along θ = 45°)

increasing maximum shear stress. Slightly higher shear moduli in the $\theta = 0°$ direction shown in Fig. 1 might be due to depositional fabric along $\theta = 0°$ direction prior to shearing along various θ directions.

Since the shear modulus had been normalized to eliminate the dependence on the void ratio and the confining stress, the reason for the initially different behavior from 0° to 90° might be because of an initial fabric introduced by the sample preparation. When a sample was sheared along different θ directions, the total number of contacts decreased or increased as dictated by the initial fabric. With an increase of maximum shear stress, the initial fabric might be erased to a certain extent and induced fabric built up in the direction of the principal stress axis. Thereafter, the normalized shear modulus decreased for all θ directions due to the decrease in the number of contacts.

The effect of initial fabric on dynamic shear modulus can be seen clearly in Fig. 2 for Series 1b experiment during which specimens were presheared along $\theta = 45°$ once. Those specimens are expected to have a preferred fabric orientation along $\theta = 45°$. As expected, the shear modulus was highest in the $\theta = 45°$ direction and lowest in the $\theta = -45°$ direction. A slightly different response for $\theta = 0°$ and 90° might indicate that those preshearing processes could not totally erase the depositional fabric, which had a preferred direction along $\theta = 0°$.

Series 2. Effect of Stress Reversals

To investigate how the fabric changes with stress and strain reversal, specimens were first sheared to a certain stress level in one direction and then were unloaded back to the isotropic stress condition. Then they were resheared in the opposite direction.

Fig. 3 shows how the normalized shear modulus changes with the stress reversal when a sample was sheared initially along the $\theta = 0°$ direction. When loading, unloading, and reloading at low stress levels, the normalized shear modulus showed reversible behavior. Since the shear modulus is a function of stress and fabric, this suggests that the initial fabric could be recovered on unloading. But once the sample was sheared to certain high stress levels, the shear modulus returned to a value lower than its initial value upon unloading. The fabric probably reached a new state by particle slip. Consequently, the previous fabric could not be recovered even after unloading back to an isotropic stress condition. When anisotropic fabric formed in one direction under some high shear stress, unloading back to an isotropic stress could not bring the fabric back to isotropic. This phenomenon has been pointed out by Cundall, et al. (Ref. 4) in their numerical simulations.

The summary of three test results is plotted in Fig. 4. In Fig. 4 there seems to be an envelope that defines a limiting relationship between the normalized shear modulus and the maximum shear stress. No matter how the

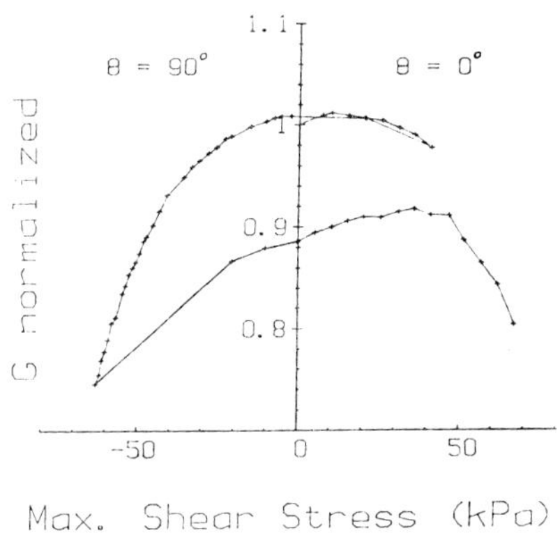

Fig. 3. Normalized shear modulus versus maximum shear stress (stress reversal)

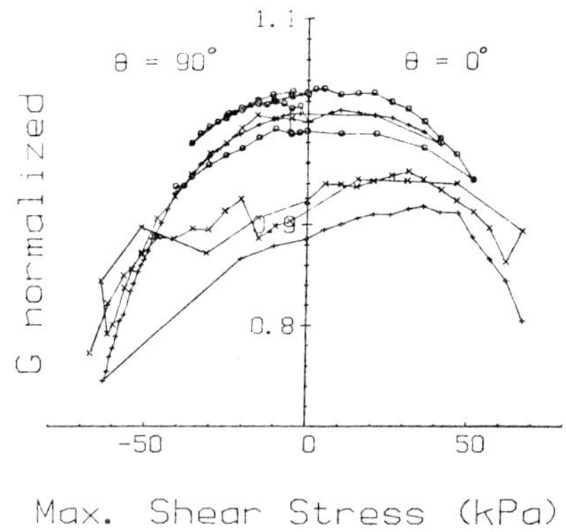

Fig. 4. Normalized shear modulus versus maximum shear stress (stress reversal - 3 data)

sample was previously sheared, once it approached this envelope its shear modulus and the stress were uniquely related. This implies that when the stress is high enough the fabric induced by the previous stress maximum in the reverse direction is totally erased. Furthermore, the new fabric induced by shearing in this direction would be nearly the same as if the sample were sheared monotonically from an isotropic condition to that stress level without any stress history.

Fig. 5 is a typical set of data showing the normalized shear modulus versus the maximum shear strain. In this figure the shear modulus decreases with an increase of maximum shear strain and increases more rapidly upon unloading. Before the maximum shear strain has returned to zero the shear modulus has reached a peak value and begun to decrease. The shear modulus continues to decrease as the shear strain is increased in the reverse direction.

Stokoe, et al. (Ref. 3) measured higher shear wave velocities in a plane of isotropy than in a plane of anisotropy under the same conditions of stress and density. This suggests that other things being equal, isotropic fabric generates the highest shear modulus. The observation that, upon unloading, the shear modulus reaches a peak value before the maximum shear strain returns to zero implies that the fabric may have returned to an isotropic state before the maximum shear strain returned to zero.

Similar test results were obtained for θ = + 45° to - 45°, θ = 15° to -75° and 60° to -30°. In all cases, envelopes which defines a limiting relationship between the normalized shear modulus and the maximum shear stress were observed and the stress was ahead of the fabric and the fabric was ahead of the strain.

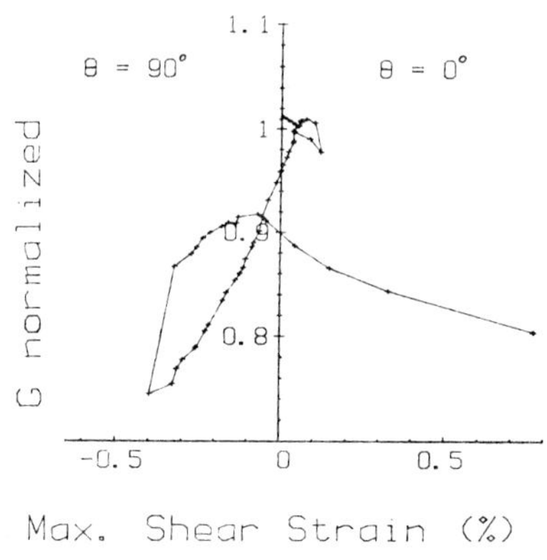

Fig. 5. Normalized shear modulus versus maximum shear strain (stress reversal)

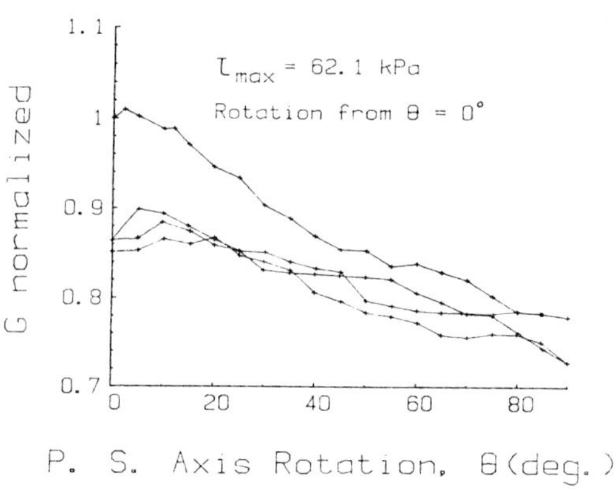

Fig. 6. Normalized shear modulus during principal stress axes rotation (initial θ = 0°)

Series 3. Effect of Principal Stress Axis Rotation

During the tests, the maximum shear stress was kept constant, while the major principal stress axis was rotated continuously from 0° to 90° and from 90° to 0°. Two maximum shear stress levels was selected as 62 kPa. The mean principal normal stress, $(\sigma_1 + \sigma_2 + \sigma_3)/3$ was kept constant as 138 kPa as mentioned earlier.

Fig. 6 (initial $\theta = 0°$) and Fig. 7 (initial $\theta = 90°$) show the normalized shear modulus as a function of the major principal stress axis direction relative to the vertical direction. The normalized shear modulus decreased for the first half cycle of stress rotation in both tests, and reduced to similar lines thereafter. The decrease of the normalized shear modulus in the first half cycle might have resulted from a decrease of contact number. After the first half cycle, the contact number presumably reached a stable condition throughout the test.

The shear moduli presented in Figs. 6 and 7 were normalized by $[(\sigma_1 + \sigma_2 + \sigma_3)/3]^{0.4}$ as previously mentioned. However, if the shear moduli were normalized by $[(\sigma_1 + \sigma_3)/2]^{0.4}$ (or $= [(\sigma_z + \sigma_\theta)/2]^{0.4}$), the replot of data, for example, from Fig. 7 becomes like Fig. 8. This newly normalized shear moduli stay nearly constant after a half cycle of principal axis rotation. Considering the fact that σ_z and σ_θ directions coincide with the direction of shear wave propagation and the particle motion and that the σ_r direction is the out of plane from the shear vibration for this experiment, this new normalization might provide a better data presentation technique. Therefore, it could be concluded that the principal axis rotation does not affect any change in dynamic shear moduli and hence over all number of particle contacts.

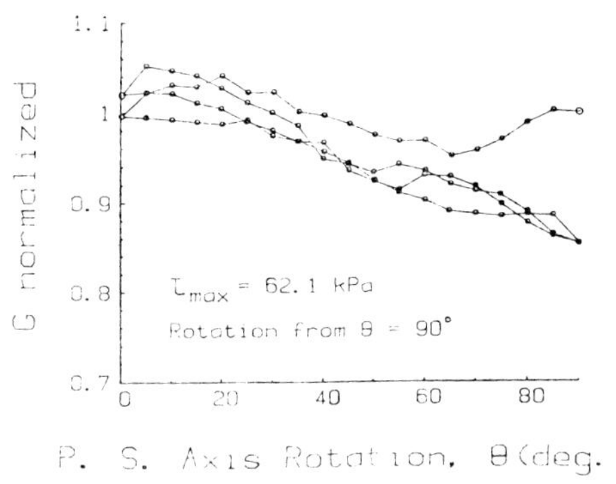

Fig. 7. Normalized shear modulus during principal stress axes rotation (initial θ = 90°)

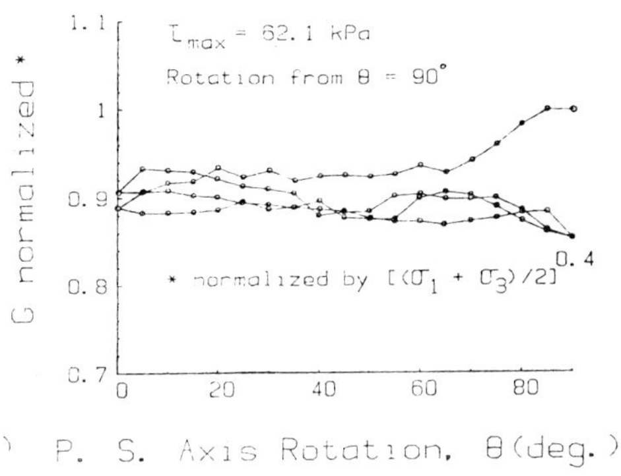

Fig. 8. Normalized shear modulus during principal stress axes rotation (normalized by $[(\sigma_1 + \sigma_3)/2]^{0.4}$)

However, the above observation also suggests that the dynamic shear moduli can detect the change of contact numbers but cannot adequately detect the directions if whole contact numbers remain the same due to its orthogonal nature.

CONCLUSIONS

Dynamic shear moduli of glass bead assemblages were measured in hollow cylindrical torsional shear apparatus for different stress paths. The effect of depositional fabric, induced fabric, stress reversal, rotation of principal stress axes were examined, and their results are summarized below.

1. There are some small depositional fabric which has a preferred contact normal distribution in the vertical direction due to the sample preparation. The above phenomenon was evidenced by a slightly higher shear moduli in the $\theta = 0°$ direction reshearing for specimens without any preshearing.
2. Application of preshearing creates a strong preferred orientation along the direction of the preshearing. The highest dynamic shear moduli were measured along the same direction of reshearing as the preshearing direction and the lowest ones along the 90° direction relative to preshearing.
3. In general the normalized shear modulus decreases with increasing maximum shear stress. It is probably due to gradual decrease in particle contact number when the maximum shear stress increased.
4. Based on the above observation and previous studies, it can be assumed that the peak normalized shear modulus might represent a stage of isotropy in

the assemblage.

5. Anisotropic fabric may develop during shear deformation due to a decrease in contacts in the minor principal stress direction and an increase in contacts in the major principal stress direction. Unloading back to an isotropic stress condition does not return the fabric to an isotropic state. With the maximum shear stress increasing in the reverse direction, the fabric passes through the isotropic state before the maximum shear strain returns to its zero value. Therefore, the stress is ahead of the fabric, and the fabric is ahead of the strain.
6. In a cycle of loading and unloading (stress reversal), the decrease in the shear modulus in the loading portion may be recovered totally or partially upon unloading, depending on the level of stress in the loading process.
7. When the principal stress axes rotate at a higher stress level (τ_{max} = 62kPa), the shear modulus decreases in the first half cycle and remains nearly constant for the rest of the test also when normalized by $[(\sigma_1 + \sigma_3)/2]^{0.4}$.
8. The above two conclusions suggest that the dynamic shear moduli can detect the change of contact numbers but cannot well detect those directions if whole contact numbers remains the same.

ACKNOWLEDGEMENTS

This research was performed as a part of National Science Foundation collaborative research project carried out at Cornell University and the University of Minnesota.

The authors wish to express their sincere gratitude to Professor J.T. Jenkins of Cornell University and Professors P.A. Cundall and O.D.L. Strack of the University of Minnesota for their enlightening suggestions and discussions throughout this research. The Foundations's financial support is greatly appreciated.

REFERENCES

1. I. Ishibashi, M. Kawamura, and S.K. Bhatia, Torsional Simple Shear Apparatus for Drained and Undrained Cyclic Testing, Advances in the Art of Testing Soils under Cyclic Conditions, ASCE Convention, Detroit, Michigan, 1985, pp. 51-73.
2. B.O. Hardin, The Nature of Stress-strain Behavior of Soils, Proceedings of the Earthquake Engineering and Soil Dynamics Conference, ASCE, Pasadena, California, Vol. 1, 1978, pp. 3-90.
3. K.H. Stokoe II, S.H.H. Lee, and D.P. Knox, Shear Moduli Measurements under True Triaxial Stresses, Advances in the Art of Testing Soil Under Cyclic Conditions, ASCE Convention, Detroit, Michigan, 1985, pp. 166-185.
4. P.A. Cundall, and O.D.L. Strack, Modeling of Microscopic Mechanisms in Granular Material, Mechanics of Granular Materials: New Models and Constitutive Relations, Elsevier, Amsterdam, Holland, ed. by J.T. Jenkins and M. Satake, 1983, pp. 137-149.

Micromechanics of Granular Materials, edited by M. Satake and J.T. Jenkins
Elsevier Science Publishers B.V., Amsterdam, 1988 — Printed in The Netherlands

A DISCRETE LIMIT ANALYSIS OF GRANULAR MATERIALS INCLUDING EFFECTS OF THE SOLID CONTACT

Norio Takeuchi[1] and Tadahiko Kawai[2]

[1]Department of Civil Engineering,Meisei University, Hino,Tokyo 191 (Japan)

[2]Department of Electrical Engineering,Science University of Tokyo,Sinjuku-ku, Tokyo 162 (Japan)

ABSTRACT

Previously the present authors have developed a discrete method of limit analysis of solids using a new element model entitled "Rigid Bodies-Spring Model" and proved its validity for collapse analysis of solids where slip failure is predominant.

In case of the granular materials,however,influence of crack initiation due to tensile load often can not be neglected.

In view of this Takeuchi developed a new algorithm recently which may be applicable to analyse the coupled failure of solids due to slippage and tensile cracking.

In this paper,this new algorithm is described with a numerical example on the anker block problem.

INTRODUCTION

In case of granular materials such as soil or sand,effects of slippage as well as tensile cracking are often considered important to understand their macroscopic behaviour.

In such a case,consideration is definitely needed on the effect of mutual contact of the solid particles.

Solid contact problems of deformable bodies are mathematically very difficult problem because of unknown contact area,local development of plastic zone as well as mutual slippage on the contact surfaces.

Finite element method has been proved to be useful in analysis of the solid contact problems using special element such as the joint elements proposed by Goodman (ref.1).

This method,however,may not be practical except in unusual cases where the contact area is apriori known.

On the other hand,the Rigid Bodies-Spring Model is proposed as a physical model especially suitable for limit analysis of solids and by which mutual slip movement of two adjacent elements can be simulated (ref.2).

Its usefulness has been duly verified by solving many collapse problems of foundation structures where slippage is considered predominant.

Using this element and adopting Yamada's incremental approach (ref.3) the present authors proposed a method of limit analysis by which collapse mechanism mainly due to slip can be treated successfully (ref.4).

Recently Takeuchi proposed an algorithm applicable to the coupled failure problems of slip and tensile cracking.

In this paper the mathematical basis of this algorithm will be described first and then feature of this method will be discussed showing a simple numerical example.

DISCRETE LIMIT ANALYSIS WHERE EFFECT OF TENSILE FAILURE IS TAKEN INTO ACCOUNT

The Rigid Bodies-Spring Model consists of a set of rigid bodies of arbitrary shape and they are connected by the spring system of two types distributed over the contact surfaces adjacent bodies.

Namely one type of spring resists the normal relative displacement while the other resists the relative shear displacement.

Calculating the strain energy of these spring system in term of the relative displacement of two adjacent elements and applying Castigliano's theorem, the stiffness matrix of this new discrete element can be derived.

Outline of the derivation is given in Fig 1 (a), (b), &(c), details of while is also given in the references listed at the end of this paper(ref.2).

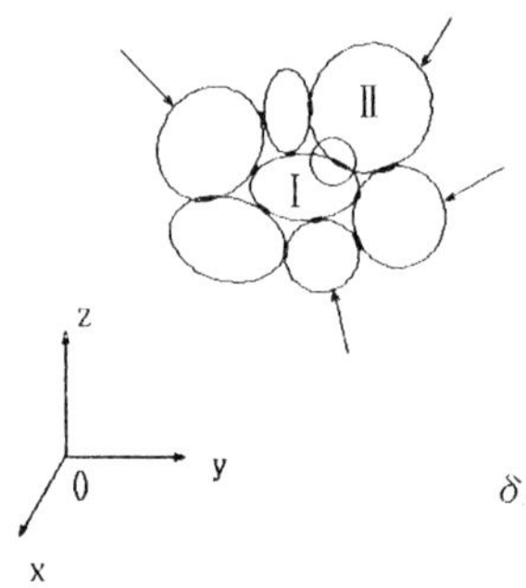

(a) General view of the RBSM system

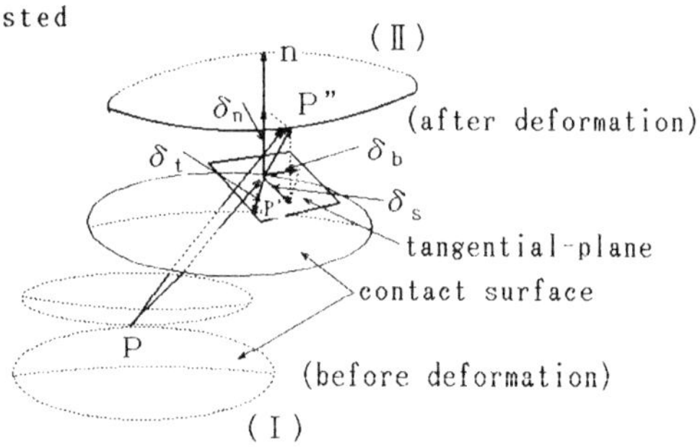

δ_n :relative normal displacements of two rigid elements (I)&(II) (dilatation) at the point P

δ_s :tangential relative displacement of elements (I)&(II)

(b) Contact boundary surface of elements (I)&(II) (enlarged figure of the domain encircled in (a))

Strain energy to be stored in the boundary spring system:

$$V = \frac{1}{2}\iint(k_n\delta_n^2+k_s\delta_s^2)\,dS = \frac{1}{2}\iint\delta^T D\delta\,dS$$

$$= \frac{1}{2}u^T K u \left(K = \iint B^T D B\,dS\right)$$

where k_n, k_S are spring constants of the normal and tangential spring system respectively and δ , $\mathbf{u}$ are displacement vectors defined as fellows:

Fig 1 (a)&(b) 3D RBSM model

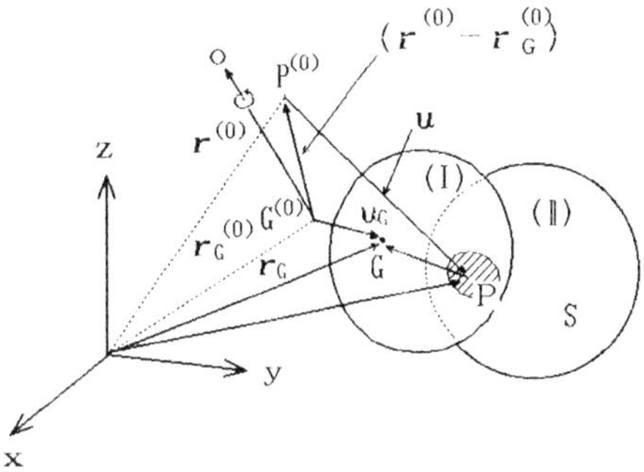

$$r = r^{(0)} + u$$

$$u = u_G + o \times (r^{(0)} - r_G^{(0)})$$

$$= u_G + o \times (r - r_G)$$

S: contact surface

superscript (0) implies

the state before deformation

Rigid body displacement field (1st element)

$$= u_{Gi} + o \times (r - r_G)_i = B_i u_i$$

$(r - r_G)$: position vector of an arbitary point with respect to the centroid

U_i : the displacement vector of an arbitrary point of a given element

u_i : rigid body displacement vector of the element centroid

$(u_i, v_i, w_i, \theta_i, \psi_i, \chi_i)$

u_G, o : translational and rotational displacements of the element centroid respectively

$$u_G = (u_i, v_i, w_i) \qquad o_i = (\theta_i, \psi_i, \chi_i)$$

B_i : shape function of the rigid body displacement field

δ : relative displacement vector on the interelement boundary

$$\delta = B u$$

where $\delta = U_{II} - U_I \quad u = (u_1, u_2)$

(c) Displacement vector in the RBSM system

Fig 1 3D RBSM model

In the RBSM element, Kawai considered that reaction stresses induced are not tensor but vector,and consequently Coulomb's condition may be most realistic constitutive law for such a discrete system representing granular materials.

As well known, Coulomb's condition can be represented by two straight lines which relates the normal stress σ and the shearing stress τ as shown in Fig 2-1.

The stress state on these straight lines represents deformation state of slippage.

If σ reaches σ_t , σ can not become greater than this value and therefore the relation between normal stress and strain can be shown as in Fig 2-2(a).

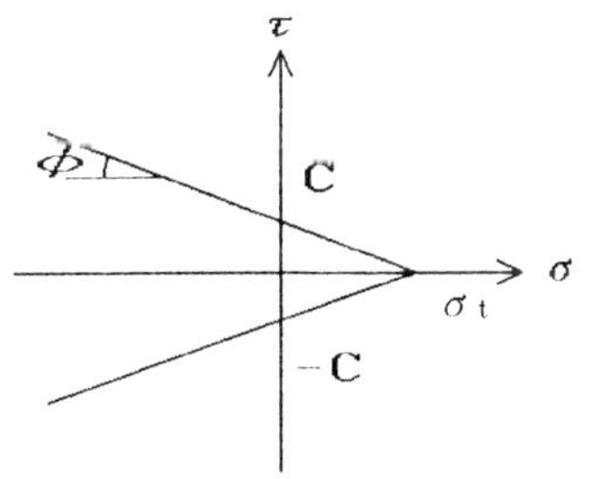

$|\tau| = C - \sigma \tan\phi$

Fig 2-1 Coulomb's condition

This is the state of plastic yielding.

In case of the granular materials like soils, it is commonly observed that σ is relieved as soon as it reaches σ_t as shown in Fig 2-2(b).

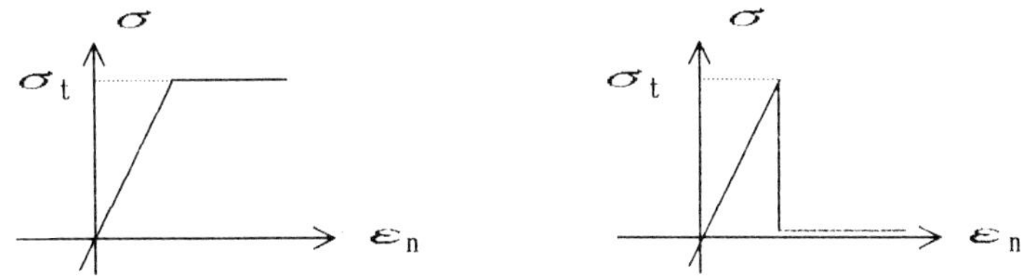

(a) tensile yielding (b) tensile fracture

Fig 2-2 Stress-strain relation at the tensile failure

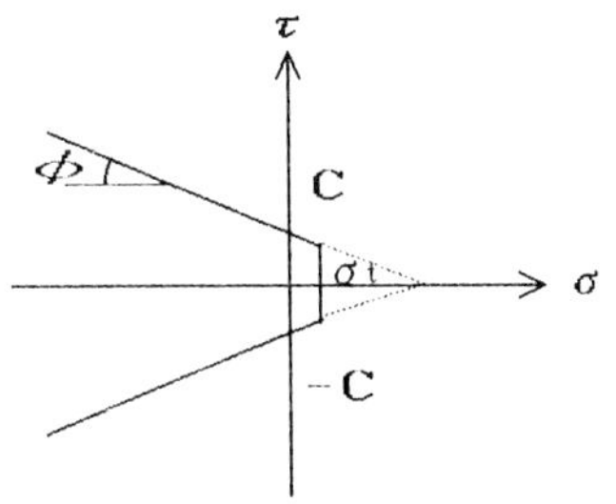

Consequently the yield condition of the soil-like materials can be modified as shown in Fig 2-3.

Regions in Fig 2-3 can be classified by the failure pattern as in Fig 2-4.

Fig 2-3 Modified Coulomb's condition where the tensile failure is considered

In this paper a new algorism is proposed by assuming the spring characteristics as shown in Fig 2-2(b) under the failure condition shown in Fig 2-4.

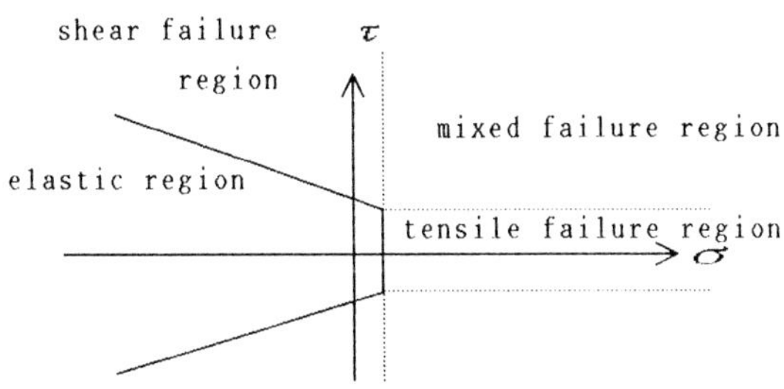

Fig 2-4 Failure patterns

A PROPOSED NEW ALGORISM

A new algorism is proposed by applying the incremental loading procedure developed by Yamada for the nonlinear problem of soil foundations whose failure condition is described in the preceding section.

In Yamada's method, necessary rate of the load increment to yield the most heavily stressed element can be calculated by stress distribution and load increment at the present stage as shown in Fig 3-1.

In this figure the point A represents the stress condition at the previous step while the point C the stress state at the present step.

Actually,however,the stress equilibium can not be attained at the point C but at the point B on the failure curve.

From this condition the required rate of the load increment r can be calculated.

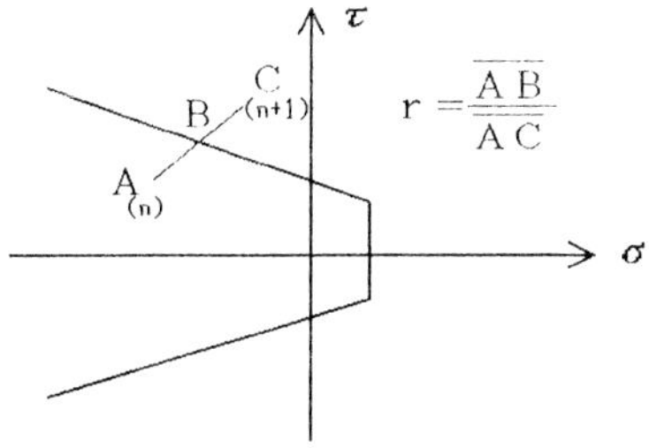

Fig 3-1 Rate of load increment in case of the shear failure

Once the stress point lies on the failure curve, it may move according to the plastic flow rule until the unloading occurs.

Similar calculation must be made in case of the tensile failure as shown in Fig 3-2.

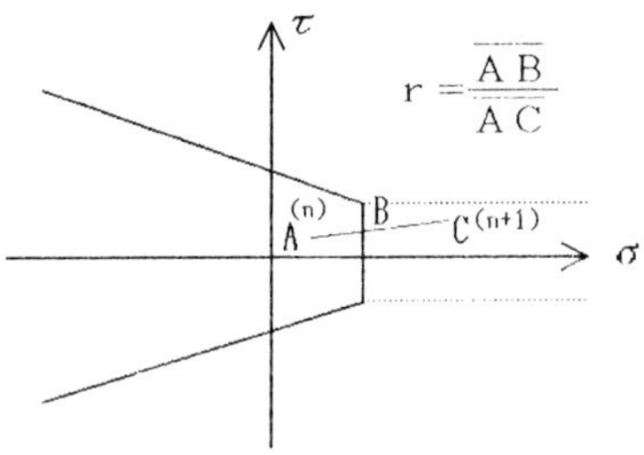

Fig 3-2

Rate of load increment in case of the tensile failure

The point B represents the tensile strength in this figure.

Namely when the stresses in some element exceeds the tensile strength, firstly the stresses should be reduced to the level corresponding to the point B using this rate of stess increment.

And then unbalance of stresses must be relieved in that element as shown in Fig 2-2(b).

Following this operation, both normal and shear springs should be cut to prevent the stress transfer on the common boundary until recontact of these elements occur.

On the other hand possibilities of the tensile (A-B-C) and shear failure (A'-B'-C') must be considered in case of the coupled failure as shown in Fig 3-3.

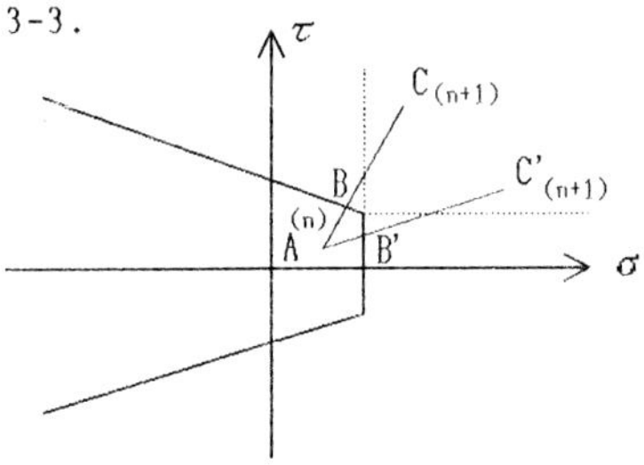

Fig 3-3

Rate of load increment in case of the coupled failure of shear and tension

In this case both rates of stress increment must be calculated and the smaller rate should be selected.

In this way true location of B or B' can be determined.

If the point B is selected, the shear failure is concluded, and the subsequent operation should be made using the flow rule of plasticity,while in case of the point B',it can be concluded that tensile failure occurs and the connection springs are cut after relief of the stress unbalance.

All the possible rates of load increment corresponding to failure patterns should be calculated in all the elements and the minimum rate of load increment must be determined at each step as the rate of load increment corresponding to the next step.

Therefore, essentially, there are only two types of failure (shear and tensile failure) although the coupled failure region exists.

It should be noted here that the spring may fracture due to tension following the plastic flow rule after the shear failure.

CALCULATION OF THE REQUIRED AMOUNT OF STRESS RELAXATION AND RATE OF LOAD INCREMENT

Stress relaxation is usually followed by the tensile failure.

If Yamada's method is applied to this stress relaxation process exactly, endless calculation should be repeated corresponding to the tensile failure which may occur continuously.

Therefore supplement of the operation as shown in Fig 4-1 is to be considered in the iteration at the present loading step.

Here r_{min} implies the cumulative rate of load increment and the calculation must be repeated until $r_{min}=1$ in each step of loading.

In this figure tensile crack initiated at the step 2 and the stress relaxation is considered at the step 3.

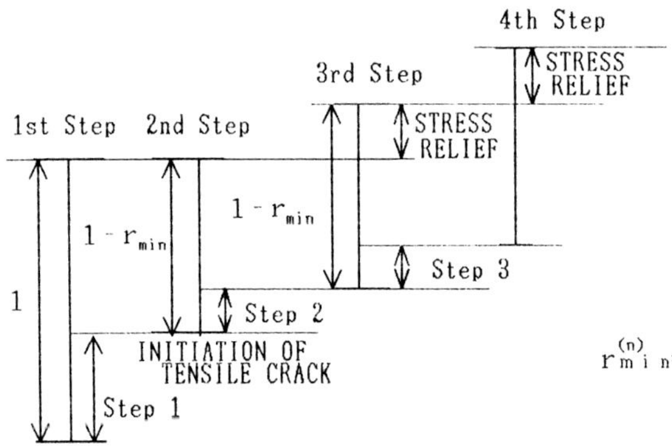

$$r_{min}^{(n)} = r_{min}^{(n+1)} + (1 - r_{min}^{(n+1)}) \times \text{Step } n$$

Fig 4-1 Concept of the rate of load increment

Calculation should be repeated for the residual load increment which is equal to the previous rate of load increment minus the amount of stress relaxation at the step 3.

As will be seen from this figure, the rate of load increment specified before calculation may change due to stress relaxation caused by crack initiation. Therefore in this case concept of r_{min} may not be correct, but the process of progressive failure may be observed.

If $r_{min}=1$ is realized, however, the iteration must be stopped and the result of calculation may be considered reliable.

Therefore if the rate of load increment is taken necessarily small, results of calculation described above may be duly justified.

Fig 4-2 shows outline of the flow diagram of the proposed algorithm.

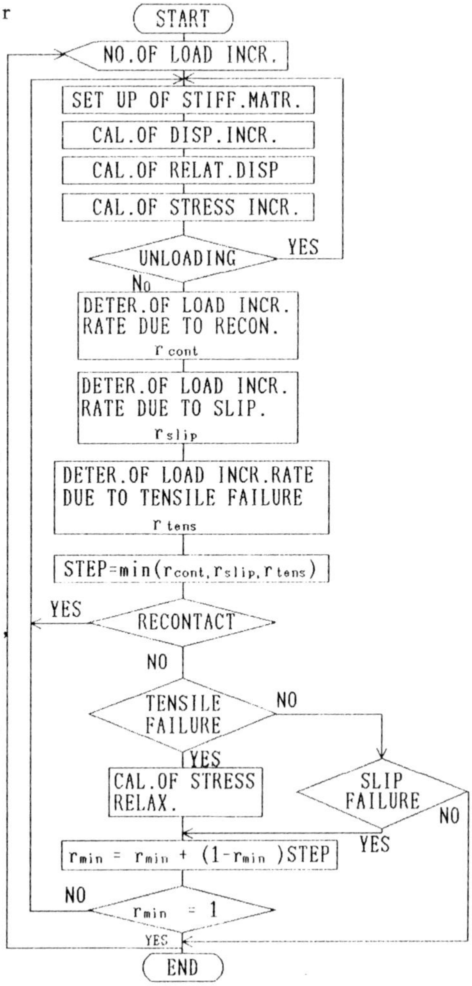

Fig 4-2 Flow diagram of the proposed new algorithm

NUMERICAL EXAMPLE

As a numerical example, behaviour of an anker block in the soil foundation subjected to a horizontal force is studied. As shown in Fig 5-1(a) tensile stress may be induced in the rear vertical wall of a given block and cavity may be produced due to tensile cracking as shownin Fig 5-1(b).

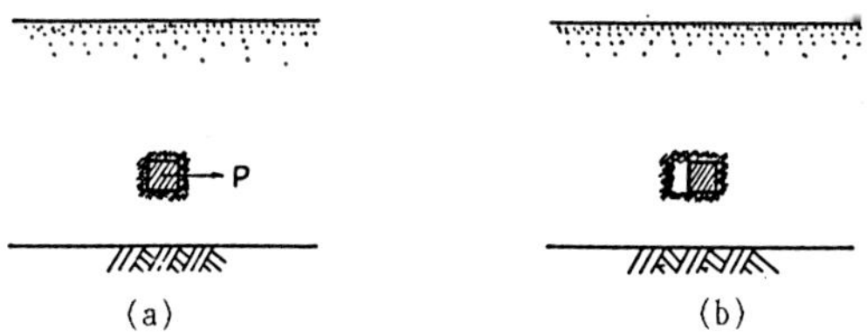

Fig 5-1 Anker block subjected to the horizontal force

Applying previous as well as the present new methods to this problem, analyses were conducted and comparison of two solutions were made.

Fig 5-2 shows the test problem to be analysed and material constants used. In the present analysis effect of the gravitational load was neglected.

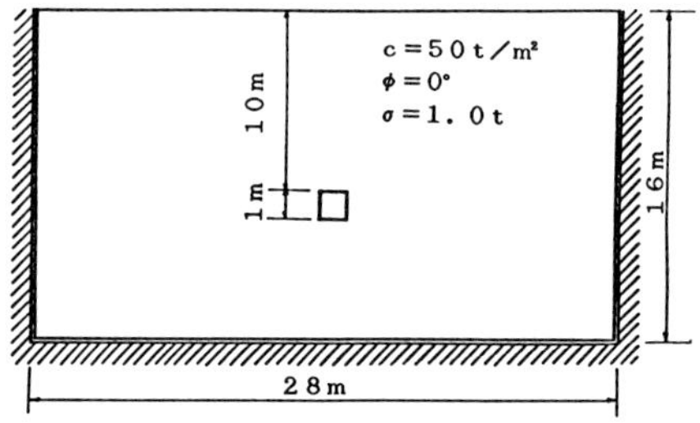

Fig 5-2 A test problem to be analysed

Fig 5-3 shows pattern of the mesh division. Number of the nodes and elements was 234,454 respectively.

And total numbers of the intelement springs and degrees of freedom were 657 and 1362 respectively.

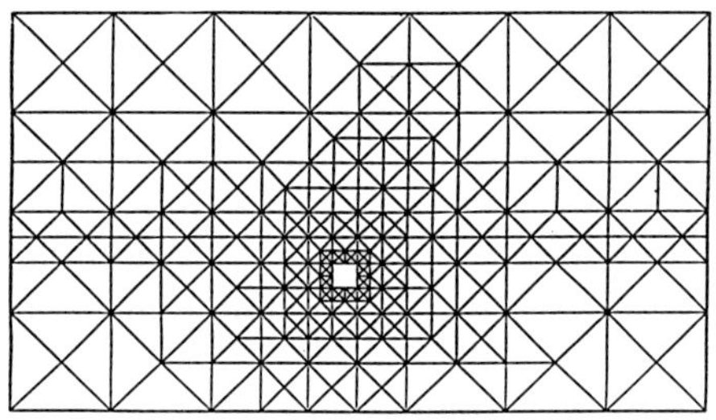

Fig 5-3 Mesh division

Figs. 5-4 and 5-5 show the slip line and displacement patterns of the solution obtained by the previous method.

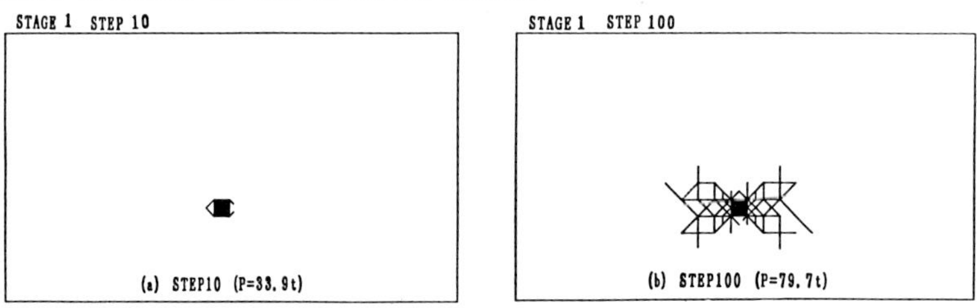

Fig 5-4 Slip line pattern of the solution obtained method

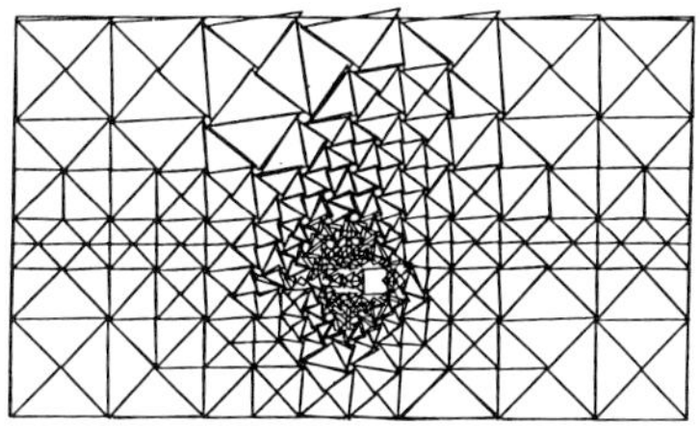

Fig 5-5 Displacement mode of the solution obtained by the previous method

As will be seen in these figures,the slip lines were developed on the rear wall of a given block and indeed the soil behind the block was stretched by the block,and these results are not considered realistic.

Actually the soil must separate from the concrete block,and this is demonstrated in the solution obtained the present method as shown in Fig 5-6(a),(b),and (c).

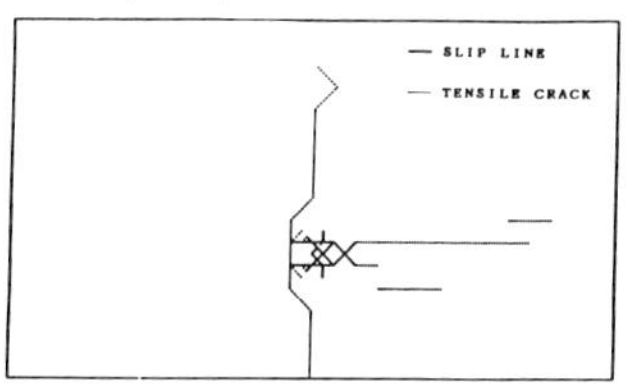

Fig 5-6(a) Slip line pattern of the solution obtained by the present method

Fig 5-6(a) shows the slip line pattern of the solution by the present algorithm at the step 5.

The displacement field and failure mode corresponding to this step are shown in Fig 5-6(b) and (c) respectively.

In this figure it can be seen that not only slip lines but also tension crack may spread on the front region of a given block and the displacement mode may be considerably different from that of the previous solution.

The load was applied in step by step manner taking the increments as 10t,10t,5t,5t,and 5t.

From this figure separation of the soil on the rear wall of the block and rise of the displacement field near the front wall can be seen.

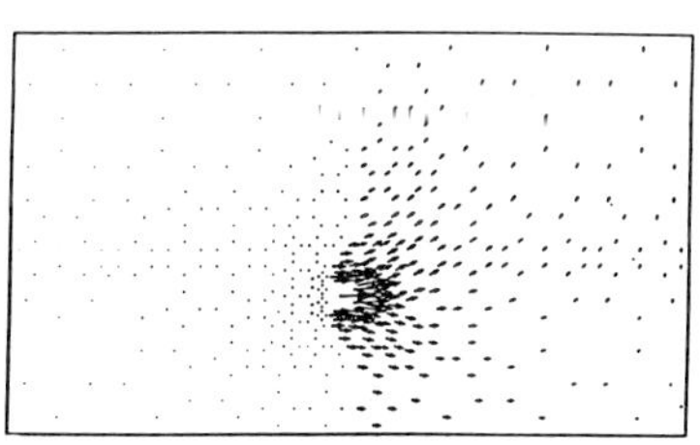

(b)displacement vectol field

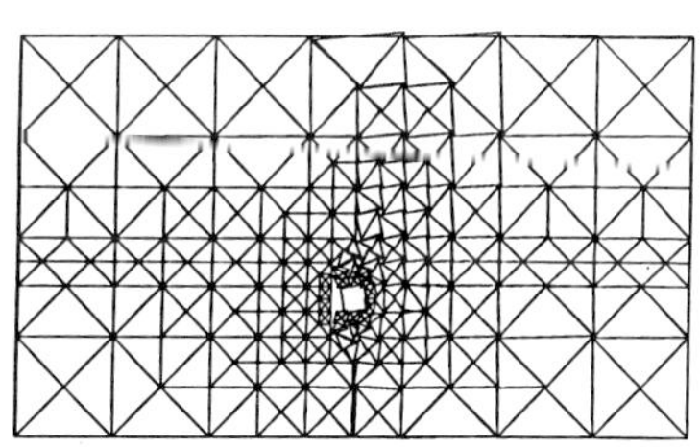

(c)the displacement mode

Fig 5-6(b)&(c)

Displacement field of the solution obtained by the present method.

Finally the load-displacement relationship is given in Fig 5-7. It can be seen from this figure that the displacement corresponding to the latter solution is greater than that of the former,while the load is smaller.

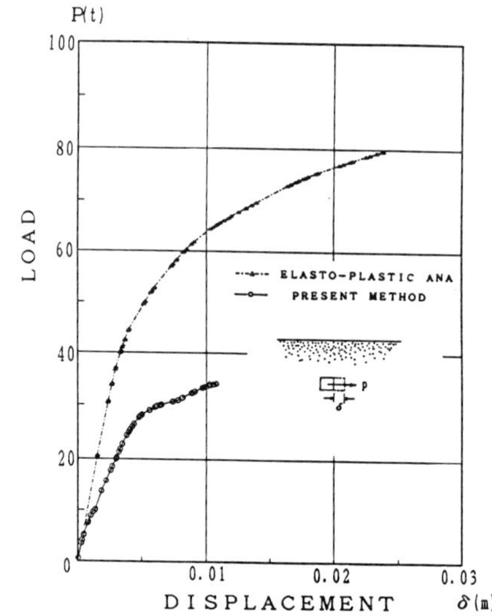

Fig 5-7 Load-deflection curve

CONCLUSION

A New algorithm was developed by which coupled failure due to shear and tensile loads can be treated in this paper.

Although a numerical example illustrated is very simple,it is believed that the present method may be useful to failure analysis of various structual problems in many other fields. It can be concluded that number of "if" sentences may be increased resulting in time-consuming caluculation to compare it with the previous solution procedure,but more realistic analysis can be expected.

REFERENCES

1 Goodman,R.E.:"Method of Geological Engineering in Discontinuous Rock", Weast Publishing (1976)
2 Kawai.T:"New Element Models in Discrete Structual Analysis",Journal of the Society of Naval Architects of Japan Vol.141 (1977)
3 Yamada.Y,Yoshimura.T,and Sakurai.T:"Plastic Stress-Stain Matrix and its Application for the Solution of Elastic-Plastic Problems by the Finite Element Method" Int.J.Mech. Sci.,10,343-54 (1968)
4 Kawai.T and Takeuchi,N:"A Discrete limit Analysis of Foundation Structures" Advances in the Mechanics and the Flow of Granular Materials Vol.II,Trans Tech Publications (1983)

Micromechanics of Granular Materials, edited by M. Satake and J.T. Jenkins
Elsevier Science Publishers B.V., Amsterdam, 1988 — Printed in The Netherlands

COMPUTER SIMULATIONS OF DENSE SPHERE ASSEMBLIES

P.A. CUNDALL
Department of Civil & Mineral Engineering, University of Minnesota, Minneapolis, Minnesota 55455, U.S.A.

SUMMARY

Results are presented from several numerical experiments done on dense sphere assemblies in three dimensions. The experiments are intended for comparison with physical experiments, described by Ishibashi and Chen (ref. 1) and the theoretical results of Jenkins (ref. 2). Periodic boundaries are used to eliminate boundary effects; the effect of sample size is studied in one series of tests, using linear contact stiffnesses. In another test series, contact stiffnesses are nonlinear, according to the Hertz theory. In general, numerical and physical results agree, except for the volume strain observed in a triaxial extension test: this discrepancy may indicate the presence of an initial fabric in the physical sample.

INTRODUCTION

Numerical experiments are done with the computer program TRUBAL, which models assemblies of spheres that interact with one another either by a linear force/displacement law or according to a nonlinear relation derived from the Hertz theory of contact. The method of calculation in two dimensions is described by Cundall & Strack (ref. 3). In three dimensions the procedure is similar (see Cundall & Strack, ref. 4), but the experiments are all done in a periodic solution space in order to eliminate the effects of boundaries. An experiment is described that is designed to show if the periodicity of the sample influences the behavior. The remaining tests are intended to reproduce the physical experiments performed by Chen et al. (ref. 5) and reported here by Ishibashi & Chen (ref. 1).

In TRUBAL, the full equations of motion for every particle are integrated numerically, using an explicit central difference formula. In order to perform quasi-static experiments, the sample is loaded at a rate that is slow enough for inertial effects to be negligible. A small amount of massproportional viscous damping is also used. Interactions between particle pairs are "soft": i.e. contact forces are related to the relative displacement vector of the two particles. The complete set of equations is given by Cundall and Strack (ref. 4).

PERIODIC SOLUTION SPACE

The solution space consists of a periodic cell in the form of a parallelepiped that has a numerical connection between opposite faces. Particles that move out of the cell, through a particular face, are re–mapped back into the cell at a corresponding location on the opposite face. The criterion for re–mapping is that the particle center lies outside of the cell. If the center is near a face, part of the particle may protrude through the face: in this case, an "image" of the particle is introduced into the opposite face, and allowed to interact with other particles in the normal way. The image is displaced from its real counterpart by a distance equal to the dimension of the periodic cell.

The assembly of particles is loaded by distorting the periodic cell, and changing its volume. This type of loading corresponds to a strain–controlled test. Numerically, the effect of a deforming space is accounted for as follows.

1. At each time step, a displacement increment due to the strain rate of periodic space is added to the displacement increment calculated from the real particle velocity:

$$\Delta u_i = \Delta u_i + \dot{e}_{ij}^{(s)} x_j \Delta t \tag{1}$$

where $\dot{e}_{ij}^{(s)}$ is the strain rate of space, which is assumed to be independent of position, and x_i is the particle coordinate.

2. The relative approach velocity of two particles (used to compute force increments) is found by summing their relative real velocities and a velocity related to the strain rate of periodic space:

$$\dot{u}_i^{(rel)} = \dot{u}_i^{(real)} + \dot{e}_{ij}^{(s)} z_j^{(diff)} \tag{2}$$

where $z_i^{(diff)}$ is the vector joining centers.

The particles are thus "carried along" by the periodic space until a collision occurs. The particles then acquire velocities relative to the space.

TEST PROCEDURE AND MEASUREMENTS

Average stresses are measured for the whole sample according to the following formula (see ref. 6):

$$\bar{\sigma}_{ij} = \frac{1}{V} \sum_p R^{(p)} \sum_c n_i F_j \tag{3}$$

where $R^{(p)}$ is the radius of particle p,

n_i is the unit normal vector at contact c,

F_i is the force vector at contact c,

V is the volume of the periodic cell,

$\sum_p$ is the summation over all particles p, and

$\sum_c$ is the summation over all contacts on particle p.

The average strain of the sample is simply the strain of the periodic cell. Since the periodic space imposes strain–control on the sample, some more stress components are held constant. A numerical servo–control is implemented, whereby the strain–rate of the periodic cell is adjusted in such a way that a measured component of the average stress tensor is held constant. For example, to perform a test at constant mean stress, the servo–control operates as follows:

$$\dot{e}_v = \dot{e}_v + G(\overline{\sigma}_o^{(desired)} - \overline{\sigma}_o^{(measured)}) \qquad (4)$$

where G is the 'gain' of the servo–control.

A typical test consists of the following stages:

1. Particles are generated, at random, with sizes corresponding to the desired distribution. There is no packing algorithm; if a particle is generated that overlaps an existing particle, a substitute is generated. The initial cell volume is large, in order to allow the required number of particles to be accommodated.
2. The periodic cell is contracted isotropically until the desired sample porosity is obtained. During part of the compaction phase, contact friction may be increased or decreased in order to modify the final porosity that will be obtained, for a given isotropic stress.
3. The numerical servo–control is activated, and the sample brought to equilibrium under the desired isotropic stress. The required loading path is then followed, by setting the strain rates of periodic space. From time to time, the effects of inertial forces are assessed by setting the strain rates to zero, and noting the changes in stress components; if the changes are too large, the whole test is repeated at lower strain rates.

NUMERICAL TESTS

Linear contact stiffnesses: effect of sample size

Tests were performed with linear springs to represent the deformability at a contact between two particles. The objectives were: a) to compare the three dimensional results wit those of previous two dimensional tests; b) to assess the influence of the periodic nature of the sample. For tests done at large strain, it is believed that the nature of the force/displacement relation at contacts is relatively unimportant, since most deformation occurs by way of sliding and geometrical re–arrangement. Three large strain tests were performed, identical except for different numbers of particles in the samples (45, 150 and 1200 spheres). Two particle sizes were used: two thirds of radius 10 units and one third of radius 15 units. The initial porosity was 0.335, and isotropic stress was 1.5×10^5 units. Normal stiffness was 2×10^8 units and shear stiffness 10^8 units. Friction coefficient was 1.0. Fig. 1 shows the results of biaxial tests done on the three samples, in which isotropic stress was held constant by means of the servo–control. The results for the three different sample sizes are quite similar, although some 'noise' is evident in the 45–particle simulation. It is concluded that the periodic nature of the sample has little effect on behavior.

It is believed that the observed strain softening is an intrinsic characteristic of the material, and not an artifice of the boundary conditions. A localized shear band is difficult to obtain in a periodic sample because a) no particular location in the sample is "special;" and b) shear bands are only kinematically possible for certain angles of band.

Tests with Hertz contact formulation

A nonlinear contact formulation based on the theory of Mindlin & Deresiewcz (ref. 7) was used in a series of small–strain experiments designed to reproduce the physical tests described in refs. 1 and 5. The normal force, F_n, at a contact between two particles is found directly from the normal displacement, u_n, using double precision arithmetic:

$$F_n = m(u_n)^{3/2} \quad , \quad m = [2(2\bar{R})^{1/2}G]/[3(1-\nu)] \quad , \tag{5}$$

where $u_n = (R_1 + R_2 - z)$,
R_1 = radius of the first particle,
R_2 = radius of the second particle,
z = distance between centers,
$\bar{R} = 2R_1R_2/(R_1 + R_2)$,
G = elastic shear modulus of both particles, and
ν = Poisson's ratio of both particles.

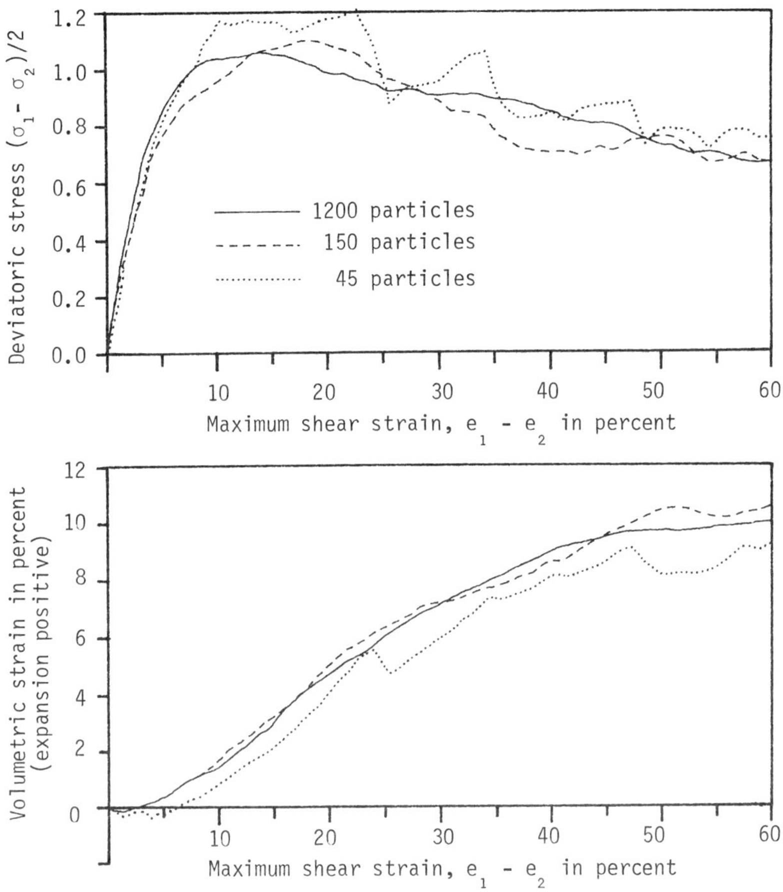

Fig. 1. Effect of sample size, for large–strain biaxial tests, using linear contact stiffnesses

The solution for shear force is more complicated, as it depends on the history of sliding and deformation at the contact. Seridi (ref. 8) presents an elastic–plastic model, involving nested yield surfaces, that produces a solution corresponding to arbitrary histories of normal and tangential forces at a contact (excluding twisting moments). The calculation is time–consuming and, in principle, requires unbounded memory at each contact. However, observations made during the tests described below suggest that it is unnecessary to represent the complete contact behavior in shear. The contribution of shear forces to the average deviatoric stress tensor is rather small in the tests reported here, even for the highest friction used: the shear partition (see ref. 6) is found to account for less than 15% of the total deviatoric stress. The major contribution to deviatoric stress comes from the unequal distribution of normal forces around the particles. Furthermore, 30 to 40% of contacts are sliding, for most of the test. Consequently, the effect of shear contact stiffness on sample behavior is likely to be small. This supposition was confirmed by numerical experiment of re–running a test with half the value of k_s given by equation (6): Fig. 3 shows that the effect is very small. In the following numerical experiments, shear force is proportional to shear displacement at each contact; the shear stiffness depends on normal force, and is equal to the initial loading stiffness, as given in refs. 7 and 8.

$$\Delta F_s = k_s \Delta u_s \ , \ k_s = 2G^{2/3}[3(1-\nu)\bar{R}\,F_n]^{1/3}/(2-\nu) \ . \tag{6}$$

(i)<u>Test conditions.</u> Conditions for the numerical experiments were designed to resemble those in the physical experiments of Ishibashi & Chen (ref. 1), in which two sizes of glass beads were used, with a weight ratio of 2:1.

Properties of glass beads:	shear modulus,	$G = 2.9 \times 10^{10} N/m^2$
	Poisson's ratio,	$\nu = 0.2$
	friction coefficient,	$\mu = 0.15$ or 0.30
Number of particles:		5 of 0.1825 mm rad.
		49 of 0.1075 mm rad.
Isotropic stress		138 KN/m^2
Initial porosity		0.367
Initial coordination number		4.44

The initial porosity in the physical tests was 0.367. It is difficult to obtain simultaneously a certain porosity and a certain isotropic stress icompaction, the assembly begins to carry load at a porosity that is partially determined by the initial random positions of the particles. It is possible subsequently to adjust the porosity, by deforming the assembly mechanicaly, but this approach was not taken, as it would bias the experiments by introducing fabric. No way has been found to achieve the desired porosity except by trial and error.

According to measurements reported by Chen et al. (ref. 5) the friction coefficient for sliding between glass beads is 0.15. However, subsequent measurements suggest a higher value approaching perhaps 0.3 (ref. 9). Results are reported below for two different values (0.15 and 0.3).

Triaxial tests are performed in which the minor and intermediate principal strains are equal. In TRUBAL, the following set of principal strain rates are applied to the periodic space.

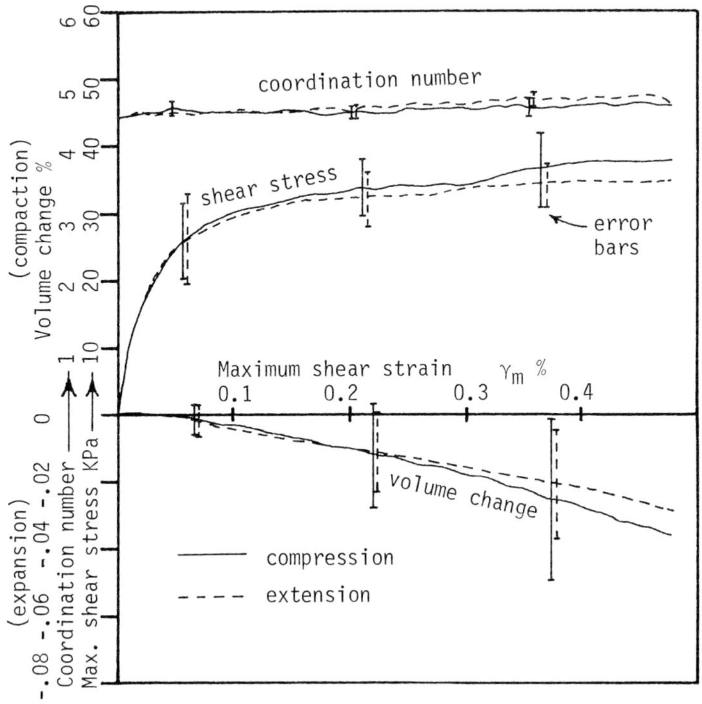

Fig. 2. 54–sphere sample: triaxial compression and extension results for friction coefficient of 0.15.

$$\dot{e}_1 = (2/3)\dot{\gamma}_m + \dot{e}_v/3$$
$$\dot{e}_2 = -(1/3)\dot{\gamma}_m + \dot{e}_v/3$$
$$\dot{e}_3 = -(1/3)\dot{\gamma}_m + \dot{e}_v/3 \quad (7)$$

where γ_m is the maximum shear strain plotted in the figures to follow.

The strain rate $\dot{\gamma}_m$ is set by the experimenter, and the volumetric strain rate, $\dot{e}_v$, is the controlled by TRUBAL so as to keep the isotropic stress constant throughout the test. In order to reduce the effect of sample variability, the results are averaged over three tests: each one corresponds to a different direction of major principal axis – in the x,y or z coordinate directions. The standard deviation of the three results is plotted as an error bar plotted on both sides of the mean curve.

(ii)Results. Fig. 2 shows results from compression and extension tests done on the 54–sphere sample for a contact friction of 0.15. Note thatcomparisons can be made with the results of Chen (ref. 5). Figs. 3 and 4 show the results from similar tests, for contact friction

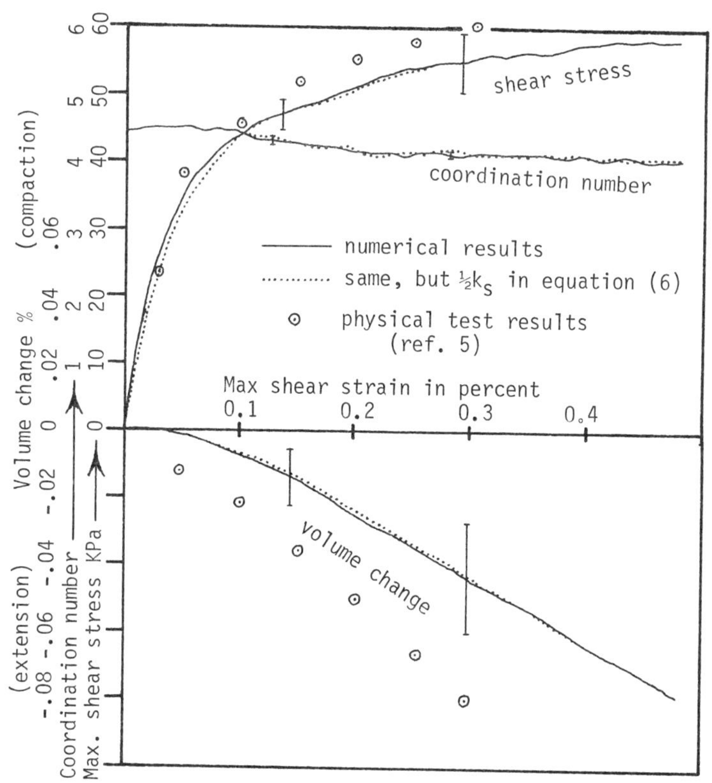

Fig. 3. 54–sphere sample: triaxial compression results for friction coefficient of 0.3; physical test results of ref. 5 superimposed.

of 0.3. Superimposed on the figures are results from the physical tests (ref. 5). The most striking difference is that the initial compaction seen in the physical extension tests is not present in the numerical tests. It is suspected that an initial depositional fabric in the physical sample is responsible. In an attempt to reproduce this effect numerically, the sample was preloaded to 0.5% shear strain, and then unloaded until the deviatoric stress was almost zero. The compression and extension tests were then repeated, starting from this new state. Fig. 5 records the results. As expected, there is initial compaction in the extension tests, but the stress–strain behavior now deviates from that of the physical test: the compression results show evidence of a yield surface that 'remembers' the pre–loading stress level. Evidently this is caused by locked–in shear forces at contacts (see ref. 6). An attempt was made to remove these locked–in forces, while retaining the anisotropic fabric, but the sample collapses to a low porosity state.

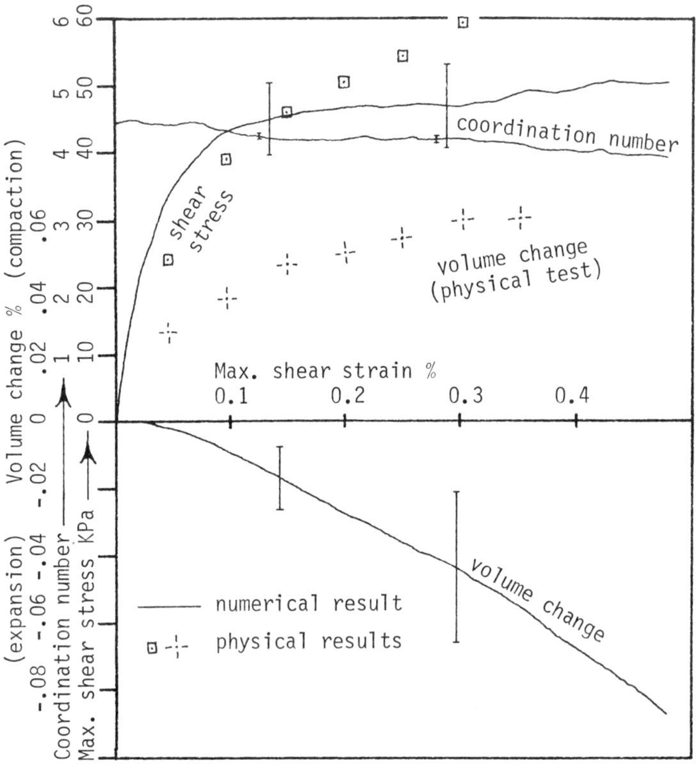

Fig. 4. 54–sphere sample: triaxia extension results for friction coefficient of 0.3; physical test results of ref. 5 superimposed.

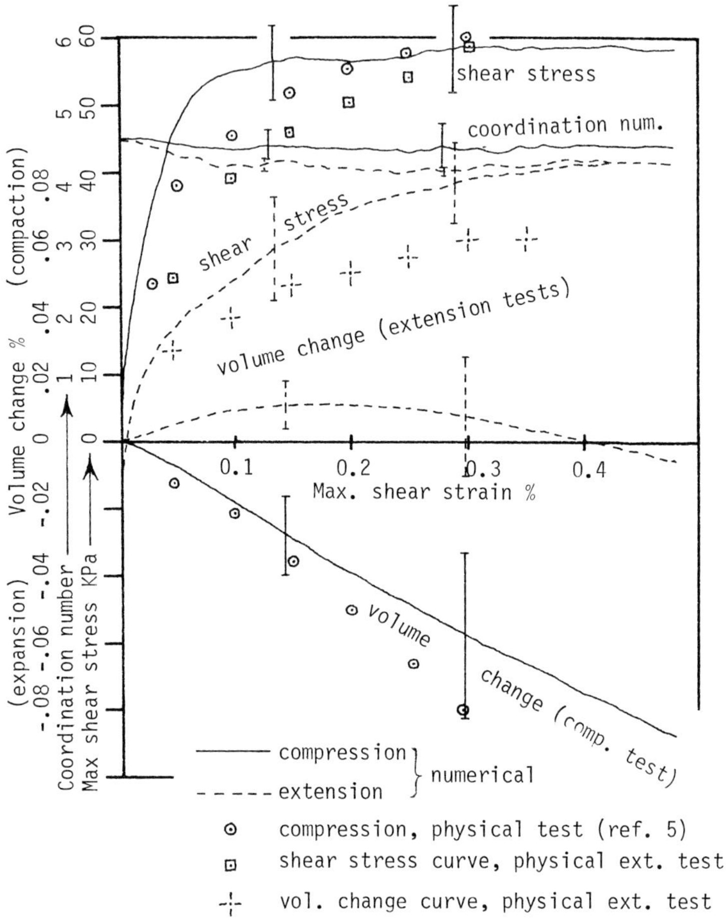

Fig. 5. 54–sphere sample: triaxial compression and extension results following initial loading and unloading cycle.

ACKNOWLEDGEMENT

Much of the development work on TRUBAL was support by the National Science Foundation under grant CEE 831–0729. The research was done in collaboration with O.D.L. Strack of the University of Minnesota, J.T. Jenkins of Cornell University, and I. Ishibashi of Old Dominion University. Their essential contribution is greatly appreciated.

REFERENCES

1. I. Ishibashi & Y–C Chen, Dynamic shear moduli and their relationship to fabric of granular materials: in this volume.
2. J.T. Jenkins, Small strain uniaxial deformations of a granular material: in this volume.
3. P.A. Cundall & O.D.L. Strack, A discrete numerical model for granular assemblies, Geotechnique, 29 (1979) 47–65.
4. P.A. Cundall & O.D.L. Strack, The distinct element method as a tool for research in granular material, Part II: report to NSF, Dept of Civil and Mineral Engineering, University of Minnesota, 1979.
5. Y.–C. Chen, I.Ishibashi, and J.T. Jenkins, Dynamic shear modulus and fabric. Part I. Depositional and induced anisotropy, Geotechnique (in press).
6. P.A. Cundall & O.D.L. Strack, Modeling of microscopic mechanisms in granular material, in: J.T. Jenkins & M. Satake (eds.), U.S.–Japan Seminar on new models and constitutive relations in the mechanics of granular material, Elsevier Scientific Publications, Amsterdam, 1983.
7. R.D. Mindlin & H. Deresiewicz, Elastic spheres in contact under varying oblique forces. J. Appl. Mech., 20 (1953), 327–344.
8. A. Seridi, An incremental elastic–plastic model for the forcedisplacement relation at the contact between elastic spheres, PhD thesis, Rensselaer Polytechnic Institute, New York, August 1984.
9. I. Ishibashi, personal communication, October 1987.

Micromechanics of Granular Materials, edited by M. Satake and J.T. Jenkins
Elsevier Science Publishers B.V., Amsterdam, 1988 — Printed in The Netherlands

GRANULAR ASSEMBLY SIMULATION FOR GROUND COLLAPSE

K. IWASHITA[1], Y. TARUMI[2], L. CASAVERDE,[3] D. UEMURA[4], K. MEGURO[5] and M. HAKUNO[6]

[1]Graduate Student, University of Tokyo, Bunkyo, Tokyo (Japan)

[2]Nihon Public Highway Cooporation, Sapporo Branch, Sapporo (Japan)

[3]Nihon Sekkei Co., Shinjuku Mitsui Building, Tokyo on leave from Catholic University, Lima (Peru)

[4]Shimizu Construction Co., Yashio City, Saitama (Japan)

[5]Graduate student, Univerity of Tokyo, Tokyo (Japan)

[6]Earthquake Research Institute, University of Tokyo, Bunkyo, Tokyo (Japan)

SUMMARY

Using Cundall's simulation method, the present work aims at study in the mechanism of slope collapse, non-linear dynamic response of structural foundation, liquefaction of loose ground under vibrational excitations, and rupture of earthquake fault.

Results achieved using the model idealized as a continuous medium, which have been widely applied in the past, differ from those gained by using the granular assembly model adopted in this study. Ignoring the influence of movements of soil particles, including subsidence and rotation of particles, in the former method seems to be the cause of such differences.

INTRODUCTION

Soil is basically noncontinuous medium of numerous particles. Some phenomena such as rock avalanche, debris flow and dilatancy of soil can be clarified only by considering the soil as a discrete granular assembly.

Tho authors modified Cundall's Distinct Element Method (ref. 1-3) by introducing a few more cooffioients, such as a restitution factor and the characteristics of cohesion. They appliod the method to several problems in the geotechnical engineering (ref. 4-7, 9) and seismology(ref. 8).

SIMULATION RESULTS

PACKING OF PARTICLES BY DROPPING (ref.6)

Fig.1 Initial Position for the Process of Packing 5040 Particles

Fig.2 Results of the First Step of the Packing Process by Dropping on the Intermediate Floors of a Box

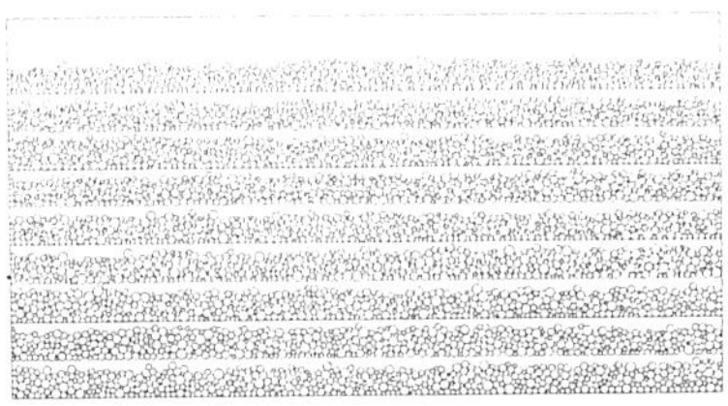

Fig.3 The Result of Approximating the Boxes and Taking off the Intermediate Floors

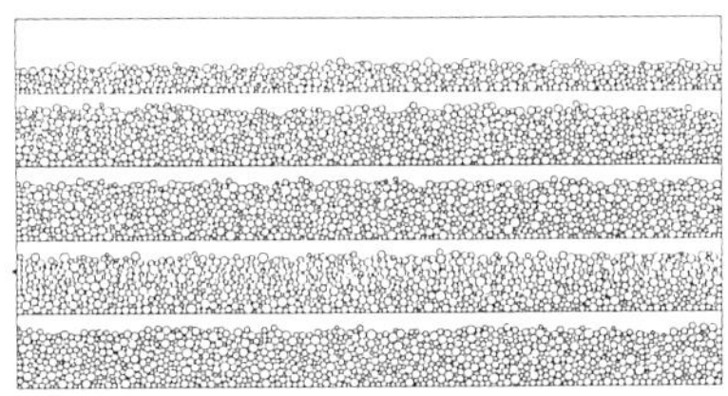

Fig.4 The Result of Packing Process

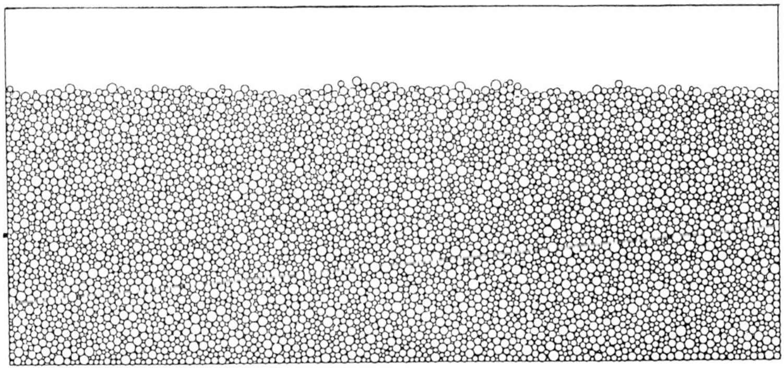

Fig.5 Result of the Packing process for 5040 Particles

Sliding of Cohesionless Cliff

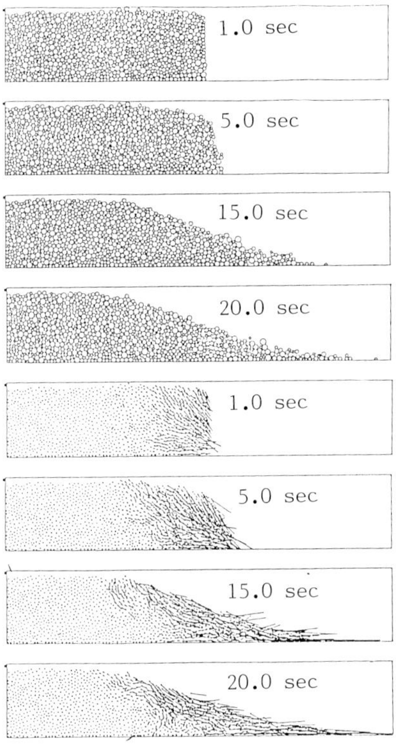

Fig.6 Particles Position and Velocity Distributions

(ref.6)

Slide of slope after 0.5 sec duration Horizontal shock

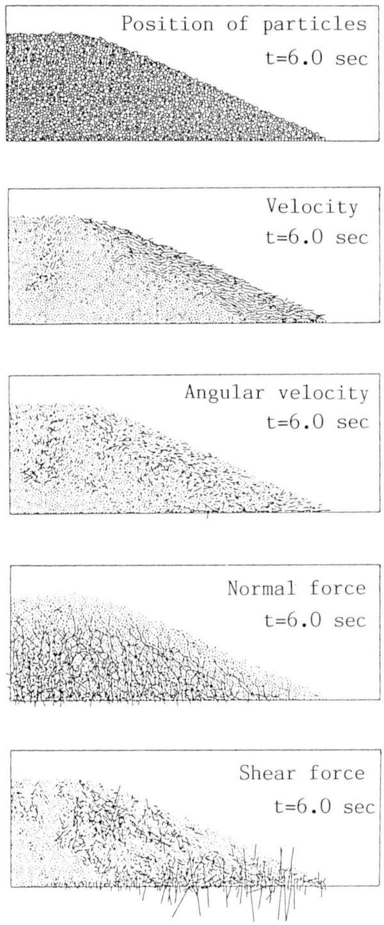

Fig.7 Slide of slope

(ref.6)

Slide of cohesive embankment due to horizontal shaking (ref.7)

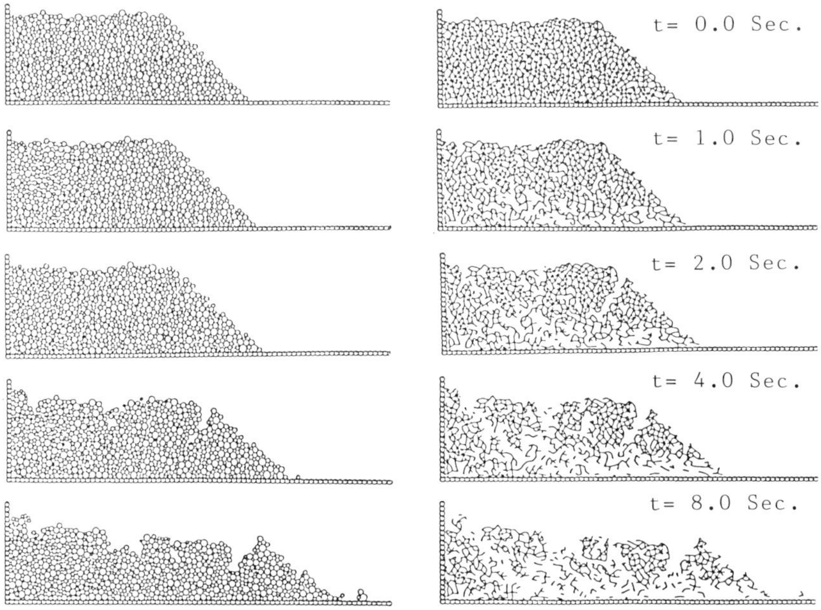

Particles location Spring between particles

Fig.8 Slope slide

(Angle of slope:45°, Shaking:2Hz, 1000Gal Sine Wave)

Slide of cliff (cohesive soil) (ref.7)

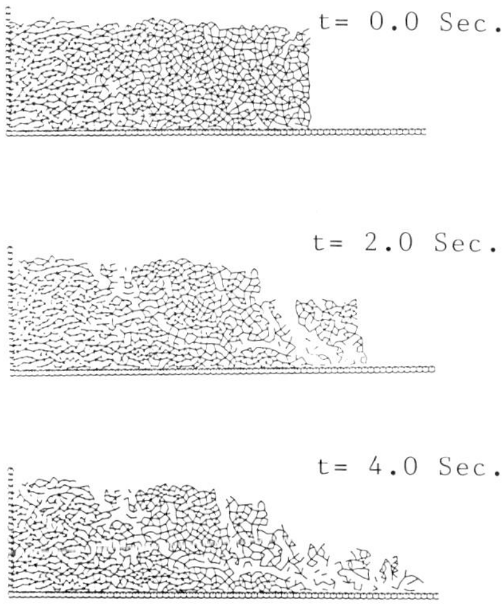

Fig.9 Slide of cliff without shaking
(Spring between Particles)

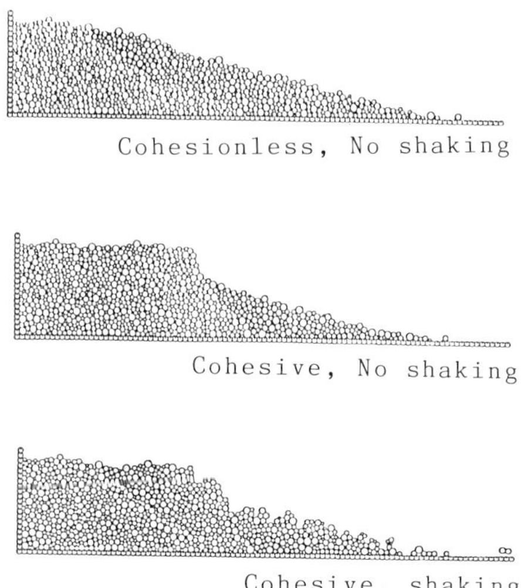

Fig.10 Slide of cliff
(t= 4.0 Sec.)

Rupture of earthquake fault (ref.8)

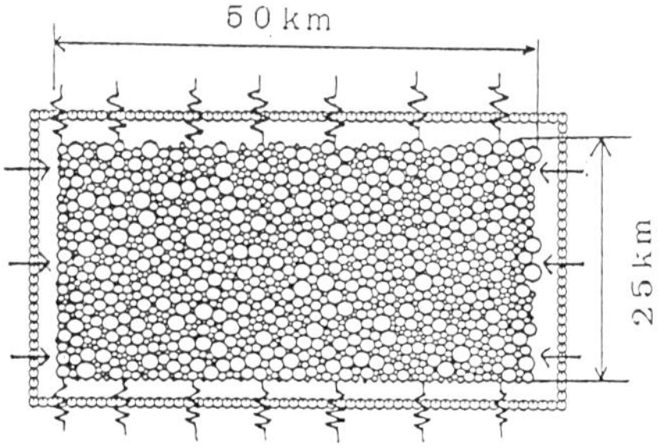

Fig.11 Ground model (1)
(Compressed in X-direction, and supported by spring in Y-direction)

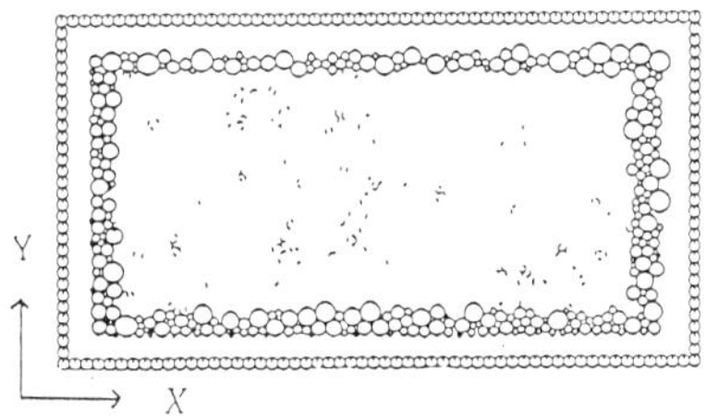

Fig.12 Rupture distribution of model (1)

Fig.13 Ground model (2)
Distribution of internal springs
(Compressed in X-direction, Dotted line is fault Zone which is weakened and thick line is ruptured point)

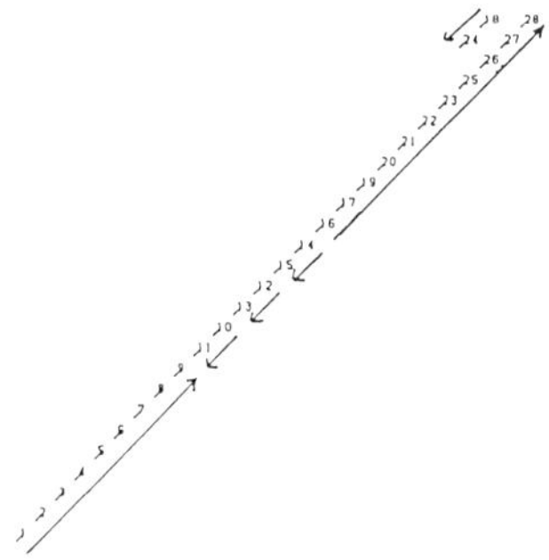

Fig.14 Repture distribution of model (3) (Number is the order of rupture and arrow shows the rupture direction

Penetration of a Pile into Ground (ref.4)

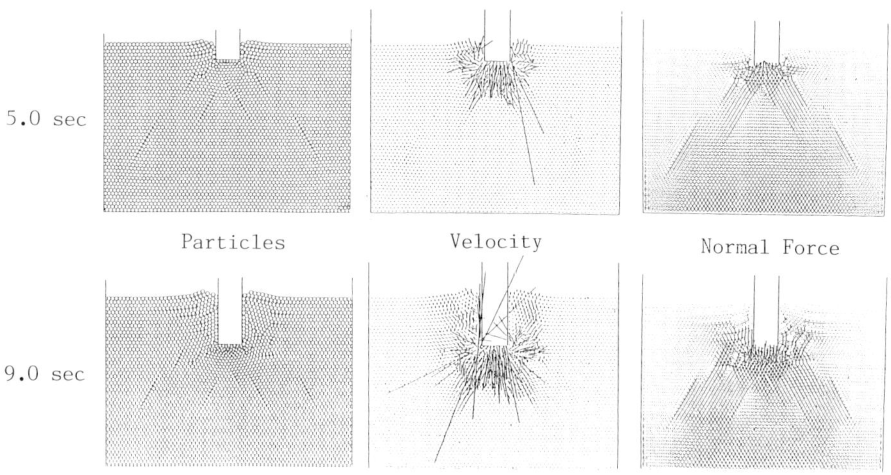

Fig.15 Penetration process of a pile

Shaking of structural foundation (ref.5)

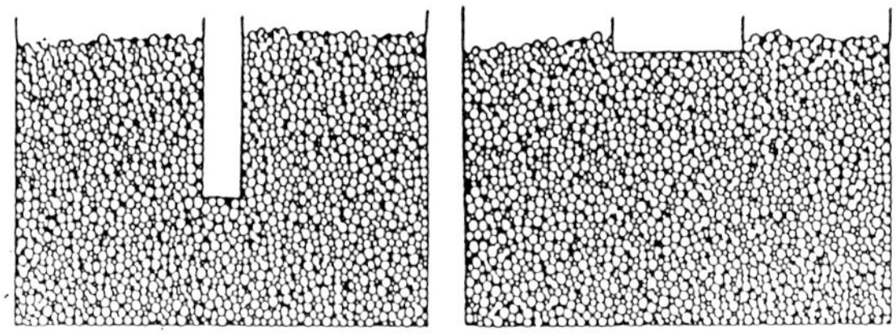

Fig.16 Assembly model

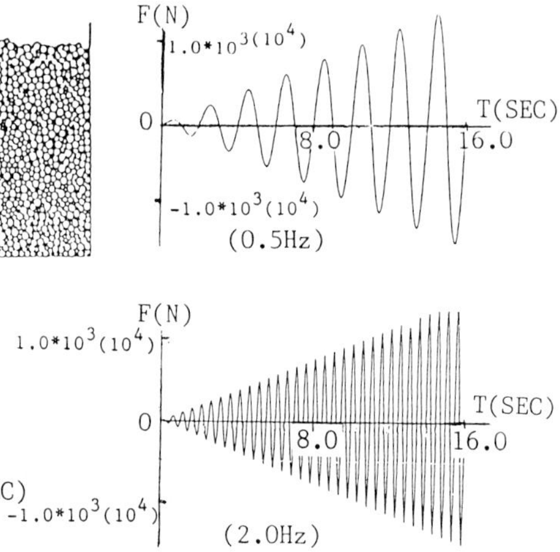

Fig.17 Horizontal Input

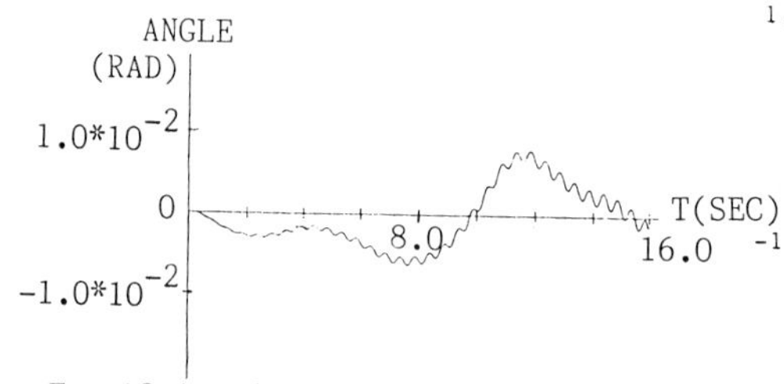

Fig.18 Angular response of a pile foundation (2Hz)

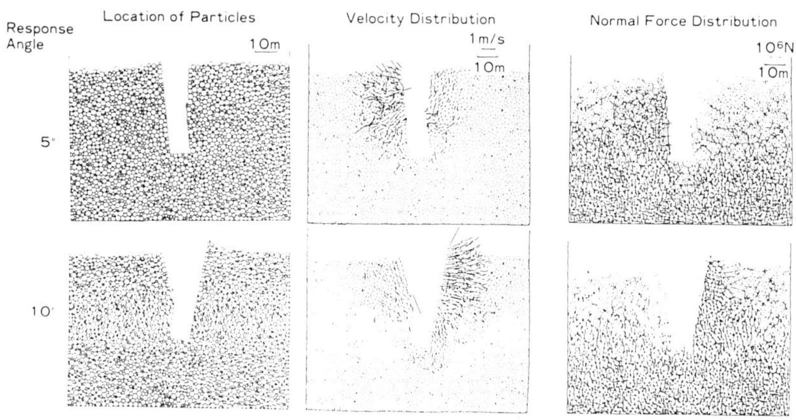

Fig.19 Response of a pile

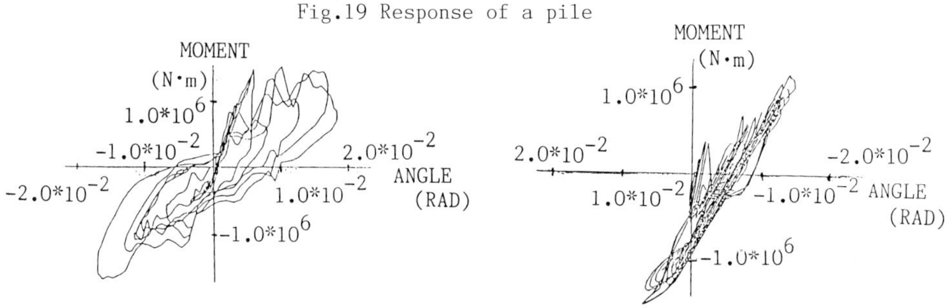

Flat foundation (0.5Hz) Pile foundation (2Hz)

Fig.20 Hysteretic relationship between moment and angular response

Liquefaction analysis of sand (ref.9)

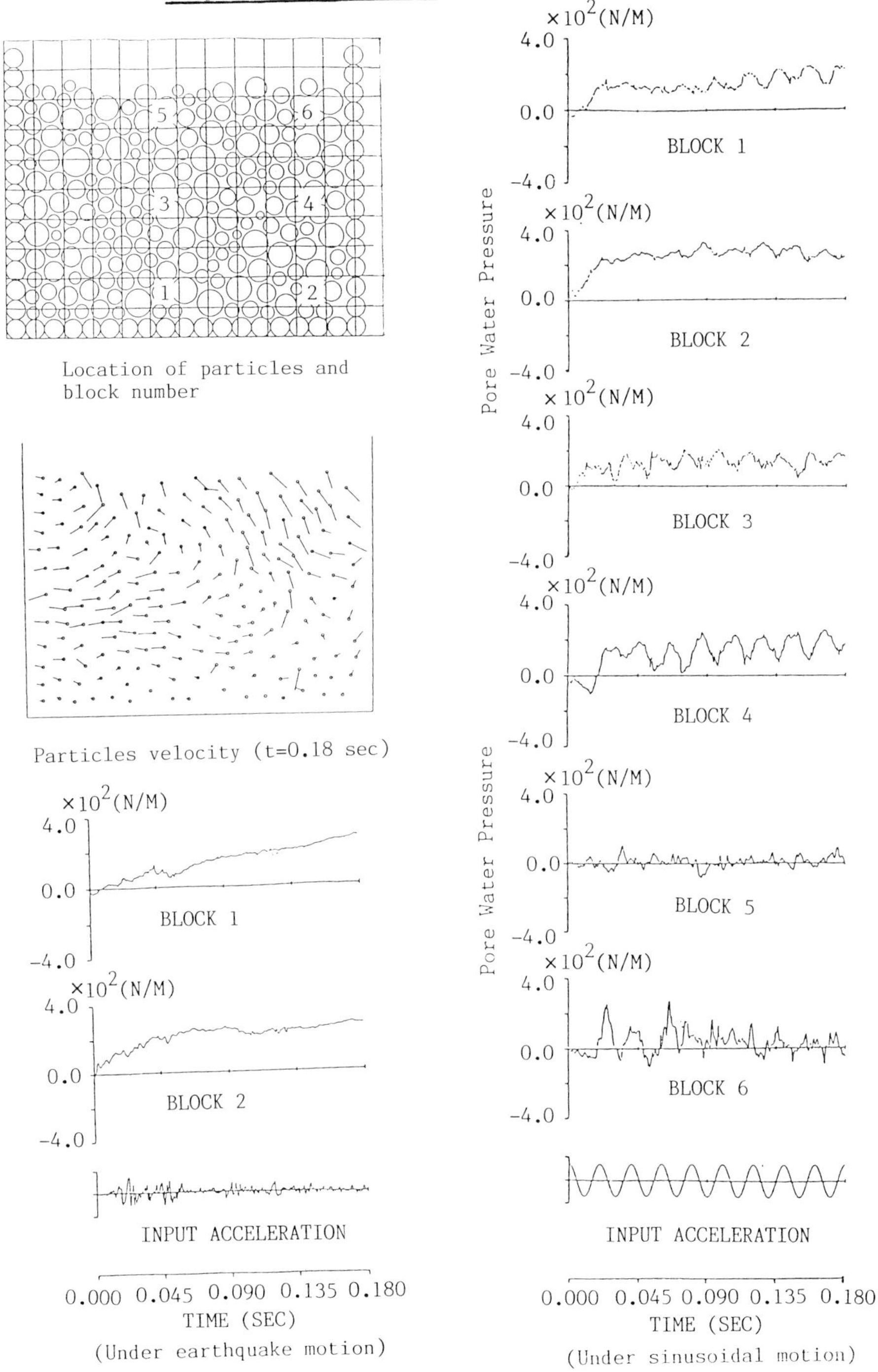

Fig.21 Time history of excessive pore water pressure

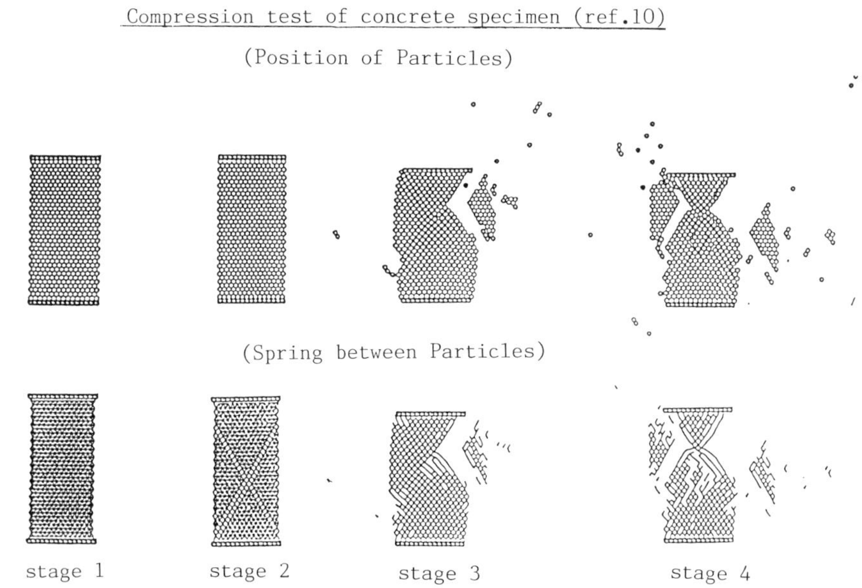

Fig.22 Fracture process of concrete specimen

REFERENCES

1. P.A. Cundall, A. computer model for simulating progressive, large scale movements in blocky rock system, Symp. ISRM, Nancy, France, Proc. 2, 1971, 129-136.
2. P.A. Cundall and O.D.L. Strack, A discrete numeral model for granular assemblies, Geotechnique, 299, 1, 1979, 47-65.
3. H. Kiyama and H. Fujimura, Analysis on gravity flow of rock granular assemblies by Cundall model, Proc. Japan Society of Civil Engineers, 1983, 333, 137-146 (in Japanese).
4. D. Uemura and H. Hakuno, Granular assembly simulation for ground collapse, Bulletin of the Earthquake Research Institute, University of Tokyo, 62, 1987, 19-59.
5. D. Uemura and M. Hakuno, Granular assembly simulation with Cundall's model for the dynamic collapse of the structural foundation, Structural Eng./Earthquake Eng. Proc. of Japan Society of Civil Engineers 4, No. 1, 1987, 155-164.
6. L. Casaverde A., Y. Tarumi, M. Hakuno, Granular assembly simulation for slope slide, Structural Eng./Earthquake Eng. Proc. of Japan Society of Civil Engineers, "submitted for publication".
7. K. Iwashita, M. Hakuno, Granular assembly simulation for collapse of cohesive materials, Structural Eng./ Earthquake Eng. Proc. JSCE" submitted for publication".
8. K. Iwashita, M. Hakuno, Granular assembly simulation for rupture of earthquake fault, Structural Eng./Earthquake Eng. Proc. JSCE "submitted for publication".
9. Y. Tarumi, M. Hakuno, Granular assembly simulation for sand liquefaction, Structural Eng./Earthquake Eng. Proc. JSCE "submitted for publication".
10. K. Meguro, M. Hakuno, Granular assembly simulation for rupture of concrete material, Structural Eng./Earthquake Eng. Proc. JSCE "submitted for publiation".

Micromechanics of Granular Materials, edited by M. Satake and J.T. Jenkins
Elsevier Science Publishers B.V., Amsterdam, 1988 — Printed in The Netherlands

APPLICATIONS OF THEORETICAL CONTACT MECHANICS TO SOLID PARTICLE SYSTEM SIMULATION

C. THORNTON and C. W. RANDALL
Civil Engineering Department, Aston University, Aston Triangle, Birmingham B4 7ET, (England)

SUMMARY

The paper considers the application of theoretical contact mechanics to computer simulation of dissipative solid particle systems. Theoretical and computational requirements are discussed, and results obtained from simulated impact experiments are presented.

INTRODUCTION

Previous computer simulations of solid particle systems based on the methodology developed by Cundall and Strack (ref 1) have used linear viscoelastic (parallel spring and dashpot) laws to model the normal and tangential stiffnesses at the interparticle contacts (refs 2, 3, 4, 5, and 6). Motivated by problems associated with the process engineering industry current implementations of the program at Aston now incorporate more realistic algorithms, based on established theories of contact mechanics to model the particle - particle interactions. A significant consequence of implementing such laws is the ability to quantify particle velocities, interparticle forces, contact stresses and the macroscopic stress state in terms of real physical units. It is also possible for certain types of problem to perform the simulated experiments in real time.

CONTACT MECHANICS

Current work is restricted to assemblies of topographically smooth elastic spheres with interparticle friction μ. It is, therefore, necessary to construct suitable algorithms, based on the theories of Hertz and Mindlin, see (ref 7), to model the normal and tangential incremental stiffnesses at the interparticle contacts.

Normal contact stiffness

For two spheres of radii R_i and elastic properties E_i, G_i and ν_i (i = 1,2) the classical theory of Hertz relates the normal force N to the relative approach α of the centres of the two contiguous spheres and the radius a of the circular area of contact by the expressions

$$\alpha^3 = 9N^2/16R^*E^{*2} \qquad (1)$$

$$a^3 = 3NR^*/4E^* \qquad (2)$$

where

$$1/R^* = 1/R_1 + 1/R_2 \qquad (3)$$

$$1/E^* = (1-\nu_1{}^2)/E_1 + (1-\nu_2{}^2)/E_2 \qquad (4)$$

and $\alpha = a^2/R^*$ (5)

Rearranging (1) we obtain

$$N = 4E^*(R^*\alpha^3)^{1/2}/3 \qquad (6)$$

from which the incremental stiffness is defined, using (5) as

$$k_n = \Delta N/\Delta\alpha = 2E^*a \qquad (7)$$

noting that on initial contact, when $a = 0$, it is necessary to use (6) with $N = \Delta N$, $\alpha = \Delta\alpha$.

Tangential contact stiffness

For a constant normal force, the effect of a tangential force $T<\mu N$ is to cause 'slip' over part of the contact area. Slip is initiated at the circumference of the circular contact area and, as T increases, an annular area of slip develops spreading radially inwards until, when $T = \mu N$, rigid body sliding occurs. The development of the slip annulus involves a dissipative process. If the tangential force is reduced then slip in the opposite direction (counterslip) spreads radially inwards from the perimeter of the contact area. In general, all load reversals initiate slip reversals that spread radially inwards from the perimeter of the contact area, rather than causing a recession of the existing slip annulus. Consequently the tangential stiffness is dependent on the loading history.

The tangential force is also dependent on the magnitude of the normal force and hence there exists an infinite set of geometrically similar force- displacement curves, each corresponding to a different value of N. Figure 1 shows two such curves diagramatically; illustrating loading, unloading and reloading conditions. A theoretical investigation of the behaviour of elastic spheres in contact under varying oblique forces was presented by Mindlin and Deresiewicz (ref 8). Several loading sequencies involving variations of both normal and tangential forces were examined from which general procedural rules were identified.

The procedure is to update the normal force and contact area radius followed by calculating ΔT using the new values of N and a. By reanalysing the loading cases considered in (ref 8) it can be shown that for loading, unloading, and reloading, the tangential incremental displacement may be expressed as

$$\Delta\delta = (1/8G^*a)[\pm\mu\Delta N + (\Delta T \mp \mu\Delta N)/\theta] \qquad (8)$$

except when

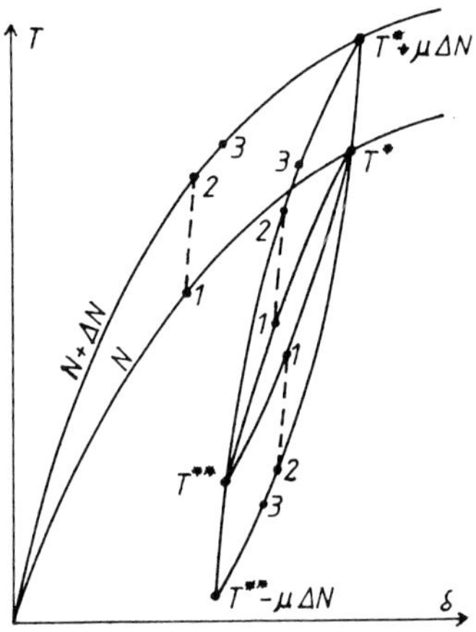

Fig.1 Tangential force - displacement relationship

$$\Delta\delta < \mu \Delta N/8G^*a \text{ and } \mathrm{sign}(\Delta\delta) = \mathrm{sign}(\Delta N) \tag{9}$$

Rearranging (8) we obtain

$$\Delta T = 8G^*a\, \theta\, \Delta\delta \pm \mu\Delta N(1 - \theta) \tag{10}$$

where

$$1/G^* = (2 - \nu_1)/G_1 + (2 - \nu_2)/G_2 \tag{11}$$

$$\theta^3 = 1 - (T + \mu\Delta N)/\mu N \qquad \text{(loading)} \tag{12}$$

$$\theta^3 = 1 - (T^* - T + 2\mu\Delta N)/2\mu N \qquad \text{(unloading)} \tag{13}$$

$$\theta^3 = 1 - (T - T^{**} + 2\mu\Delta N)/2\mu N \qquad \text{(reloading)} \tag{14}$$

and the negative sign in (10) is only invoked during unloading; the parameters T^* and T^{**} are defined in Fig 1. For a current state given by point 1 in Fig. 1 (during loading, unloading, or reloading) a tangential incremental displacement corresponding to

$$|\Delta\delta| = \mu |\Delta N| / 8G^*a \text{ with } \Delta N > 0 \tag{15}$$

will result in a new state given by point 2 on the curve corresponding to the new value of N. Larger values of $|\Delta\delta|$ will result in a state farther along the curve such as point 3. A problem occurs if the conditions given in (9) occur since point 2 is not reached and the new state does not lie on the curve corresponding to the new value of N. This case has been considered analytically by Szalwinski (ref 9) but for computer implementation the problem is solved by setting $\theta = 1$ in (10) until $8G^*a\Sigma\Delta\delta > \mu\Sigma\Delta N$.

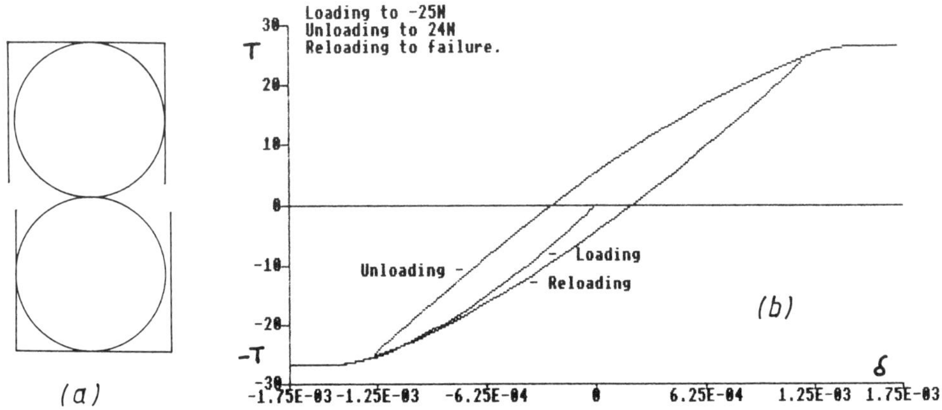

Fig. 2. Validation test results (constant normal contact force = 88N)

COMPUTER IMPLEMENTATION AND VALIDATION

To implement the above on the computer a flag is required to indicate the initial direction of the tangential displacement. It is also necessary, as indicated in Fig. 1, to store T^* and T^{**} and, after using (10), to update $T^* = T^* + \mu\Delta N$ and $T^{**} = T^{**} - \mu\Delta N$ in preparation for the next timestep. Obviously much more complex loading histories than that shown in Fig. 1 are possible during a computer simulated experiment. However, this would require all load reversal points to be stored in the memory map and since the number of possible reversals is not known a priori this would be impractical. Consequently, it has been decided to model according to theoretical contact mechanics the loading, unloading and reloading behaviour as illustrated in Fig. 1. However, if during reloading further reversals occur within the range T^{**} to T^* the behaviour is treated as nonlinear elastic using (10) and (14).

In order to check the coding for the new interaction laws a series of simple test simulations were devised. Two identical spheres were created with the following properties: R = 10 mm, μ = 0.3, ρ= 7800kg/m^3, E = 210 GPa, ν = 0.3. The relative displacement of the two spheres was achieved by kinematically controlled frictionless walls with the same elastic properties as for the spheres. The initial configuration is shown in Fig. 2a. In the first test a normal force of 88N was developed between the two spheres and then, with the normal force held approximately constant, relative tangential displacement was simulated so as to produce loading, unloading and reloading behaviour. The tangential force-displacement curves obtained are shown in Fig. 2b; the difference between the simulated results and theory is indistinguishable. Simulation of more complex loading histories confirmed the validation.

TIMESTEP CONSIDERATIONS

As a result of the contact mechanics interaction laws it is possible to simulate experiments using real time units. For dynamic processes, such as impact problems, which are of short duration we need to consider the elastic wave propagation across particles and the time for transfer of load from one particle to another. If a dynamic load is applied to an elastic sphere then three types of wave motion occur (ref. 7) The first two wave types are dilational (or pressure) waves and distortional (or shear) waves which travel through the body of the sphere with velocities c_P and c_S respectively, where

$$c_P{}^2 = 2(1 - \nu)\, G/(1 - 2\nu)\rho \qquad (16)$$

$$c_S{}^2 = G/\rho \qquad (17)$$

The third type of wave are Rayleigh waves which propagate along the surface of the sphere with a velocity

$$c_R = \beta\, c_S \qquad (18)$$

where β is obtained (ref. 7) from

$$(2 - \beta^2)^4 = 16(1 - \beta^2)[1 - (\beta c_S/c_P)^2] \qquad (19)$$

For the spring-dashpot analogy normally employed the energy radiated by wave motion corresponds to the energy 'dissipated' by the dashpot. It has been show by Miller and Pursey, see (ref 7), that the Rayleigh waves account for 67% of the radiated energy in comparison with dilational (7%) and distortional (26%) waves. It is not unreasonable, therefore, in attempting to simulate dynamic problems to assume that all of the energy is transferred by the Rayleigh waves. Given this assumption the time for transfer of load from one particle to another is dependent on particle size. Consequently, in order to keep the simulation simple a timestep is chosen to correspond to the time for transfer of load between contiguous particles based on the average particle size. Thus, the timestep is defined as

$$\Delta t = \pi\langle R\rangle/c_R \qquad (20)$$

Using the above definition of the timestep, even for millimetre sized particles the transfer time is in microseconds. Hence one million calculation cycles would be required to span one second of real time. For gravity free systems such as those normally employed in quasi-static shear simulations a real time simulation may be achieved by scaling up the particle densities. This has the effect of reducing the elastic wave velocity and increasing the timestep, thereby permitting long test durations to be accommodated on current computers.

IMPACT EXPERIMENTS

A problem readily amenable to real time simulation is that of impact due to the small timescale of the events. In a preliminary study of simulated impact problems the impact of a

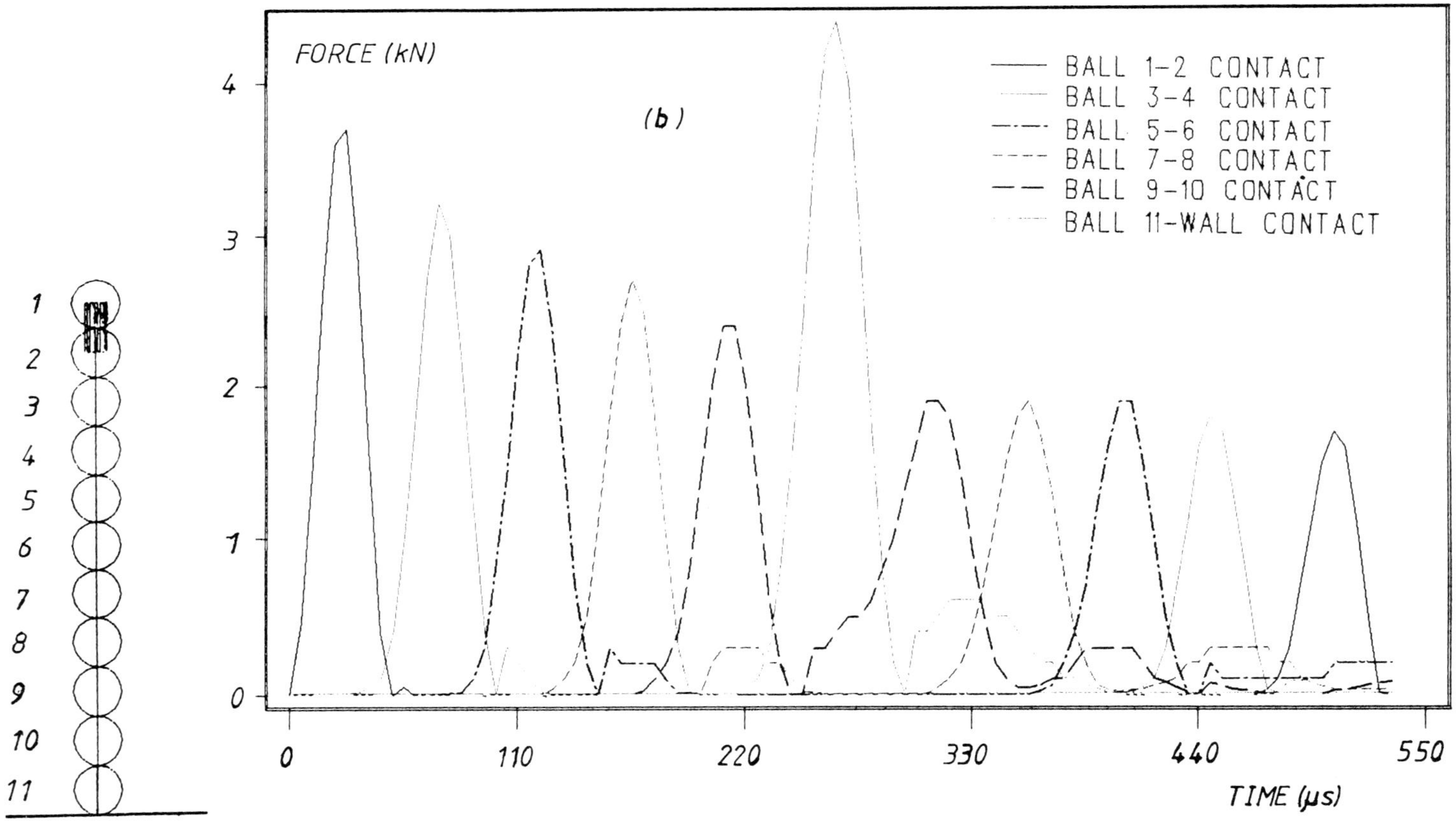

Fig. 3. Force transmission through a column of spheres

(a)

single sphere ($R = 10mm$, $\mu = 0.0$, $\rho = 7800\ kg/m^3$, $E = 210$ GPa, $\nu = 0.3$) falling under gravity onto a horizontal stationary frictionless wall, (same elastic properties as the sphere), was simulated in order to validate the code for the Hertzian contact law.

Having confirmed the interaction law a second simulation test was performed in which the same sphere impacted upon a column of 10 similar spheres contained within frictionless walls. The arrangement is shown in Fig. 3a and the evolution of the force transmission is illustrated in Fig. 3b. From Fig 3b it can be seen that the force transmitted through the column of particles attenuates such that on returning to the original impacting sphere the force has halved in magnitude. The results shown in Fig. 3b are for zero damping thus demonstrating that force attenuation through particulate material is primarily due to the discrete nature of the material and the interactive behaviour at the contacts.

It was noted during the simulated experiment that the initial impact resulted in disruptive contact by separation of particles due to inertia effects. It has been found from experiments on photoelastic discs that this contact release is usually followed by rearrangement and repositioning of individual particles and gives rise to a so called "rearrangement wave", Rossmanith and Shukla (ref 10). That a small secondary wave propagates through the column can clearly be seen in Fig 3b. Detailed comparison of the simulated data with results from dynamic photoelasticity experiments (ref 10) are encouraging.

To demonstrate the potential of the program the same initial sphere was impacted upon a previously prepared bed of particles. The particle bed consisted of 700 spheres of 6 different sizes which had been deposited under gravity until an equilibrium condition was attained. Results of the initial stage of the impact process are shown in Figs. 4 and 5. Figure 4 shows the particles with the velocity vectors superimposed Fig. 5 illustrates the corresponding force distributions (the thickness of the lines scaled to the current maximum force). From the velocity diagrams the development of the disturbed zone can be seen clearly and the force distribution figures nicely illustrate the propagation of the stress wave through the assembly.

The above example is merely illustrative and does not properly account for plastic deformation due to impact. The necessary code to realistically account for a permanent change in the geometry of the contact surface and the non-Hertzian pressure distribution would be complex and costly in computer time. However, the energy dissipation due to plastic deformation of the particles can be modelled by adjusting the contact damping. The results shown in Figs 4 and 5 are for 0.3 critical damping. The correspoonding values for maximum contact force and maximum particle velocity for zero damping were F_{max}= 1292N, 476N, 246N and V_{max}= 1.681 m/s, 0.854 m/s, 0.658 m/s.

INTERGRANULAR STRESS

Use of contact mechanics interaction laws results in the ability to quantify all parameters

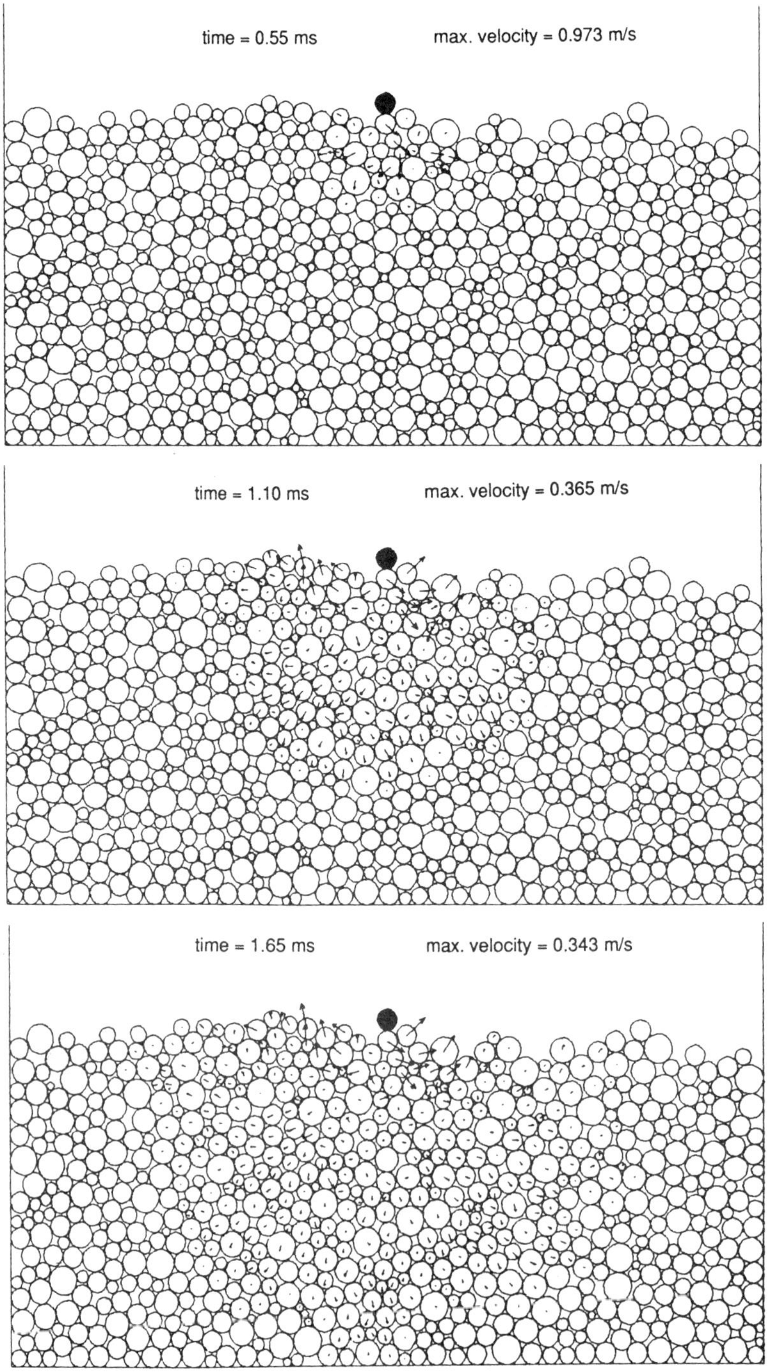

Fig. 4. Particle impact on a particle bed (velocity field)

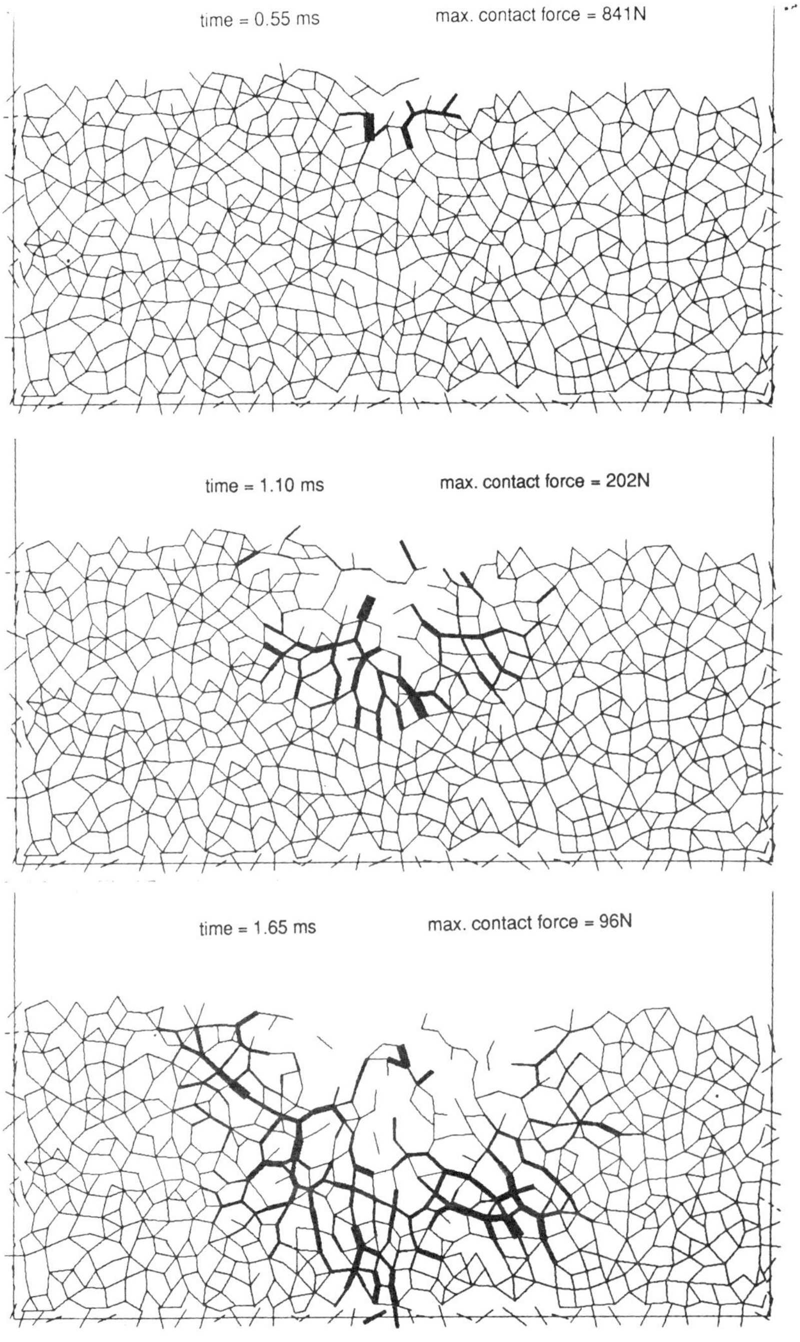

Fig. 5. Particle impact on a particle bed (force transmission)

in terms of real physical units. To illustrate this, an assembly of 743 spheres of 14 different sizes was randomly generated within a circular area and then subjected to an increasing applied isotropic stress. The radii of the spheres varied from 0.3mm to 1.0mm, other properties were identical for all spheres ($E = 70$ GPa, $\nu = 0.25$, $\mu = 0.35$, and $\rho = 2650$ Kg/m^3). The applied stress was incrementally raised to 25, 50, 75, and 100 kPa. At the end of each stage calculation cycles were continued until an equilibrium condition was reached. The mean values (for the assembly) of various contact parameters are given below for different applied stress states.

applied stress	(kPa)	25	50	75	100
normal contact force	(mN)	44	86	128	170
contact radius	(μm)	6.27	7.90	9.03	9.92
average contact normal stress	(MPa)	298	375	429	471
average contact tangential stress	(MPa)	47.3	57.6	64.1	70.5

If additional particle properties such as compressive yield stress and crushing force are available then the type of data shown above can provide useful information related to the extent of plastic deformation of particles and particle breakdown. This type of information may be more useful to the process engineering industry than to geotechnical engineering. However, with further improvements in computer power, the realistic application of the program to boundary value problems such as footings and retaining walls will become attainable. For such problems the reliable determination of interface pressures will require interaction laws based on contact mechanics theory.

REFERENCES

1 P. A. Cundall and O. D. L Strack, Geotechnique, 29 (1979) 47 - 65.
2 P. A. Cundall, A. Drescher and O. D. L. Strack, in P. A. Vermeer and H. J. Luger (Eds.) Deformation and Failure of Granular Materials, Balkema, Rotterdam, 1982, pp 355 - 370.
3 P. A. Cundall and O. D. L. Strack, in J. T. Jenkins and M. Satake (Eds.), Mechanics of Granular Materials: New Models and Constitutive Relations, Elsevier, Amsterdam, 1983, pp 137 - 149.
4 O. R. Walton, in J. T. Jenkins and M. Satake (Eds.), Mechanics of Granular Materials: New Models and Constitutive Relations, Elsevier, Amsterdam, 1983, pp 327 - 338.
5 O. R. Walton, Int. J. Engng. Sci., 22 (1984) 1097 - 1107.
6 C. Thornton and D. J. Barnes, Acta Mechanica, 64 (1986) 45 - 61.
7 K. L. Johnson, Contact Mechanics, Cambridge University Press, Cambridge, 1985, 452 pp.
8 R. D. Mindlin and H. Deresiewicz, J. Appl. Mech., 20 (1953) 327 - 344.
9 C. M. Szalwinski, J. Appl. Mech., 52 (1985) 62 - 66.
10 H. P. Rossmanith and A. Shukla, Acta Mechanica, 42 (1982) 211 - 225.

Micromechanics of Granular Materials, edited by M. Satake and J.T. Jenkins
Elsevier Science Publishers B.V., Amsterdam, 1988 — Printed in The Netherlands

DISC MODEL ANALYSIS OF GRANULAR MEDIA

Y. KISHINO

Department of Civil Engineering, Tohoku University, Aoba, Aramaki, Sendai 980 (JAPAN)

SUMMARY

A new simulation method has been developed to study the micromechanics of granular materials. In this paper the method is called the disc model analysis, because it is explained specifically for disc assemblies in two dimensional space. It is assumed that the contact force is produced by the action of springs when two discs overlap each other. The Coulomb's law is assumed for the slippage between discs. The iteration process adopted in this method consists of the succesive releasing of fixing vectors which is achieved through the calculation of contact stiffness matrices. Through the applications to element tests, the efficiency of the disc model analysis is shown especially for the definite study on the statistical characteristics of granular materials.

INTRODUCTION

The micromechanical approach to the mechanics of granular materials should be supported by observations of the fabric and other micromechanical quantities in granular assemblies. However it is very hard to achieve this in real experiments. On the other hand, numerical experiments in computers are expected to offer various kinds of detailed informations in the aspect of micromechanical study.

The distinct element method developed by Cundall has been successfully applied to analyses of granular assemblies (ref.1). The fundamental idea of the method is based on the Newton's second law and problems are treated as dynamical ones. Thus it may be important to regulate the time effect for utilizing the method to the static problems. Generally, the movement of grains is not co-axial with the acceleration vector because of the constraints by neighboring grains and/or boundaries. However, in the distinct element method, the direction of movement is simply determined by the acceleration calculated from the resultant force and the moment. This simplification may affect results for dense assemblies of grains.

A new method explained in this paper has been developed for the precise analyses of statical behaviors of granular materials. As shown in the examples including a liquefaction test, the method has capabilities as a tool for the micromecanical studies on granular materials.

DISC MODEL ANALYSIS

In this paper, the term 'disc' is used instead of 'grain'. However, the fundamental idea of this analysis may be applied to the three dimensional assemblies of arbitrary-shaped grains under quasi-static loading.

The method presented here is based upon the following assumptions: a) The elastic behavior of a disc is replaced by the action of springs attached virtually at the contact point in the normal and tangential directions, and the disc itself is displaced and rotated rigidly. b) Discs can overlap each other, and only at that time the contact force is produced between them. Usually, the amount of overlapping is small enough compared with the radii of discs. c) For the slippage between discs, the Coulomb's friction law is assumed, and the tangential component of contact force can not exceed the limit determined by it. d) To attain the equilibrium state of the disc assembly, each disc is displaced and rotated iteratively according to the contact stiffness matrix defined from the locations of neighboring discs. e) The boundary can be either strain-controlled or stress-controlled, and the interaction between a boundary element and a disc is specified in a similar way as the interaction between discs.

Force-displacement law between discs

In the following, the usual tensor notations in the Cartesian co-ordinate system are adopted. As shown in Fig.1, let two discs A and B, whose centers are located at x_i and x_i' $(i=1,2)$ respectively, are in contact with each other. The normal and tangential components of the contact force p_i which is acting on the disc A are denoted as $p_n = -p_i\, n_i$ and $p_t = -p_i\, t_i$ respectively, where n_i and t_i are the normal and tangential unit vectors on the disc A. The increment of relative displacement is given by the components

$$\Delta u_n = (\Delta x_i - \Delta x_i')\, n_i \quad (1)$$

$$\Delta u_t = (\Delta x_i - \Delta x_i')\, t_i + r\,\Delta\omega + r'\Delta\omega' \quad (2)$$

where r and r' are the radii and ω and ω' are the rotations of the discs A and B respectively.

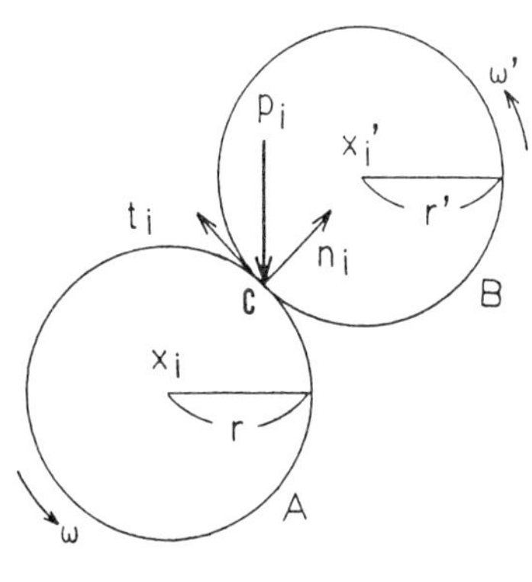

Fig.1. Contact force.

Between the contact force and the relative displacement, the following incremental force-displacement law is assumed:

$$\Delta p_n = k_n\, \Delta u_n \quad (3)$$

$$\Delta p_t = k_t\, \Delta u_t \quad (4)$$

where k_n and k_t are stiffnesses.

Fixing vector

The disc model analysis is performed through an iterative calculation, which is repeated until the resultant force and the moment of each member in the disc assembly become sufficiently small. Thus, when the configuration of disc assembly in the iterative process is required to be fixed, virtual forces and moments should be applied to the discs.

The fixing vector F is defined as a vector which consists of the force f_i and the moment m required for fixing a disc at its center, and it is calculated in terms of the contact force $P_c = (p_n, p_t)_c^t$ applied at the contact point C and the body force vector $B = (b_1, b_2, 0)^t$ (see Fig.2) as

$$F = (f_1, f_2, m/r)^t = \Sigma_c T_c P_c - B \tag{5}$$

where Σ_c represents the summation over the contact points of a disc and

$$T_c = \begin{bmatrix} \cos\theta & -\sin\theta \\ \sin\theta & \cos\theta \\ 0 & 1 \end{bmatrix}_c \tag{6}$$

is the transformation matrix.

Contact stiffness matrix

The contact stiffness matrix is defined as a set of additional fixing vectors required to give unit movements of a disc while other discs being fixed. From Eq.(1) through Eq.(4), the incremental movement of a disc

$$\Delta X = (\Delta x_1, \Delta x_2, r\Delta\omega)^t \tag{7}$$

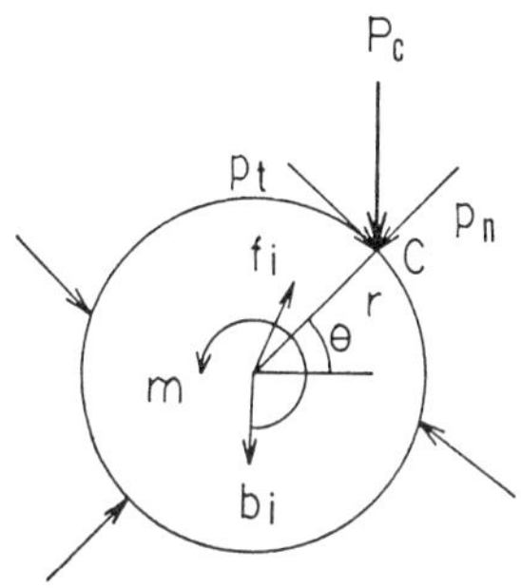

Fig.2. Fixing of disc.

produces the incremental contact force

$$\Delta P_c = S_c T_c^t \Delta X \tag{8}$$

where

$$S_c = \begin{bmatrix} k_n & 0 \\ 0 & k_t \end{bmatrix}_c \tag{9}$$

Thus, the increment of fixing vector is expressed as follows:

$$\Delta F = S\Delta X \tag{10}$$

where

$$S = \Sigma_c T_c S_c T_c^t \tag{11}$$

is the contact stiffness matrix. In the process of iteration, the contact stiffness matrix keeps changing with the relative movements of discs. However, it is kept constant during each iteration step.

As far as only one disc is concerned, the equilibrium condition for the disc is to release its fixing vector by the equation $F + \Delta F = 0$, which leads to $\Delta X = -S^{-1}F$ for the case $\det S \neq 0$.

Releasing of fixing vectors

The equilibrium state of a disc assembly under a prescribed boundary condition is obtained by releasing the fixing vectors. An iterative calculation as shown in Fig.3 seems a profitable way to do this, because some of the disc stiffness matrices may be singular and the independent variables themselves will be changed by the touching and separation between discs. However, the releasing of a fixing vector in the way explained in the previous section causes the alteration of other fixing vectors, and the order of releasing affects the result of analysis. To avoid this, all of the discs are moved simultaneously by use of the contact stiffness matrices determined with respect to the configuration in the last iteration step. Instead the amount of movement calculated through the contact stiffness matrix needs some modification.

A disc, for which $\det S \neq 0$, is displaced and rotated by

$$\Delta X = -\alpha S^{-1} F \qquad (12)$$

where $\alpha < 1$ is the modification factor. If the number of contact points of a disc is less than two, then S becomes singular. For a disc which has no contact point, the movement is given by

$$\Delta X = \beta B / k_n \qquad (13)$$

where $\beta > 1$ is a constant. If a disc has only one contact point, ΔX is determined under the assumption $\Delta u_t = 0$ as

$$\Delta X = \begin{pmatrix} \cos\theta & -\sin\theta \\ \sin\theta & \cos\theta \\ 0 & -1 \end{pmatrix} \begin{pmatrix} -\alpha' \; f_i \; n_i / k_n \\ \beta' \; b_i \; t_i / k_n \end{pmatrix} \qquad (14)$$

where α' and β' are constants.

The amounts of additional contact forces are calculated by Eqs.(3) and (4). If p_n becomes less than zero, we put $p_n = 0$. If p_t exceeds the limit of the Coulomb's law, we put

$$|p_t| = p_n \tan\phi + c \qquad (15)$$

where ϕ is the friction angle and c is the cohesion.

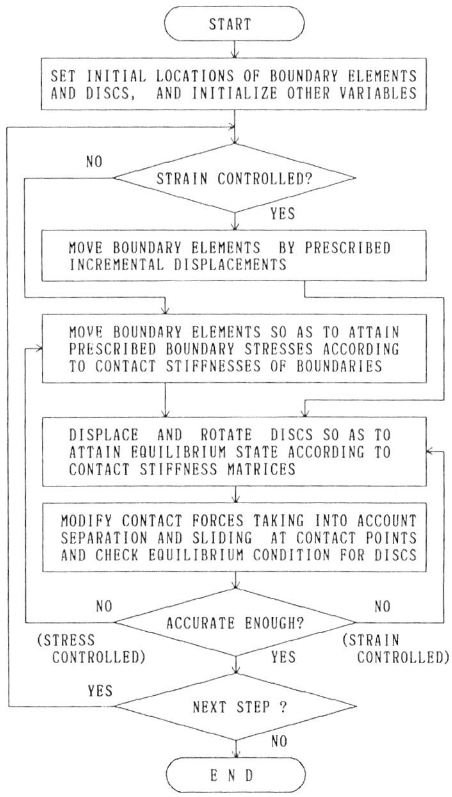

Fig.3. Flow chart of analysis.

Boundary elements

In this analysis the boundary is given by a set of elements. In usual bi-axial problems, four straight walls may be used as boundary elements. In other cases, a set of discs with the same range of radii as those of the inner discs may be used for a precise control of the disc assembly. The interaction between a boundary element and a disc is represented by virtual springs in a similar way as the interaction between two discs. The fixing vector of boundary element consists of the external force and the external moment, and the contact stiffness of boundary element is defined as a force or a moment required for the unit movement of the boundary element while the neighboring discs being fixed.

For the stress-controlled tests, the fixing vector of boundary element is prescribed for each loading step. The difference between the current fixing vector and the prescribed one is eliminated by moving the boundary element according to the contact stiffness of boundary element. The releasing of the fixing vectors of inner discs causes changes in the fixing vectors of boundary elements, so that the above procedure is repeated until the disc assembly reaches its equilibrium state. It should be noted that the boundary condition given only by stresses is insufficient to determine the rigid motion of the disc assembly without applying proper constraints to the movements of boundary elements.

For the strain-controlled tests, a part of movements of boundary elements is constrained so as to maintain the strain conditions, while the other part is determined by the stress condition in the similar way as described above. The undrained condition is similar to the constant volume condition. However, the volume (area) changes slightly, according to the bulk modulus of water. If the time effect of the seepage can be omitted, the pore-pressure is determined directly from the area change of disc assembly. The forces applied to discs is determined by the effective stress which is obtained by subtracting the pore-pessure from the external stress.

Flow of analysis

The flow chart of the disc model analysis is shown in Fig.3. A disc assembly, for which numerical tests are performed, must fulfil the equilibrium condition as well as the geometrical condition. To derive the expected configuration from an artificial packing, the disc model analysis itself is utilized. Even if the artificial packing does not fulfil the equilibrium condition, the iteration process leads to a proper configuration provided that discs and boundary elements do not overlap extensively. Generally, limited movements of discs and boundary elements are required for the stable convergence in each iteration step.

The loading process corresponds to the successive steps of incremental movements of boundary elements. Movement of boundaries are controlled in two ways as shown in Fig.3. Sometimes, each loading step is devided into several sub-steps so that the boundary conditions are applied separately. For example, the first sub-step for the undrained condition may be the movements governed by the increment of shear stress, and then the second one corresponds to the volume change calculated through the effective stress.

The iterative calculation is repeated until the equilibrium condition for each grain is fulfilled with sufficient accuracy.

CALCULATION OF DISSIPATION ENERGY

The tangential component of the relative displacement at contact point C in Fig.1 is given by Eq.(2). The elastic and plastic parts of Δu_t are determined through Eq.(4) as follows:

$$\Delta u^e_t = \Delta p_t / k_t \tag{16}$$

$$\Delta u^p_t = \Delta u_t - \Delta u^e_t \tag{17}$$

Thus, the increment of dissipation energy at contact point C is given by

$$\Delta D_C = p_n \tan \phi \quad | \Delta u^p_t | \tag{18}$$

and the increment of total dissipation energy in a disc assembly is

$$\Delta D = \Sigma_D (\Sigma_C \Delta D_C) \tag{19}$$

where Σ_D represents the summation over the discs.

DEFINITION OF STATISTICAL TENSORS

The fabric tensor for granular assemblies is defined through the summation of the dyad products of contact normal n_i (ref.2). However, the definition of fabric tensor is not unique. The freedom to choose the weight of summation and the denominator in the equation makes it possible to define various kinds of statistical tensors. In the following, two types of tensors are presented. The first one will be called the co-ordination tensor, which is defined as

$$C_{ij} = \{ \Sigma_D (\Sigma_C n_i n_j) \} / N \tag{20}$$

where N is the number of discs in the disc assembly. The trace of the tensor reduces to the so-called co-ordination number which represents the average number of contact points per disc. The second one is defined as

$$E_{ij} = \{ \Sigma_D (\Sigma_C e_C n_i n_j) \} / A \tag{21}$$

where e_C is the elastic energy produced in a disc by the contact with a neighboring disc and A is the area of disc assembly. The trace of this tensor represents the elastic energy density, and it will be called the energy tensor. In the disc model analysis, the amount of e_C is calculated easily as one half of the elastic energy stored in the normal and tangential springs between two discs or between a disc and a boundary element.

EXAMPLES

A bi-axial shear test and an undrained cyclic test were performed on a disc assembly of Fig.4. The disc assembly consists of three types of discs with different diameters and they have the same total area. The void ratio of the initial state was 0.26. In Fig.4, each contact force vector is expressed by a bar whose thickness is proportional to the intensity of the contact force. In each test, the boundary elements are moved symmetrically with respect to the central axes of disc assembly.

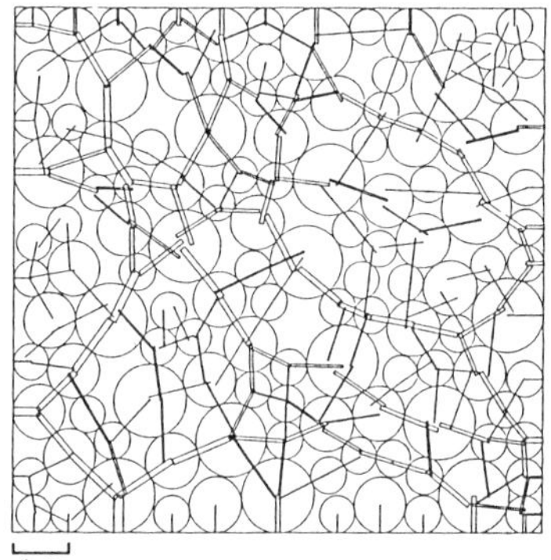

Fig.4. Distribution of contact forces at initial state.

The constants used in the tests are as follows (depth = 1 cm):
diameter of disc = 6,9,12 mm; weight of disc in water = 1500 dyn/cm^2;
stiffness (k_n, k_t) = 1 $\times 10^8$, 7 $\times 10^7$ dyn/cm;
friction angle (ϕ) = 25°(between discs), 0°(between disc and wall)
cohesion (c) = 0 dyn; confining stress (σ_0) = 1.5$\times 10^5$ dyn/cm;
bulk modulus of water = 1$\times 10^{10}$ dyn/cm^2;
modification factor α = 0.75, α' = 0.5, β = β' = 20;
accuracy (maximum absolute value of fixing vector) = 100 dyn.

In the following, suffixes 1 and 2 of the principal tensor quantities correspond to the vertical and horizontal directions respectively.

Bi-axial shear test

The drained test was assumed. During the test, the shear component of loading was controlled by the strain, while the isotropic confining stress $\sigma_0 = (\sigma_1 + \sigma_2) / 2$ was kept constant. The increment of shear strain $\Delta(\varepsilon_1 - \varepsilon_2)$ for each loading step was 0.05% .

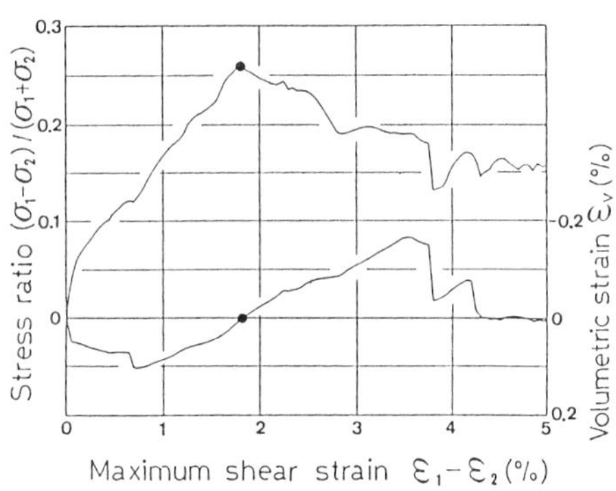

Fig.5. Stress-strain relationship in bi-axial shear test.

Fig.5 shows the stress-strain relationship which is very similar to test results on loose sands. Sudden drops observed in the curves are due to local collapses of the fabric in disc assembly. It seems that local

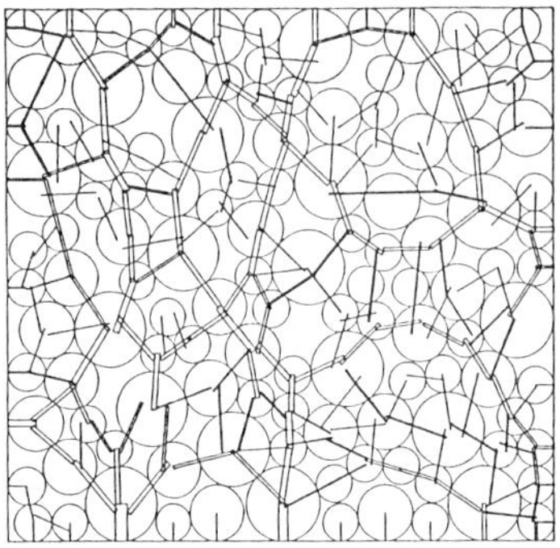

Fig.6. Distribution of contact forces at peak state.

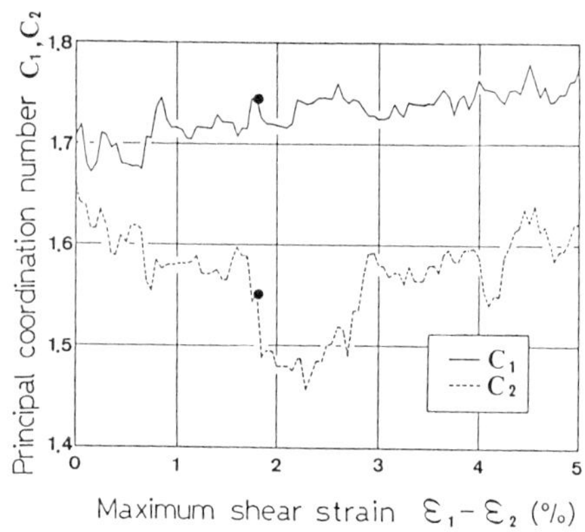

Fig.7. Change of co-ordination tensor in bi-axial shear test.

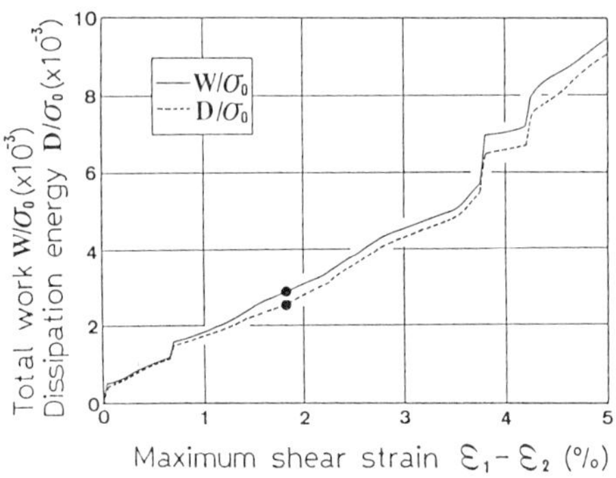

Fig.8. Accumulation of dissipation energy in bi-axial shear test.

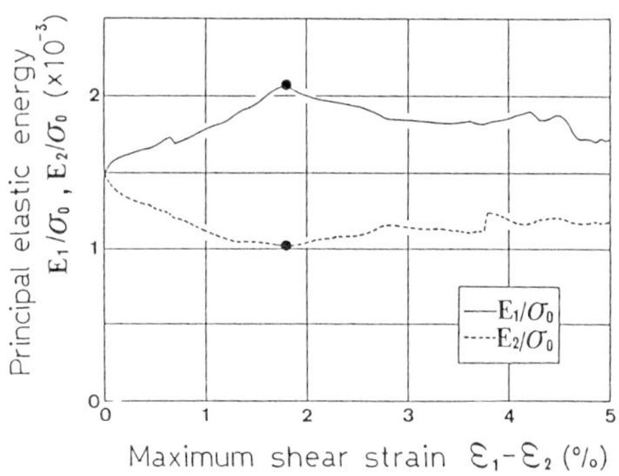

Fig.9. Change of energy tensor in bi-axial shear test.

changes in fabric are significant in such a small scale model as the example. In the following figures, dots indicate the peak state of shear stress. The distribution of contact forces in the peak state is given in Fig.6 where a concentration of forces in the vertical direction is observed.

Fig.7 illustrates the relation between the shear strain and the principal values of co-ordination tensor (Eq.(20)). It is seen that the development of anisotropy depends mainly on C_2 which decreases abruptly at the peak state. The abrupt change may be originated by an eigenvalue problem.

Calculated results of the external work and the dissipation energy (Eq.(19)) are shown in Fig.8. The difference between these quantities should be equal to the elastic energy, which is given separately in Fig.9 in the form of principal values of energy tensor (Eq.(21)). All these values are given in densities per unit area. It is noted that the dissipation energy changes almost lineary with the shear strain.

Undrained cyclic test

While the external confining stress σ_0 was kept constant, the shear stress $(\sigma_1 - \sigma_2)/2$ was increased or decreased by the steps of one-tenth of the amplitude ($\pm 2.5 \times 10^4$ dyn/cm). The number of cycles was four.

The stress path and the stress-strain relationship are shown in Figs.10 and 11 respectively. The numbers attached in these figures represent the loading steps. Though uneven changes are observed locally, the overall behavior of the disc model is very similar to that of real sands. In Fig. 10, it is observed that the friction angle of disc assembly is less than the friction angle between indivisual discs.

As shown in Figs.4, 12 and 13, the intensity of contact forces decreases with the increase of pore-pressure. Up until the step 109 which is just before the remarkable shear takes place, rather large pore-pressure has developed, but the most of the discs have scarecely moved. Even up to the final stage 160 of the cyclic test, no drastic change in fabric is observed.

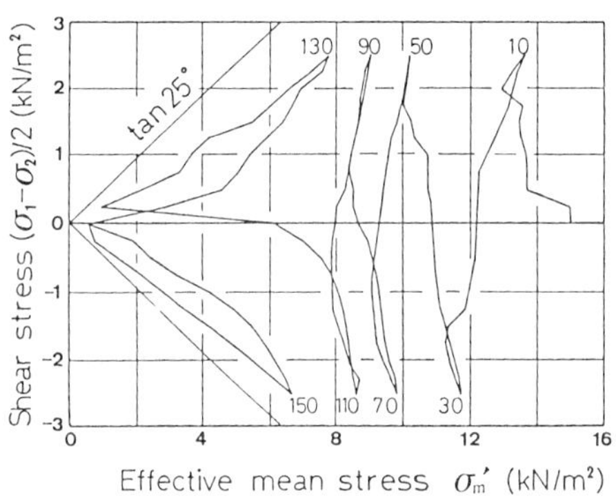

Fig.10. Stress path in undrained cyclic test.

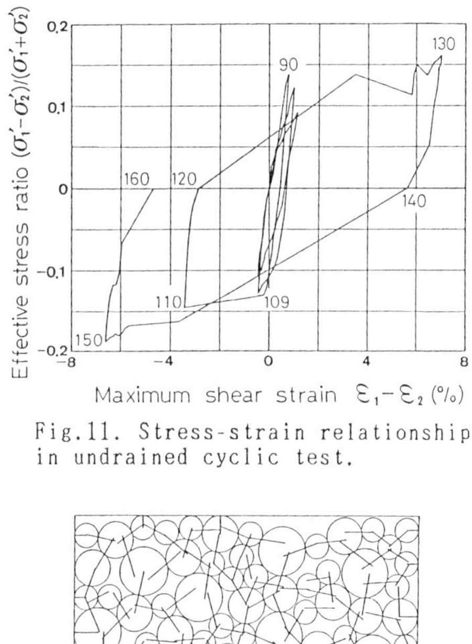

Fig.11. Stress-strain relationship in undrained cyclic test.

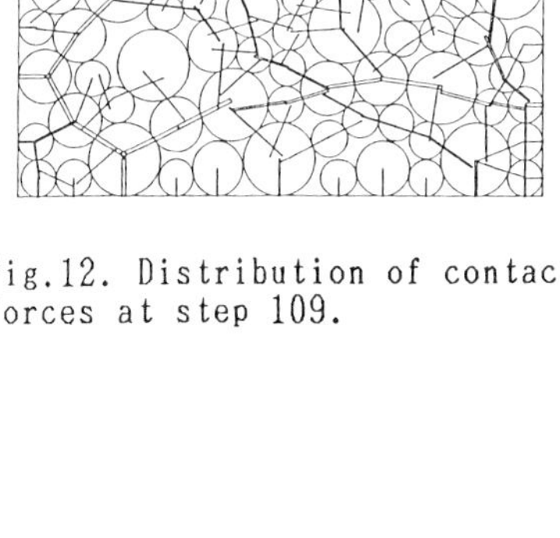

Fig.12. Distribution of contact forces at step 109.

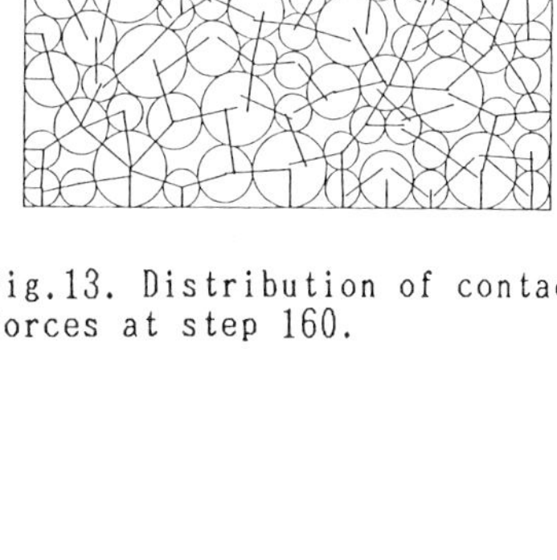

Fig.13. Distribution of contact forces at step 160.

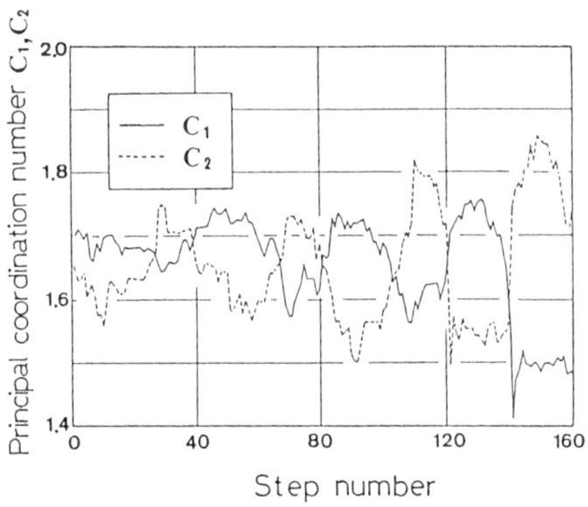

Fig.14. Change of co-ordination tensor in cyclic test.

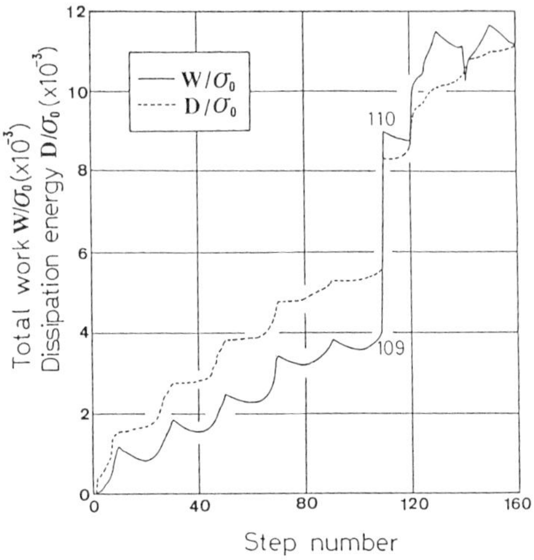

Fig.15. Accumulation of dissipation energy in cyclic test.

Fig.14 shows the change in the principal values of co-ordination tensor. The larger deviation of principal value from the initial mean value (1.68), is observed in the negative direction.

The external work done by the effective stress and the dissipation energy are shown in Fig.15. Until the step 109 the dissipation energy exceeds the external work. The reason is that the elastic energy stored in the disc assembly is consumed as shown in Fig.16. It can be said that the liquefaction is a process of loosing elastic energy.

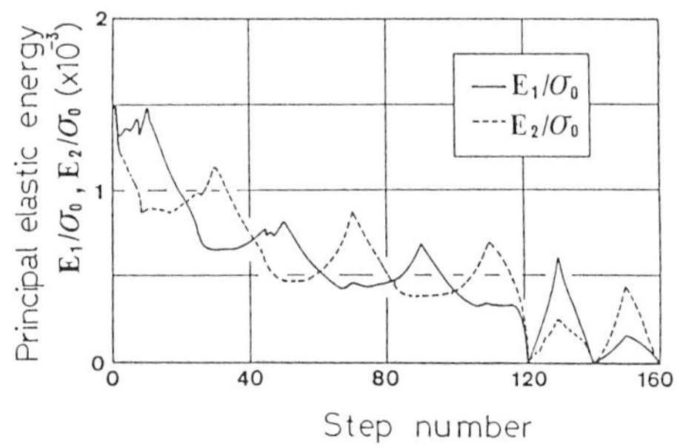

Fig.16. Change of energy tensor in cyclic test.

CONCLUDING REMARKS

As shown in the examples, the disc model analysis presented here is a useful tool to study the micromechanics of granular materials. The computer used for the analyses is a 32-bit microcomputer with 4MB memories and 80MB disks, which seems sufficient to perform the 'element tests'.

REFERENCES

1 P.A. Cundall and O.D.L. Strack, A Discrete Numerical Model for Granular Assemblies, Geotechnique, 29-1 (1979) 47-65.

2 M. Satake, Constitution of Mechanics of Granular Materials through Graph Representation, Theor.Appl.Mech., 26, Univ.Tokyo Press, (1978) 257-66.

Micromechanics of Granular Materials, edited by M. Satake and J.T. Jenkins
Elsevier Science Publishers B.V., Amsterdam, 1988 — Printed in The Netherlands

PARTICLE-DYNAMICS CALCULATIONS OF GRAVITY FLOW OF INELASTIC, FRICTIONAL SPHERES*

O. R. WALTON, R. L. BRAUN, R. G. MALLON, and D. M. CERVELLI
Lawrence Livermore National Laboratory, Livermore CA 94550, USA

SUMMARY

Three-dimensional discrete-particle computer models that calculate the motion of each individual grain in assemblies of hundreds of particles in steady shearing flows with either periodic or real boundaries have been modified to simulate gravity flow of particles through arrays of cylindrical horizontal rods and down inclined chutes. The particle interaction models reproduce experimentally measured recoil trajectories for colliding frictional particles, including rotation effects. Laboratory measurements of the flow of glass beads cascading down through an array of horizontal cylindrical rods correlate well with gravity flow calculations of inelastic, frictional spheres falling through a similar rod array. Less elastic particles are found to cascade through the array faster than nearly elastic particles. Likewise, smaller particles are found to flow faster than large ones. Model simulations of nearly two-dimensional inclined chute flow tests of $6\,mm$ diameter cellulose-acetate spheres flowing over a rough surface between parallel vertical glass plates, result in particle velocities that are considerably higher than values measured in similar laboratory tests at UCLA; however, inclusion of approximate air drag effects in the calculational model eliminates most of the discrepancy producing both density and velocity profiles that are close to the measured values.

BACKGROUND

The flow behavior of granular solids is of considerable technological importance. Yet, despite considerable engineering and scientific attention, this behavior is not well understood. During the past several years there has been a significant resurgence of interest and corresponding flurry of activity in developing a microstructural understanding of granular solids. New optical measuring techniques, statistical mechanical theories for inelastic particles, and new numerical models based on tracking individual particle motions have all been applied to flowing granular solids. These new efforts are yielding new insight into the microstructural details of granular flows. The discrete particle computer models utilized in this study of particles flowing under the influence of gravity, have previously been applied to determining the rheologic properties of assemblies of inelastic, frictional spheres and disks under conditions of rapid shearing flow (refs. 1-4). These steady shearing flow models use non-equilibrium molecular-dynamics methods that have previously been applied to simulation of molecular motions (refs. 5, 6). The major difference for macroscopic particles being the inclusion of inelastic, frictional forces and the accompanying particle rotations.

*Work performed under the auspices of the U.S. Department of Energy by the Lawrence Livermore National Laboratory under Contract W-7405-Eng-48

The steady shearing flow calculations with these models generally agree with laboratory measurements, when such data are available, and they agree with theories when comparable approximations are made in the models (such as assuming frictionless and nearly elastic particles). Calculational studies with these models have confirmed, for example, that stresses in systems of inelastic particles generally vary as the square of the shear rate and the square of the particle radii (refs. 1, 2), agreeing with measurements (*e.g.*, ref. 7), and confirming theories (*e.g.*, refs. 8, 9). The calculations also show that the stresses depend strongly on the degree of inelasticity and void volume, and less strongly, but still significantly, on the interpartiulate friction coefficient. For the series of calculations described in this paper, fixed 'particles' in the form of fixed spheres, cylinders or planes were added to our three-dimensional spherical particle model, 3DSHEAR, while retaining the capability of having periodic boundaries in each of the three Cartesian directions. An approximate air-drag model, with a velocity squared dependent drag force, was also added. We used this new model to calculate gravity flow of inelastic, frictional spheres in two configurations. First, was a series of calculations of particles cascading down through an array of horizontal cylindrical rods, directly simulating laboratory measurements of glass beads flowing through a similar array. The second set of calculations (still in progress) is a direct simulation of nearly two-dimensional inclined chute flow tests of spherical particles flowing between two vertical glass plates. As described below, the simulation calculations correlate well with the laboratory measurements in both of these situations.

MEASUREMENTS AND SIMULATIONS

Gravity Flow Through Periodic Array of Horizontal Cylindrical Rods

New equipment being designed to cool retorted oil shale from a pilot scale oil shale retort at Lawrence Livermore National Laboratory (LLNL), includes a delayed-fall granular-solids mixer, consisting of a stationary array of horizontal rods through which crushed shale will flow (ref. 10). We conducted a series of laboratory tests of $4\,mm$ diameter glass beads, crushed oil-shale, and 0.6 - $0.8\,mm$ diameter glass ballotini falling through the rod array, to determine the relationship between the mean vertical velocity of particles and the total mass flux. The rod array consisted of $0.625\,in$ diameter stainless steel rods located in horizontal rows with $1.125\,in$ center-to-center spacing (*i.e.*, $0.50\,in$ gap between rods). All rods were parallel and located so the rods in each layer had their centers under the gaps in the layer above. Vertical spacing between layers was $0.975\,in$ center-to-center, resulting in $0.50\,in$ gaps between rods in alternate layers. A series of constant mass flux tests were conducted with $4\,mm$ glass beads, with the mass flux ranging from near zero, for a single particle cascading down through the array, up to a maximum mass flux of $125\,kg/m^2s$ (corresponding to a mean solids loading of approximately $\nu = 0.24$ in the space between the rods). Each test consisted of flowing material at a constant mass flow rate through the rod array and then 'capturing' the material in a fixed height of the rod array by simultaneously closing horizontal sliding doors at two different heights. The mass flow rate was controlled by flowing through previously calibrated fixed orifices, and the mass of the 'captured' material was obtained by opening the bottom door, removing the material, and weighing it. The mean particle velocity was obtained by dividing the known mass flow rate by the measured mass per unit height captured between the sliding doors. The mean velocity of the 4 mm beads varied from just under $0.2\,m/s$ to over $0.27\,m/s$ as the solids loading in the system was increased. When smaller glass beads were tested we found that they traveled as much as a factor of two faster than the large beads with the mean vertical velocity of the particles first increasing then decreasing as the particle loading (and thus, the mass flux) was increased. A distribution of particle sizes (*i.e.*, -6 +50 *mesh*) existed in the crushed oil shale which also exhibited a maximum in its particle velocity *vs* particle loading relationship, with a peak velocity of approximately one half metre per second. These results are summarized in Figure 1 which shows the relationship between measured mean vertical velocity and the mass flux through the rod array.

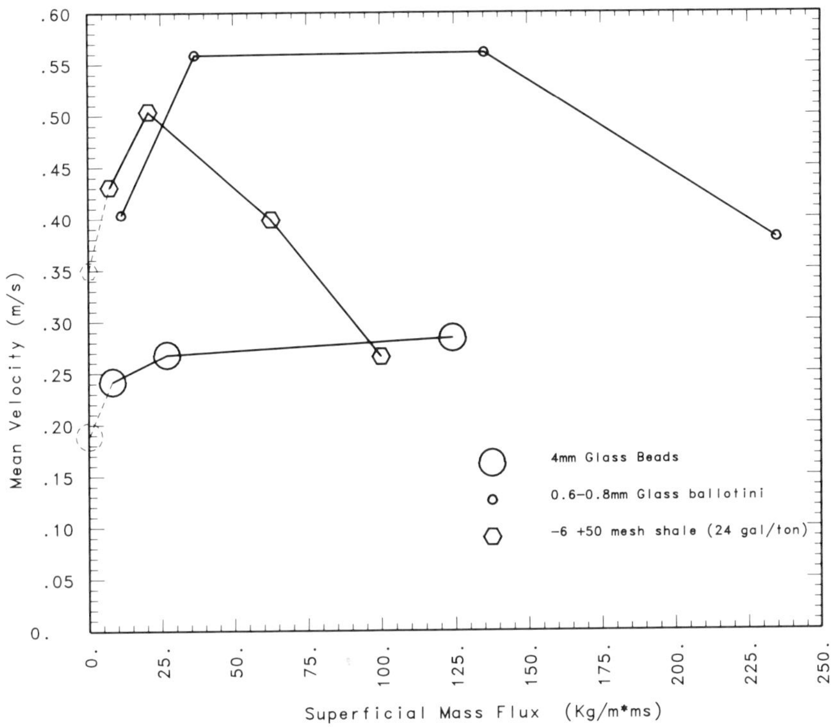

Figure 1

Measured mean velocity of particles falling through a horizontal rod array as a function of superficial mass flux through the array for 4 *mm* glass beads, 0.6-0.8 *mm* glass ballotini, and -6 +50 *mesh* crushed oil shale.

Various physical arguments were proposed to explain the observed flow behavior. The observation that small particles travel much faster through the array than large particles was hypothesized to be due to the larger effective diameter of the rods seen by the larger particles (the *effective diameter* being the sum of the rod and colliding particle diameters). The initial increase in particle velocity with increasing solids packing in the rod array was thought to be caused by the increased likelyhood of a particle rebounding from a collision with a rod, having its next collision with other particles moving down through the array, instead of with another stationary rod. The argument being that, at low to moderate solids loadings, particle-particle collisions that occur in between particle-rod collisions cause increased net velocity, but, at extremely high solids loadings, the entire system tends to become jammed, and flow velocities again decrease. In order to test these arguments and to determine how the flow velocity depends on the physical properties of the particles flowing through the array, we modified the 3DSHEAR model to simulate this rod geometry and performed a series of flow simulation calculations.

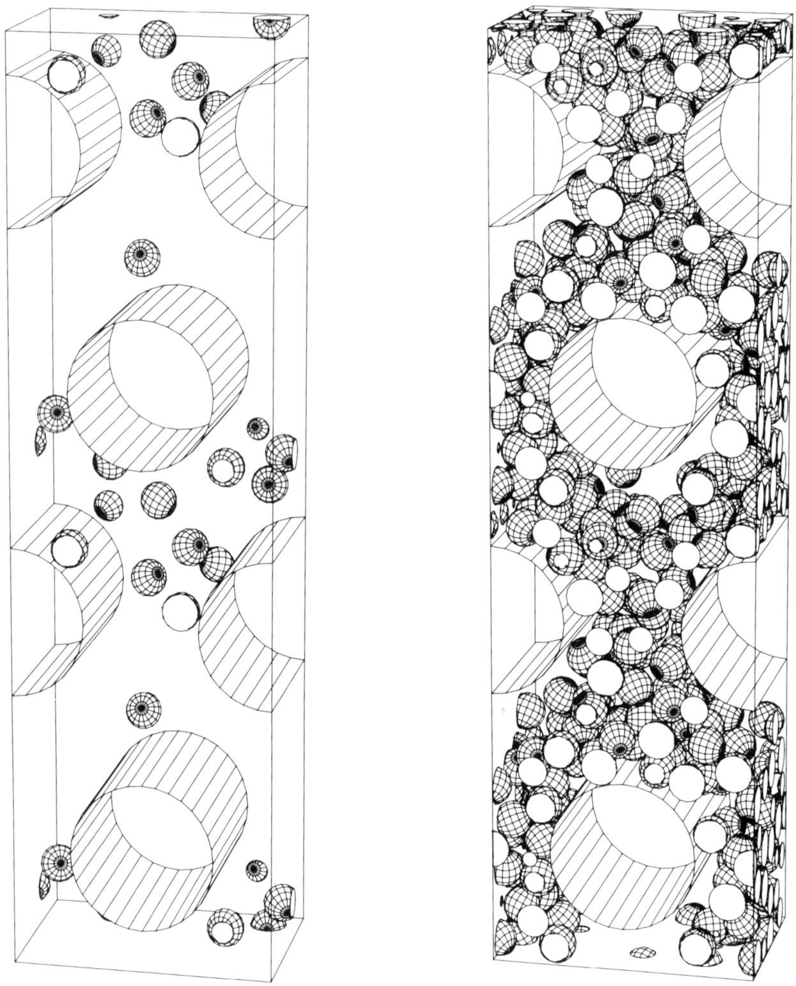

Figure 2

Perspective views of two simulation calculations of 4 *mm*, inelastic, frictional spheres falling through a regular array of horizontal cylindrical rods (*a*) mean solids fraction, $\nu = 0.0186$ (11 spheres per calculational cell), (*b*) mean solids fraction, $\nu = 0.279$ (165 spheres per calculational cell).

An initial set of simulation calculations examined the effects of coefficient of restitution and solids concentration on the mean vertical velocity of the spheres falling through a rod array. Fixed 'particles' in the form of cylindrical rods, were added to the 3DSHEAR model in order to directly simulate the rods of the oil-shale mixer used in the laboratory tests. Periodic boundaries were used on all sides of a basic cell containing two horizontal cylindrical rods, making the simulation equivalent to an infinite array. Figure 2 shows perspective views of two such simulation calculations with different solids concentrations of 4 *mm* spherical particles. Previous experimental studies with glass beads have obtained effective coefficients of restitution in the range 0.8 to 0.95 for collisions of glass beads on glass. Through simple drop tests we determined that the coefficient of restitution, e, for glass beads on steel plates was in the range $0.6 < e < 0.75$. In the simulation calculations we used spheres with a coefficient of friction of $\mu = 0.5$, and various values for the coefficient of restitution. A partially-latching-spring normal force model, with a linear loading and somewhat steeper linear unloading curve, was used to model the inelastic normal direction force, and a three-dimensional incrementally-slipping friction force, modeled after the theory of Mindlin and Deresiewicz (ref. 11), was used for the tangential direction forces (this an extension of the model previously used for two-dimensional flow calculations, ref. 1). Previous comparisons with individual collision measurements have confirmed that this contact force model captures the essential features of inelastic, frictional impacts, including proper recoil trajectories and particle rotations. The rod-array flow calculations indicated that, generally, less elastic particles flow through the array faster than more elastic ones. The flow velocity *vs* solids loading relationship for spheres with a constant coefficient of restitution of $e = 0.8$ for all collisions is shown in Figure 3 where we see that such particles fall at nearly the same rate as the glass beads in the laboratory tests. As is also shown on this figure, slightly better agreement was obtained between the calculations and the experiments when the simulated particles' restitution coefficient was set to $e_1 = 0.9$ for sphere-sphere collisions and to $e_2 = 0.6$ for sphere-rod collisions. In addition to the calculations shown on Fig. 3 we also varied the size of the rods and the spheres. We found that simply reducing the diameter of the rods, so that the sum of the rod plus 4 *mm* sphere diameters was the same as the sum of the full sized rod diameter plus the ballotini diameter, was insufficient to cause the 4 *mm* simulated beads to flow as fast as the measured 0.7 *mm* ballotini velocities. We found that this *effective diameter* argument was able to account for less than half of the factor of two increase in velocity seen in going from the large to the small glass particles. We conclude that the remainder of the increase must be due to the extra particle-particle collisions that occur in the systems of small particles.

We also conducted a few simulations for systems of small spheres (*i.e.*, $\sigma = 0.7mm$) at low solids loadings. We were unable to simulate the entire range of solids loading tested with the small ballotini because of the large number of particles required (*i.e.*, thousands); however, we did find that the dilute limit velocity for the small spheres was only slightly higher than the large sphere velocities. We also found that the velocity *vs* solids loading curve was increasing rapidly with solids loading, approaching the values measured in the tests with 0.7 *mm* ballotini.

Previous calculational models of particles cascading through arrays of horizontal rods have only dealt with the dilute limit of a single particle randomly bouncing its way through a rod array (*e.g.*, refs. 12, 13). To the authors' knowledge these are the first simulations that have included the effects of collisions with other particles simultaneously traveling through a rod array. These simulation calculations of spheres falling through horizontal rods were also the first comparisons between our particle-dynamics computer models and laboratory tests involving gravity flow and real boundaries. The good agreement achieved gives us added confidence that this calculational approach can correctly simulate the important features of most granular solids flows. Further calculational studies of flow through horizontal rod arrays are planned.

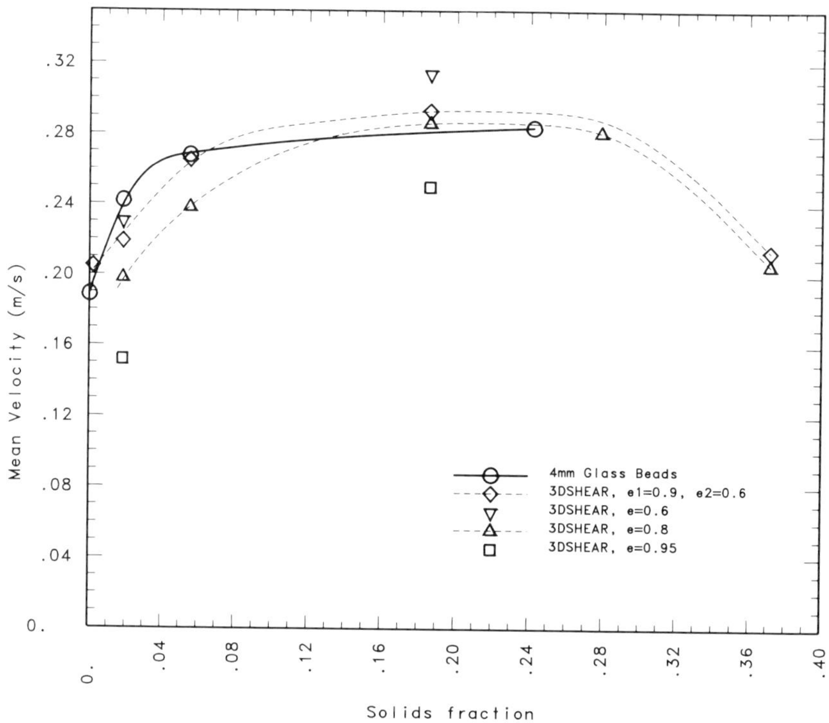

Figure 3

Mean vertical velocity for spherical particles falling through regular array of horizontal rods as measured in laboratory tests and as obtained in simulation calculations with various choices for the coefficient of restitution.

2-D Chute Flow Simulations

As part of a collaborative research project with R. Shreve and T. Drake at UCLA (partially funded by the Institute for Geophysics and Planetary Physics, IGPP) we are using our discrete particle models to simulate inclined chute flow tests of identical plastic spheres confined to a single vertical layer by parallel glass sidewalls. The UCLA researchers observe and document the motions of individual grains within the flows by means of high-speed motion pictures, and then digitize the films frame-by-frame to obtain the density, velocity, kinetic energy and collision information desired (ref. 14). They have only fully analyzed one of several flow test movies to date. It is for a chute inclined at an angle of 42.75^o from the horizontal, with a measured flow rate of approximately 1300 particles per second. The average number of particles per unit length along the chute bed in this test was 0.5568 *particles/mm*. The flow depth, mass flow rate and number density of particles per unit length all appeared to be at, or very near, steady state after traveling just over a meter down the nearly two meter long test chute.

We conducted simulation calculations of this test in which we modeled a fixed length (42.5 *mm* or 85 *mm*) of the inclined chute containing 5 or 10 boundary and 24 or 47 active particles (*i.e.*, 0.5647 or 0.5529 *particles/mm*) respectively. Fixed boundary particles were randomly placed in the simulation chute bed according to the same spacing rules as used for placing the fixed bed particles in the laboratory tests. The active particles were initially placed randomly above the incline within 20 particle diameters of the bed surface. These particles were assigned an initial mean velocity parallel to the incline and also given an initial random deviatoric velocity. Initial calculations with a two-dimensional disk model did not include the glass side walls. These simulations produced flows that were similar in character to the UCLA test results after a simulation time of approximately $0.8\,s$ except that the simulated particles continued to accelerate, reaching velocities up to three times the measured values after flowing for $2.5\,s$. Full three-dimensional simulations, including the frictional interactions with the vertical side walls, were performed with the 3DSHEAR model. These simulations also resulted in mean flow velocities well above the measured flow velocities (*e.g.*, the simulated flow rate was approximately 3000 *particles/s* at a time of $4\,s$, which is more than twice the measured 1300 *particles/s* flow rate). In order to obtain lower flow velocities in the simulation calculations we added an approximate air drag, in the form of a velocity squared drag term, to the forces acting on the particles. Initial results with an approximate air drag term are much closer to the measured results. Figure 4 shows snapshots of particle positions during two such simulation calculations. Figure 5 shows the velocity profile obtained in the simulation calculations and the corresponding measured velocity profile (ref. 14). Measurements of the true air drag effects in the chute flow apparatus are planned for next year as are additional measurements of the frictional forces between the plastic beads and glass surfaces at low normal loads. We expect that these new measurements will provide independent determination of input parameters for further simulations. Additional calculations are planned to determine the reason for the velocity differences observed near the bed surface and to simulate other flow conditions. These will be reported in the near future (ref. 15).

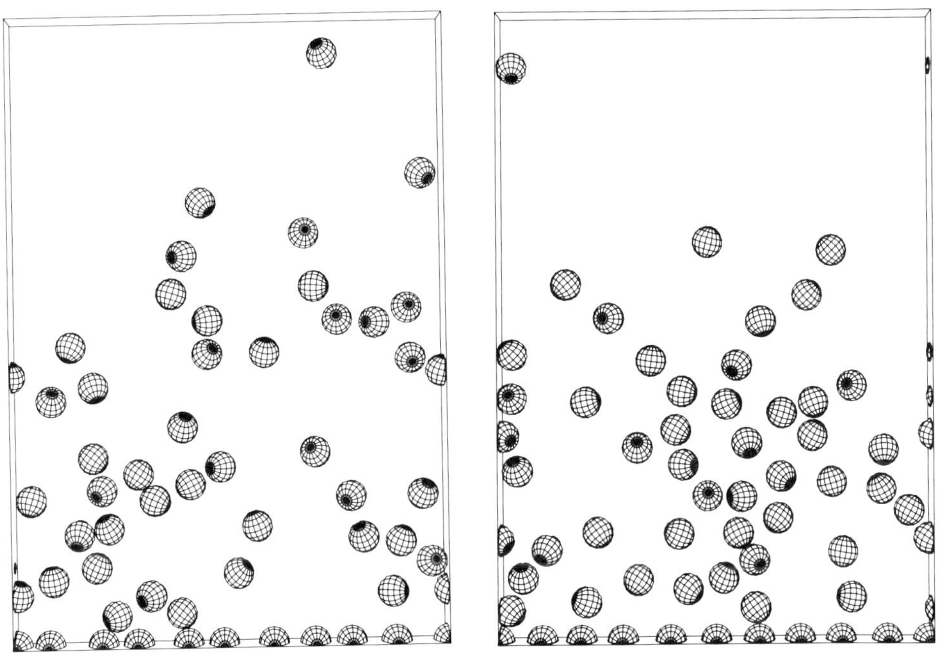

Figure 4

Perspective views of particle positions in two 3DSHEAR model simulation calculations of $6\,mm$ inelastic, frictional spheres flowing down a chute inclined 42.75^o from horizontal, with fixed particles on lower boundary and vertical planes, $6.25\,mm$ apart, in front and in back of the mono-layer of spheres.

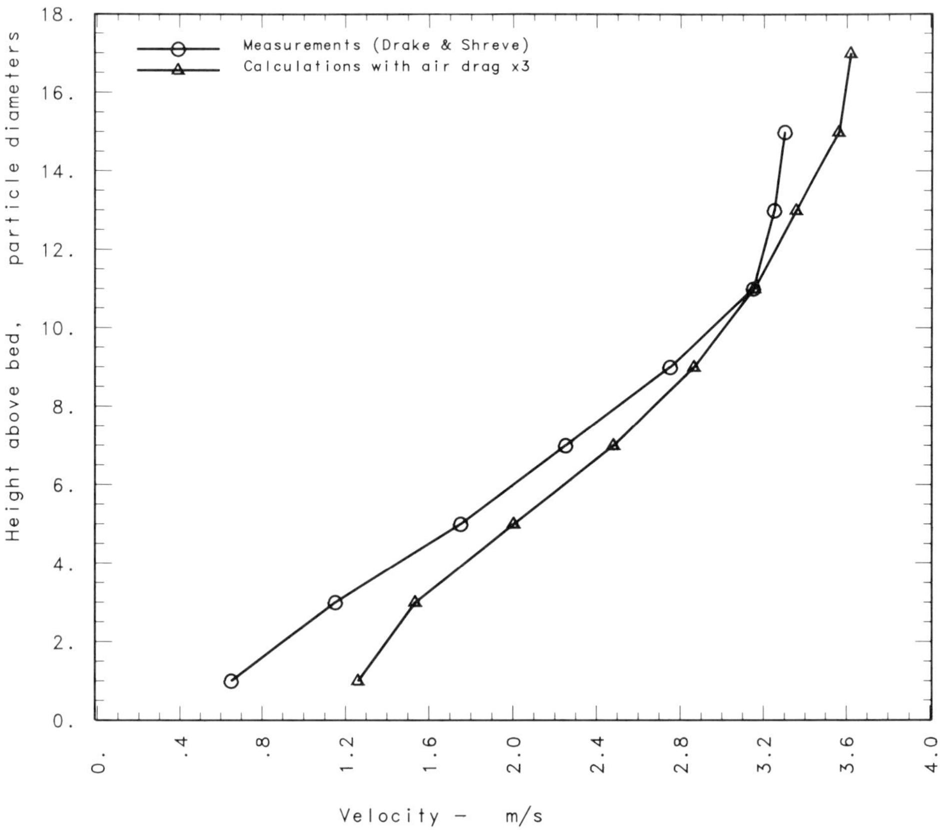

Figure 5

Velocity profiles obtained in laboratory chute flow tests at UCLA (ref. 14) and in 3DSHEAR model calculations including side walls and an enhanced air drag equal to three times the single-sphere drag coefficient for a 42.75° incline with nominally 1300 *particles/s*.

ACKNOWLEDGEMENTS

This work was supported under the Solids Transport Program of the US-DOE Office of Fossil Energy, Advanced Research and Technology Development, through the Pittsburgh Energy Technology Center. Partial support for the collaborative work with UCLA was also received from University of California's Institute for Geophysics and Planetary Physics. The authors are especially grateful for the experimental measurements and assistance provided by T. Drake and R. Shreve at UCLA, and U. Tüzün at University of Surrey, England.

REFERENCES

1 Walton, O.R., and R.L. Braun "Viscosity and Temperature Calculations for Assemblies of Inelastic Frictional Disks," *J. Rheology*, **30**(5), 1986, 949-980.

2 Walton, O.R., and R.L. Braun "Stress Calculations for Assemblies of Inelastic Spheres in Uniform Flow," *Acta Mechanica*, **63**, 1986, 73-86.

3 Walton, O.R., and R.L. Braun "Particle Dynamics Calculations of Binary Mixtures of Inelastic, Smooth and Frictional Spheres in Rectilinear Shear Flow," presented at ASCE/EMD Specialty Conference, May 20-22, 1987, Buffalo, NY (in preparation).

4 Walton, O.R., and R.L. Braun "Molecular-Dynamics Calculations of Inelastic Frictional Spheres in Uniform Shear," presented at March 1987 Mtg. APS, New York, NY (in preparation).

5 Hoover, W.G. *Lectures on Molecular Dynamics*, (to appear) Springer-Verlag, 1987.

6 Evans, D.J. "On the Representation of Orientation Space," *Molec. Physics*, **34**, 1977, pp.317-325; also Evans, D.J. and Murad, S., "Singularity Free Algorithm for Molecular Dynamics Simulations of Rigid Polyatomics," *Molec. Physics*, **34**, 1977, 327.

7 Savage, S.B., and M. Sayed "Stress Developed by Dry Cohesionless Granular Materials Sheared in an Annular Shear Cell," *J. Fluid Mech.*, **142**, 1984, 391-430.

8 Bagnold, R.A. "Experiments on a Gravity Free Dispersion of Large Spheres in a Newtonian Fluid Under Shear," *Proc. R. Soc. London, A*, **225**, 1954, 49-63.

9 Lun, C.K.K., S.B. Savage, D.J. Jeffery, and N. Chipurniy "Kinetic Theories for Granular Flow: Inelastic Particles in Couette Flow and Slightly Inelastic Particles in a General Flowfield," *J. Fluid Mech.*, **140**, 1984, 223-256.

10 Walton, O.R., A.E. Lewis, R.G. Mallon, and R.L. Braun "Cascading-Bed Process for Retorting Oil Shale," presented at ASME Energy-Sources Technology Conference, February 17-18, 1987, Dallas, TX (LLNL misc. pub. 4418).

11 Mindlin, R.D. and H. Deresiewicz "Elastic Spheres in Contact Under Varying Oblique Forces," *J. Appl. Mech., Trans. ASME*, **20**, 1953, 327.

12 Moran, B., and W. G. Hoover "Diffusion in a Periodic Lorentz Gas," *J. Stat. Phys.*, **48**, 1987.

13 Cook, M. H., and J. Bridgwater "The Simulation of a Particle Disperser," *Powder Technology*, **33**, 1982, pp.238-247.

14 Drake, T. G., and R. L. Shreve "High-speed motion pictures of nearly steady uniform, two dimensional, inertial flows of granular material," *J. Rheology*, **30**(5), 1986, 981-993.

15 Walton, O.R., and Drake, T.G. "Granular Flow on Inclined Two-Dimensional Chute: A Comparison of Measurements and Numerical Simulation," to be presented at ASCE/EMD Specialty Conference, Blacksburg, VA, May 1988, (in preparation).

Micromechanics of Granular Materials, edited by M. Satake and J.T. Jenkins
Elsevier Science Publishers B.V., Amsterdam, 1988 — Printed in The Netherlands

Boundary Interactions for Two-dimensional Granular Flows: Asymmetric Stresses and Couple Stresses

Charles S. Campbell
University of Southern California
Los Angeles, Ca. 90089-1453 USA

SUMMARY

This paper reports the results of a computer simulation study of the effects of solid boundaries on two-dimensional granular flows. In particular, two-types of smooth wall boundaries are explored: one, the Type A boundary, represents a fully rough wall, in the sense that there is no slip between the surface of the particle and the wall on departure following a collision; the other, the Type B condition, in an attempt to approximate a classic no-slip condition, assumes that there is no slip between the center of the particle and the wall on departure after a collision. The major difference between the two is that the Type A condition applies a large torque to the particles while the Type B applies no torque to the particle. The Type A condition is associated with significant local asymmetries in the stress tensor with corresponding large couple stress gradients in the region near the wall while the Type B condition has a nearly symmetric stress tensor and only small couple stress gradients.

INTRODUCTION

The challenge in granular material flows has been to derive macroscale continuum equations of motion for the entire granular mass starting from the basic microscale knowledge of how individual particles interact. But, until recently, appropriate boundary conditions to use with these equations have remained largely unexplored. Basically, the problem is the same: to start with a microscale level understanding of how particles interact with solid surfaces and from them to derive how on the macroscale level, the entire granular mass interacts with the surface. A further problem is to decide what continuum field quantities are necessary and sufficient to describe a boundary. From this point of view, conventional fluid mechanics has been blessed with the "no-slip condition" reflecting that, on the macroscale, there is insignificant slip (relative tangential velocity at the location of the boundary) between the molecules of the fluid and the molecules of the wall except at extremes of high velocity and/or small density. However, in a granular material, where particles may simply roll along a flat surface, slip will be the norm. In fact much of the problem facing experimentalists is to control the slip at the boundaries. For example, much of the rheological knowledge of granular flows comes from shear cell tests that use roughened walls to produce no-slip and hopefully a uniform velocity gradient within the material. However, how the wall is roughened significantly affects the results of the rheological tests. Both Hanes and Inman [5] and Savage and Sayed [7] have performed nearly identical shear cell tests on similar granular materials, yet the stresses measured by Hanes and Inman were often double those measured by Savage and Sayed. The major difference between the two studies is that Savage and Sayed roughened the walls of their shear cells with sandpaper, which had closely packed granules significantly smaller than the test material, and Hanes and Inman glued on particles from their test material. This indicates that there is a significant impact of the nature of the boundary on the entire behavior of the material. This dicrepancy has led to recent studies [4,6] on the effects of roughened walls. In particular, a previous computer simulation study, Campbell and Gong [4], studied the effects of roughening particle size and spacing on two dimensional granular flows.

Another view of the effects of boundary conditions can be found in the early computer simulation studies of Campbell and Brennen [1,2]. They studied two different boundary characterizations. The first, called the Type A boundary, assumed that both the particle and walls were fully rough in the sense that on departure after a collision, there would be zero tangential slip between the surface of the particle and the wall. Thus the boundary applied a force at the particle surface, off the center of mass of the particle, and therefore applies a large torque to the particle. Another consequence of this type of boundary characterization is that it allows for a large degree of slip between the particle center and the wall. In a naive attempt to control the slip and thus to control the shear rate, Campbell and Brennen employed another boundary characterization, called the Type B boundary, which forced the particle center to have the same velocity as the wall and did not change the particle's state of rotation. The results differed dramatically from one boundary condition to the other. With the Type A boundary, large velocity gradients were observed in the neighborhood of the solid bounding surfaces with almost no velocity gradient further away, and these regions of large velocity gradient correspond to regions of small particle density and large granular temperature. (The granular temperature is a measure of the energy contained in the random velocities of the granules in direct analogy to the temperature of a gas reflecting the energy contained in the random motion of the molecules.) With the type B boundary, the velocity gradient, density and granular temperature were all nearly uniform across the channel. (As will be shown later, this is not entirely correct.) Thus the nature of the boundary strongly affected the entirety of the flowfield. Furthermore, as the roughness of the sandpaper used by Savage and Sayed [7] is much smaller than a typical particle size, the induced flowfield may more closely correspond to the Type A boundary while the large glued particles used by Hanes and Inman [5] may be enough to effectively halt the particles near the wall and be much closer to the Type B boundary. (This conclusion is supported by Campbell and Gong [4].) The difference in the results presented by Savage and Sayed from those found by Hanes and Inman may then be explained by the smaller velocity gradient induced in the bulk of material by the Type A boundary than by the Type B boundary. The purpose of this paper is to study the Type A and B boundaries and to try and isolate the continuum properties of the system that best characterize the nature of the boundaries.

COMPUTER SIMULATION

This study is performed in a computer simulation that is nearly identical to that used by Campbell and Brennen [1] and so will only be briefly described here. The study considered a Couette flow of two dimensional discs, of mass m and radius R, trapped in a control volume which is shown schematically in Figure 1. The particles interact by instantaneous collision. Energy is dissipated in the collision through a coefficient of restitution, ϵ, (ϵ=0.8 is assumed throughout) and through particle surface friction. In this paper, the particle surface friction is always assumed to be large enough to halt any relative motion between the particles on departure after a collision. (This is equivalent to an assumption of an infinite surface friction coefficient.) The control volume is bounded on the top and bottom by solid walls, separated in the y-direction by a distance H, one of which is in motion in the x-direction with a velocity U relative to the other to induce a shear flow. In the direction of flow, the control volume is bounded by periodic boundaries; when a particle exits the downstream periodic boundary, it reenters the upstream with exactly the same y-coordinate and velocity with which it left. Once started, the simulation is allowed to progress

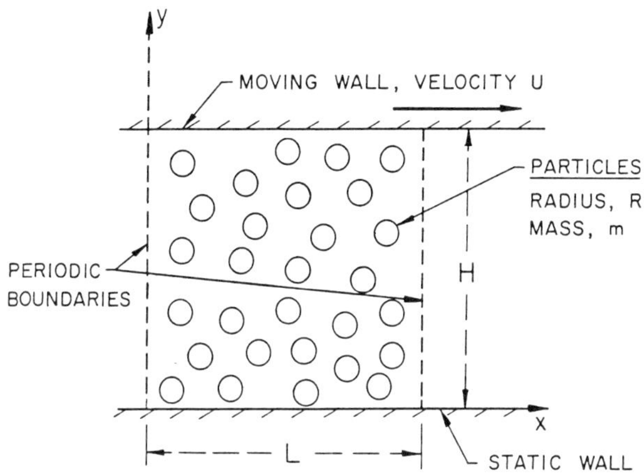

Figure 1: Schematic of the simulation control volume

until it reaches a steady state. From that point on, the properties of the system are averaged to obtain whatever information is required.

GENERAL FEATURES OF THE FLOWFIELD

Figure 2 is a composite of the velocity (plotted as $<u>/U$, where the symbol $< >$ is used to represent an average so that $<u>$ is the averaged value of the particle velocity in the x-direction), density (plotted as the solid fraction $\nu = \rho/\rho_p$, where ρ is the density of the bulk material and ρ_p is the density of the solid material alone), rotational velocity, (plotted as $-<\omega>H/U$, where $<\omega>$ is the average angular rotation rate of the particles), and the granular temperture, (plotted as $(<u'^2>/(RU/H)^2$ where $<u'^2>$ is the average value of the squared fluctuating velocity of the particles). The figure shows the results for an average solid fraction of $\nu \simeq 0.55$ for both the Type A and Type B boundary conditions. Note here that the difference between the two cases is not nearly as dramatic as those presented in Campbell and Brennen [1] and especially if that data is replotted in the manner shown in Campbell and Gong [4]. These discrepancies lead the author to suspect that the previous studies were not full converged but instead represented a slowly evolving system that could not reach full convergence and was evolving too slowly to make the non-convergence apparent with the computer facilities available to that project. (The current runs are some 30 times longer than those in Campbell and Brennen [1].) Still most of the general features are the same although the current results show that the Type B results retain more of the features of the Type A than was reported previously. The Type A velocity profile shows slip at the wall and regions of somewhat higher shear in the neighborhood of the wall (although not as marked a difference as in Campbell and Brennen [1]); both of these are missing from the Type B profiles. (Although at larger density, high shear high shear regions also appeared in the Type B profiles.) The Type A density profiles exhibit regions of low density near the boundaries and higher density near the center of the channel; much the same behavior, though to a much lesser degree is seen in the Type B profiles. (Again the low density regions were absent from the Type B profiles published by Campbell and Brennen.) The granular temperature profiles (fourth panel) show very large temperatures near the boundaries and substantially reduced temperatures near the center of the channel for the Type A boundary. The large temperatures near the boundary can be directly

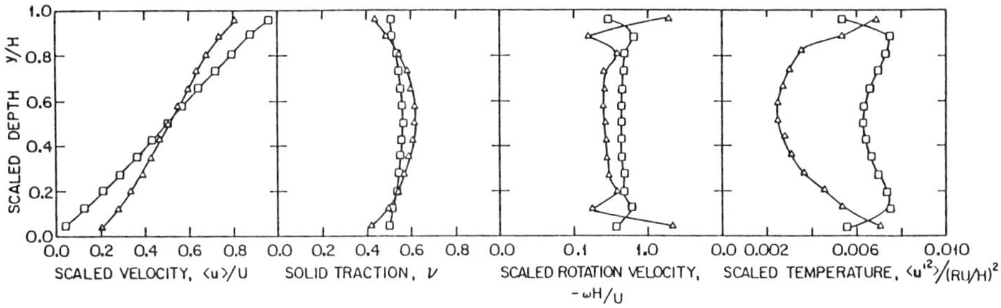

Figure 2: General features of the Type A and Type B Couette flows. △ Type A boundary, □ Type B boundary

related to the boundary condition: The type A boundary imposes a large torque and hence a large rotation rate on the particles that collide with it; on susequent collisions this large rotational velocity is converted through particle surface friction into random translational velocities and consequently larger temperatures. The Type B boundary shows much the same behavior, but to a much smaller degree; and, right next to the boundaries, the Type B profile shows reduced temperatures, but this should be expected as the particle assumes the walls velocity on collision and thus loses any random velocity components.

The third panel shows the average rotational state of the particles and, in many ways, holds the key to the entire story. (For what is to follow, it will be important to know that the points in Figure 2 are separated in the vertical direction by very close to a particle diameter.) The Type A profiles show a large rotation rate at the boundaries which is to be expected because of the large torque the boundary applies to the particles. Near the center of the channel the rotation rate is nearly uniform and for the few points near, but not next to, the walls, the rotation rate oscillates about the mean value. This behavior can be explained as a competition between two meachanisms of torque generation which are illustrated in Figure 3. First of all, (Figure 3a), in any shear flow, a particle will primarily collide either with faster moving particles from above and behind or with slower moving particles in front and below. Both types of collision will induce a clockwise rotation of the particle and Campbell and Gong [3] have shown that the mean rotation rate will be approximately one half the shear rate except at extreme values of the mean flow density. (Note that in the center of the channel, the average rotation rate is also about 1/2 the shear rate.) The second process is illustrated in Figure 3b. On colliding with a wall, however, the particle will pick up a large rotational velocity which is proportional to the slip velocity. On subsequent collisions, this will induce a counter-rotation in the slower rotating particles in the next layer further out from the wall. Thus, the particles in the second layer out from the wall have a rotation rate that is smaller than the value found near the center of the channel. These particles, in turn, induce a counter rotation in the next layer out from the wall whose particles have a rotation rate slightly larger than that at the channel center. The rotational interchange continues until the wall effect (Figure 3b) finally gives way to the mean shear effect (Figure 3a).

The rotational velocity profile for the Type B boundary condition is much more uniform across the channel, although, it still exhibits the same kind of jagged shape near the walls. Note, however, that the jags are exactly out of phase with the Type A curve. This is a bit strange because, unlike the Type A

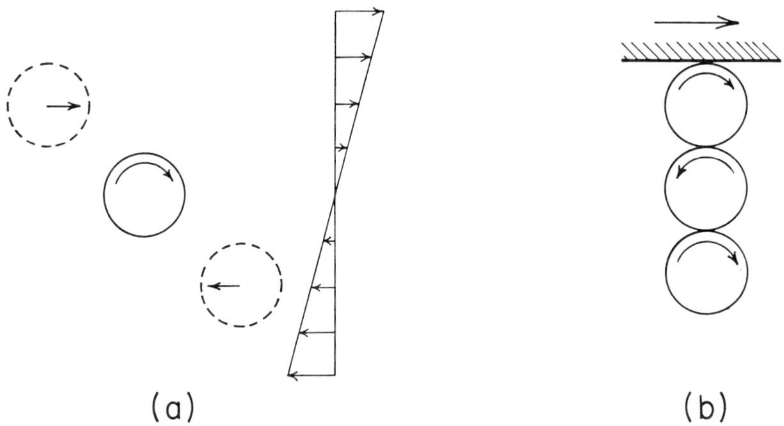

Figure 3: The two mechanisms of rotation inducement. (a) inducement by a mean shear flow, (b) inducement near a boundary

boundary, the Type B boundary exerts no torque to change the rotational state of the particles. But, at the same time, this behavior may be understood in terms of the discussion above. As shown in Figure 3a, a particle far from the walls will collide with faster moving particles from above and slower moving particles from below, both of which will induce clockwise rotation. The boundary will block collisions on one side or the other for particles right next to the boundary. Thus the internal state of torque (which is described by couple stresses and will be discussed later) is smaller for particles right next to the boundary than for those further away. This results in a smaller average rotation rate right next to the boundary. Then, following exactly the same logic as above, this will induce a countertorque and hence a slightly larger rotational velocity in the particles one layer away from the wall. This, in turn, will induce a slightly lower rotational velocity in the next layer out and so on. Thus, even though the Type B boundary does not transmit any torque, it has much the same effect as it causes a discontinuity in the internal state of torque in the material.

INTERNAL STRESS DISTRIBUTION

There are two mechanisms of internal momentum transport by which stress is transmitted through a granular material. The first, or streaming mode, denoted τ_s, reflects the momentum that is carried by particles as they move through the bulk material with their random velocity. In appearance, the streaming stress tensor is exactly the same as the Reynold's stress tensor in turbulent flow:

$$\tau_s = -\rho_p \nu \langle u'u' \rangle$$

where $\langle u'u' \rangle$ is the average of the dyadic product of the fluctuating velocity vectors, u'. Obviously, the magnitude of the streaming stresses should be related in some way to the magnitude of the granular temperature. Note that τ_S is necessarily symmetric. Also, when two particle collide, momentum is transferred from the center of one particle to the other. This is the collisional mode of momentum transfer and results in the collisional stress tensor, τ_c:

$$\tau_c = 2R\langle \mathbf{Jk} \rangle \eta$$

where $\langle \mathbf{Jk} \rangle$ is the average of the dyadic product of **J**, the vector impulse exerted by a collision, k, is the unit vector connecting the particle centers at the time of collision and η is the collision rate per unit volume. This reflects the fact that every collision causes the transport of momentum **J**, the distance 2R in the direction **k**, between the particle centers. Obviously, the collisional mode will be more important at large density where a particle cannot move far between collisions and the streaming mode is dominant at low densities where collisions are infrequent and particles move a large distance between collisions. The complete stress tensor, τ, is found by summing these two contributions. For more information on how the stress tensor is found, the reader is reffered to Campbell and Gong [3]. Here, as there, the plotted stresses are scaled by dividing the actual stresses by $\rho_p R^2 (U/H)^2$. Note that the plots show absolute value of the stresses; for the cases shown here, the normal stresses are all negative and the shear stresses are positive.

Previous computer simulation studies to measure the stress tensor in granular flow have purposely chosen conditions for which the state of stress is uniform across the control volume. In this study, however, the boundaries will induce nonuniformities in stress. For this reason, the distribution of stress across the channel was examined by dividing the control volume into strips parallel to the wall and averaging the state of stress within the strips. For the streaming stresses, this procedure is exactly the same as described in Campbell and Brennen [1]. But for the collisional stresses, there is the possibility that the colliding particles may be in different strips; thus the fraction of $\langle \mathbf{Jk} \rangle$ attributed to each strip is weighted by the transport distance through that strip.

The internal distribution of stress is plotted for the Type A boundary in Figure 4a and for the Type B boundary in Figure 4b. All four of the stress components, τ_{xx}, τ_{xy}, τ_{yx} and τ_{yy} are plotted as well as all of the complete, streaming and collisional stress tensors. (As τ_s is necessarily symmetric, only the τ_{xy} and not the τ_{yx} component is plotted in that frame.) The complete stress levels for the type B condition are significantly larger than for the type A which illustrates that a difference in boundary condition could explain the larger stresses measured by Hanes and Inman [5] over those measured by Savage and Sayed [7]. Note that, at the rather large density (ν=0.55) shown here, that the collisional mode of momentum transport is much more important than the streaming. As would be expected, the streaming stresses are noticeably larger in the lower density and high granular temperature regions near the boundaries and this is accompanied by a slightly smaller collisional stress tensor. Furthermore, in the Couette flow configuration, the τ_{xx} and τ_{yx} stresses cannot be applied by the boundaries; hence these stresses are self-equilibrated and the values they assume are completely determined by the flow conditions. On the other hand, the τ_{xy} and τ_{yy} stresses can be applied by the boundaries and the stress exerted at one boundary can only be balanced by the other. Hence, these two stresses must be transmitted across the channel and therefore the value of the complete stresses must be uniform throughout the interior and equal to the value applied at the wall. This is supported by the complete stress tensor results in Figure 4 except for the points right next to the wall - a discrepancy which is a useful byproduct of an error in the sampling procedure. All of the collisional stresses plotted as open points are calculated from particle-particle collisions. When a particle collides with a wall, momentum is also carried from the center of the particle to the wall, crossing the

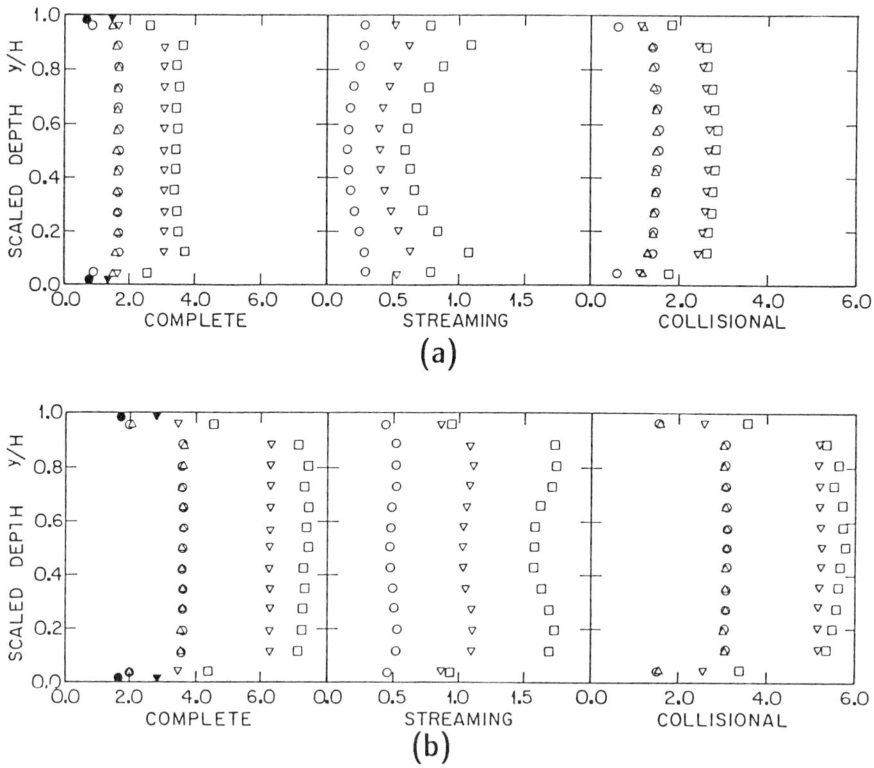

Figure 4: Internal stress distribution. □ τ_{xx}, ○ τ_{xy}, △ τ_{yx}, ▽ τ_{yy}; ● τ_{xy} correction, ▼ τ_{yy} correction; (a) Type A boundary, (b) Type B boundary

distance of a particle radius within the strip. Thus, to fully assess the momentum transport within the strip, this extra momentum transport must be assessed and added to the strip. This portion is labeled the "correction stresses" and are separately plotted as solid points in Figure 4. As the wall can only apply τ_{xy} and τ_{yy} stress, there are only corrected values for these two components. Note that when the correction stresses are added, the total τ_{xy} and τ_{yy} stress next to the boundary will be the same as the rest of the channel. In order to distinguish particle and wall effects, the correction stresses are plotted separately and not added to the particle-particle stress next to the wall.

In the few layers nearest the walls of the Type A boundary distribution, the stress tensor is assymetric, i.e. $\tau_{xy} \neq \tau_{yx}$. This is a bit surprising since in elementary continuum mechanics, there is a simple proof that the stress tensor must be symmetric. The proof involves relating the angular momentum change of the material within a contour to the contour integral of the stress; on shrinking the contour, the moment of inertia disappears faster than the contour integral implying that the value of the integral must be zero which, in turn, implies that the stress tensor is symmetric. This breaks down for a granular material because for a continuum model to hold, any such contour must contain several particles. Hence, such a contour cannot be shrunk smaller that a particle. In this light, it is reasonable to associate the assymmetry in the stress tensor with torques on the particles, an association which is supported by integrating the surface traction force over the surface of a particle. Note then that in the layer nearest the walls, $\tau_{yx} > \tau_{xy}$ while in the next layer out, $\tau_{xy} > \tau_{yx}$, and, (although it is barely perceptable,) in the next layer further out, $\tau_{yx} > \tau_{xy}$ again. This reflects the varying torques on the particles (which should be anticipated from the mechanism illustrated in Figure 3b) that lead to the jagged behavior of the $-\omega H/U$ curve in Figure 2. Now if $\tau_{yx} > \tau_{xy}$, as in the layer nearest the boundaries, the granular mass exerts a counterclockwise torque to resist the clockwise torque exerted by the wall.

However, if there is an unbalanced torque on the particles, their angular velocity would continually accelerate. Yet, these are steady flows so the average angular velocity must be constant which means that the torque that is implied by the asymmetric stress tensor must somehow be balanced. This in turn indicates that the stress tensor alone is inadequate to explain the transport of angular momentum within the granular mass. In the process of averaging the microscopic details of the collisions to obtain the macroscopic stress tensor, much of the information about the moments has been lost; for example, as long as there is surface friction on the particles, there are infinitely many different collisions exerting infinitely many different torques that will yield the same dyadic product **Jk**. In a macroscopic continuum sense, then, the transport of angular momentum is independent of the transport of linear momentum even though, on the microscopic level, both result from the same collisions. Thus another field variable is required to describe the transport of angular momentum which is the "couple stress tensor". This is a direct analog to the Cauchy stress tensor and reflects the transport of angular momentum both by streaming and collisional action. The streaming portion of the couple stress tensor, $\mathbf{M}_s$, is given by:

$$\mathbf{M}_s = -I\langle u'\omega' \rangle$$

where I is the particle moment of inertia and $\langle u'\omega' \rangle$ is the average of the dyadic product of the fluctuating linear and angular velocities. Also, each collision causes a tranport of angular momentum $R\mathbf{J}\times\mathbf{k}$, a distance 2R (between the particle centers) in the **k** direction. Thus the collisional contribution to the couple stress tensor is given by:

$$\mathbf{M}_c = 2R^2\langle \mathbf{J}\times\mathbf{k}\mathbf{k} \rangle \eta$$

where, once again, η is the collision rate per unit volume. The complete couple stress tensor is found by summing the streaming and collisional contribution to the couple stress tensor. Each component, M_{ij}, of the couple stress tensor may be read as "the surface torque component in the i direction exerted on a surface with outward pointing normal unit vector in the j direction. Performing an angular momentum balance on a control volume in steady flow, one obtains the expression:

$$M_{ij,j} + \epsilon_{ijk}\tau_{jk} = 0$$

Thus the asymmetry in the stress tensor is balanced by the divergence of the couple stress tensor. In two-dimensions, the particles are forced to rotate within the plane, and, as in Couette flow all gradients are in the y-direction, this expression becomes:

$$M_{zy,y} + (\tau_{xy} - \tau_{yx}) = 0$$

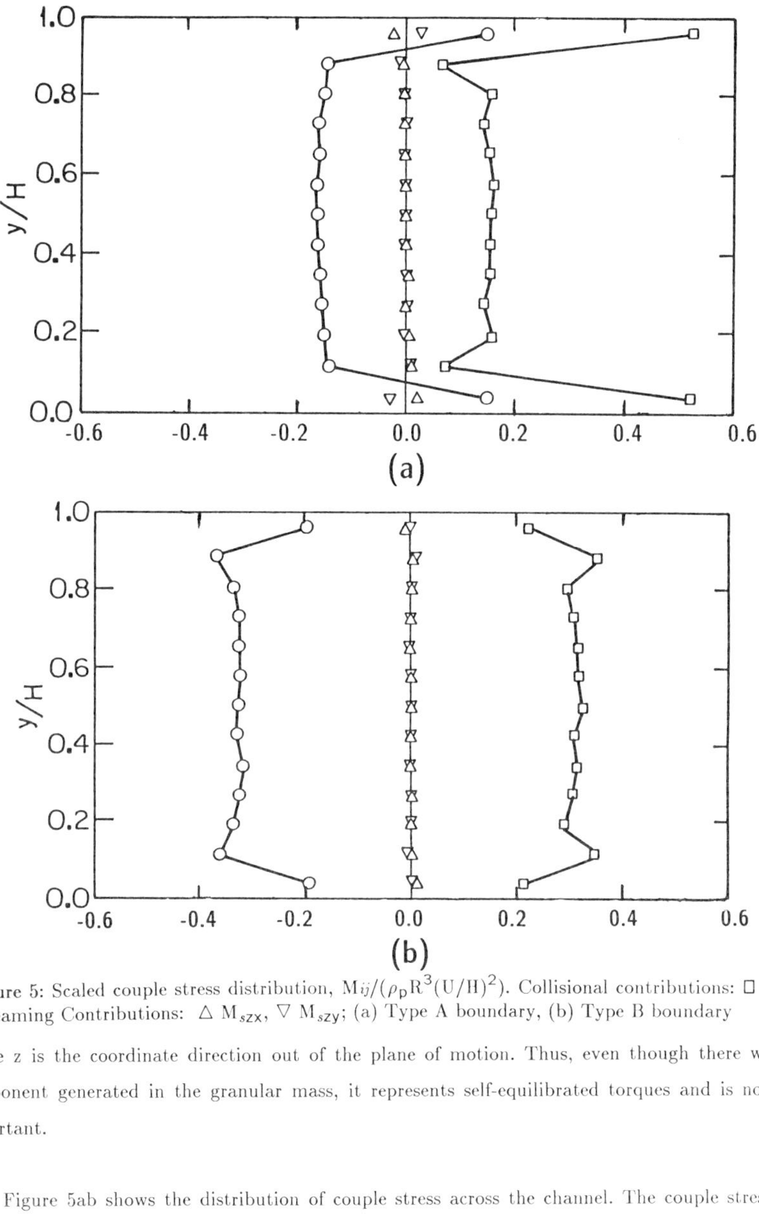

Figure 5: Scaled couple stress distribution, $M_{ij}/(\rho_p R^3(U/H)^2)$. Collisional contributions: □ M_{czx}, ○ M_{czy}; Streaming Contributions: △ M_{szx}, ▽ M_{szy}; (a) Type A boundary, (b) Type B boundary

where z is the coordinate direction out of the plane of motion. Thus, even though there will be an M_{zx} component generated in the granular mass, it represents self-equilibrated torques and is not dynamically important.

Figure 5ab shows the distribution of couple stress across the channel. The couple stresses have been scaled by dividing by $\rho_p R^3(U/H)^2$. Here, unlike the Cauchy stresses plotted in Figure 4, the actual (and not the absolute) values of the couple stresses are shown because the M_{zy} component changes sign across the control volume for the type A boundary condition. Both the streaming and collisional components are plotted on the same graph. Note that the streaming contribution is negligible compared to the collisional contribution and is only non-zero right next the the walls. This indicates that there is little correlation between the fluctuating linear, u', and rotational, ω', velocities. Thus the complete couple stress tensor is not plotted as it does not vary significantly from the collisional contribution. It can be clearly seen, at least qualitatively, that the gradients of couple stress do balance the Cauchy stress tensor assymetries according the above prescription; however, as the gradients change sign over distances on the order of a particle diameter, it is not possible to obtain an accurate quantitative evaluation of the couple stress gradients and thus not possible to demonstrate a quantitative agreement. Note that the M_{zy} component of the Type A distribution becomes positive near the walls reflecting the large positive torque exerted by the granular mass to resist the negative torque imposed by the boundary. (Once again, the graph shows only the torque exerted by particle-particle interactions.) This is not true for the Type B boundary where the generated torques are gentler and the gradients of couple stress are much smaller. Far from the boundaries, for both cases, the M_{zy} component of the couple stress is nearly constant and negative; the sign reflects the torque induced by the velocity gradient which generates the clockwise rotation of the particles. Note, however, that in this central region the M_{zx} component is positive and of nearly the same magnitude as the M_{zy} component. As the couple stress becomes nearly uniform, it is of no dynamic importance far from boundaries.

CONCLUSIONS

This paper has examined the effects of two different wall boundaries on a granular Couette flow. The Type A boundary assumes that on departure, after collision with a solid wall, there is no slip between the particle <u>surface</u> and the wall and the Type B condition assumes that there is no-slip between the particle <u>center</u> and the wall with no change in the rotational velocity. Both conditions are somewhat fictitious; the Type A because it assumes an infinite surface friction coefficient and the Type B because there is no way to apply a force directly to the center of the particle (although Campbell and Gong [4] show that similar behavior may be had by roughening the wall with glued particles). A major difference between the two is that the Type A condition applies a torque to the particles while the Type B does not.

This apparently small change in boundary condition produces a large change in the behavior of the system as a whole. Most obvious is that the Type A condition allows the material to slip at the wall which, in turn induces a large rotational velocity in the particles there. This rotation is converted by successive collisions into granular temperature in the region nearest the walls which is accompanied by a reduced density in the same area. The torque applied by the wall is balanced by a countertorque from the granular mass which appears as a locally asymmetric stress tensor. In steady flow there can be no unbalanced torque on the particles so the torque implied by the asymmetric stresses is balanced by gradients in a couple-stress tensor. Far from a boundary, the stress tensor is symmetric and the couple stresses are uniform, implying that, for the Couette flow problem at least, couple stresses are only important near boundaries. Surprisingly, many of the same effects are apparent for the Type B boundary (though to a much smaller degree) even though such a boundary exerts no torque on the particles. This is because the boundary, by exerting no couple stress, disturbs the internal couple stress field.

An essential feature separating a Type A boundary from a Type B boundary is how the two transmit couple stress to the bulk material. Thus, couple stresses, which so far have been pretty much ignored in granular flow theories, appear to be vitally important in the coupling of a granular flow to a boundary, even though they appear to have little dynamic importance further away. Furthermore, how a boundary transmits couple to the material is an important parameter in the characterization of that boundary.

ACKNOWLEDGEMENTS

The author would like to thank Professor Joseph Goddard for his instruction reguarding couple stress, Dr. Otis Walton for suggesting the idea of examining the distribution of collisional stresses and Ailing Gong for her help in preparing the simulation program and the figures. This work was supported by the National Science Foundation under grants MEA-8352513 and CBT-8615160, the Air Force Office of Scientific Research under grant AFOSR-87-0284 and IBM for with the author is most grateful.

REFERENCES

[1] Campbell, C.S. and Brennen, C.E., Computer Simulation of Granular Shear Flows, J. Fluid Mech., 151, (1985) 167-188

[2] Campbell, C.S. and Brennen, C.E., Chute Flows of Granular Material: Some Computer Simulations, J. App. Mech., 52, (1985) 172-178

[3] Campbell, C.S. and Gong, A., The Stress Tensor in a Two-Dimensional Granular Shear Flow, J. Fluid Mech. 164, (1986) 107-125

[4] Campbell, C.S. and Gong, A., Boundary Conditions for Two-Dimensional Granular Flows, Proceedings of the International Symposium on Multiphase Flows, Hangzhou China, August 1987, vol. 1, 278-283

[5] Hanes, D.M. and Inman, D.L., Observations of Rapidly Flowing Granular-Fluid Flow, J. Fluid Mech., 150, (1985) 357-380

[6] Jenkins, J.T. and Richmond, M.W., Boundary Conditions for Plane Flows of Smooth, Nearly Elastic, Circular Discs, submitted to J. Fluid Mech.

[7] Savage, S.B. and Sayed, M., Stresses Developed by Dry Cohesionless Granular Materials in an Annular Shear Cell, J. Fluid Mech. 142, (1983) 391-430

Micromechanics of Granular Materials, edited by M. Satake and J.T. Jenkins
Elsevier Science Publishers B.V., Amsterdam, 1988 — Printed in The Netherlands

DESCRIPTION OF CYCLIC DEFORMATION BEHAVIOR OF SAND BY AN ANISOTROPIC HARDENING MODEL

E. YANAGISAWA[1] and Y. TOBITA[2]

1 Dept. of Civil Eng., Tohoku University, Sendai 980, Japan

2 Dept. of Civil Eng., Hachinohe Institute of Technology, Hachinohe 031, Japan

ABSTRACT

A simple anisotropic hardening model for granular material is proposed to describe cyclic behavior of sands during earthquake excitations. Special emphasis was placed on the formulation of unloading and reloading behavior of the model. Some numerical results of response of a sand element under repeated loading condition are presented in order to demonstrate some features of the present model. Considerations are made on the calculation results by comparing to the laboratory test results and possible alternative formulation of the constitutive model is suggested.

INTRODUCTION

There has been a large number of papers published on the behaviors of sands under cyclic loading conditions, and a great deal of effort has been devoted to constitutive formulation of the behavior. In these studies constitutive models based on the elasto-plastic theory are often used to simulate the cyclic behavior of sands. In this paper, a simple anisotropic hardening model is proposed for the prediction of the cyclic behavior of cohesionless soils. Typical examples of the anisotropic hardening model would be ones proposed by Ghaboussi and Momen (ref.3), Mroz and Norris (ref.4) and Poorooshasb snd Pietruszczak (ref.5). The constitutive model proposed here with the property of rate independent can explain some specific features of the deformation of sands subjected to cyclic loading as will be demonstrated in the following numerical examples. Special attention was paid on the mathematical simplicity of the constitutive model by introducing some assumptions. It should be noted, however, that the simplicity of constitutive models for geomateials can be achieved on the sacrifices of the applicability to wider situations. This can be attribute to the fact that some assumptions introduced for idealization may ignore in some cases the fundamental micromecanical processes of the materials. It is therefore understood that micromechanical consideration on the macroscopic constitutive models seems to be of more importance to understand the applicabity and limitation of the constitutive models. The apparent shortcomings of the proposed model will be briefly discussed and possible ways of modification will be suggested later. It will be found that any refinement of the model will lead to considerable mathematical complexity.

MATHEMATICAL FORMULATION OF ANISOTROPIC HARDENING MODEL

In this section we briefly follow the development of the proposed anisotropic hardening model. Two distinct surfaces in stress space are employed to predict deformation characteristics of sands in virgin loading and cyclic loading. For virgin loading, isotropic loading surface which is defined by a function of stress state and strain history is assumed. The loading surface also defines equivalent plastic modulus surface along which plastic modulus has an equal value h_A. For cyclic behavior, loading surface is defined as a function of current stress state, stress history and strain history.

The isotropic hardening surface is given as

$$F(\sigma_{ij}, \lambda_t) = 0 \quad \text{and} \quad \hat{F} = (\partial F/\partial \sigma_{ij})\dot{\sigma}_{ij} > 0 \tag{1}$$

As far as Eq.(1) is satisfied, mechanical behavior of sand is assumed to obey non-associative flow rule in virgin state. The flow rule for the virgin state is given by following set of equations:

$$\dot{e}^p_{ij} = (1/h)(\partial F/\partial \sigma_{ij})'\hat{F} \quad \text{and} \quad \dot{\varepsilon}^p_{kk} = \Omega(\lambda_t)\dot{\lambda}_t \tag{2}$$

where e_{ij} denotes strain deviator; i.e., $e_{ij} = \varepsilon_{ij} - (1/3)\varepsilon_{kk}\,\delta_{ij}$, δ_{ij} : the Kronecker's delta with 1 for same indices and 0 for others. ()' denotes deviatoric part of the enclosed quantity. λ_t is an equivalent shear strain defined as $\lambda_t = \int_0^t \sqrt{(2/3)\dot{e}_{ij}\dot{e}_{ij}}\,ds$. $\Omega(\lambda_t)$ is the dilatancy factor which defines the ratio of dilatancy rate with respect to equivalent shear strain rate. Using the consistency condition of the equivalent plastic modulus surface, for the virgin loading we have

$$\dot{F} = (\partial F/\partial \sigma_{ij})\dot{\sigma}_{ij} + (\partial F/\partial \lambda_t)\dot{\lambda}_t = 0 \tag{3}$$

We can obtain (1/h) in an explicit form as a function of inner product between the gradient of F regarding stress components and stress rate and λ_t. For the isotropic loading surface we use

$$F = \sqrt{3J_2'} - R(\theta)k(\lambda_t)p' = 0 \tag{4}$$

where J_2' denotes the second invariant of stress deviator; namely $J_2'=(1/2)s_{ij}s_{ij}$ and $s_{ij}=\sigma_{ij}-(1/3)\sigma_{kk}\delta_{ij}$. $R(\theta)$ denotes the effect of the intermediate principal stress on the shape of the isotropic loading surface, and p' denotes the effective mean stress, i.e. $p'= (1/3)\sigma_{kk}$. For the function form of $k(\lambda_t)$ we employ the following hyperbolic relation:

$$k(\lambda_t) = (\lambda_t)/(a + b\lambda_t) \tag{5}$$

For the cyclic behavior of sands within the equivalent plastic modulus surface, we use the following loading function;

$$f = \sqrt{3J_2^*} - R(\theta)\sigma_o(\lambda_s)p' = 0 \tag{6}$$

where $J_2^* = (1/2)(s^*_{ij}s^*_{ij})$ and $s^*_{ij} = s_{ij} - \alpha_{ij}$.

α_{ij} denotes the back stress tensor indicating the position of the center of the loading surface and $\sigma_o(\lambda_s)$ is the isotropic hardening parameter indicating the degree of isotropic hardening during cyclic deformation of sands. The schematic representation of the loading surface in principal stress space is shown in Fig.1. $\sigma_o(\lambda_s)$ was assumed to have a functional form

$$\sigma_o(\lambda_s) = b_o(1 - \exp(a_o\lambda_s)) \quad \text{with } b_o > 0 \text{ and } a_o < 0 \tag{7}$$

The following flow rule is assumed to derive the plastic strain deviator;

$$\dot{e}^p_{ij} = (1/h)(\partial f/\partial\sigma_{ij})'\hat{f} \tag{8}$$

where $\hat{f} = (\partial f/\partial\sigma_{ij})\dot{\sigma}_{ij}$ and $()'$ denotes the deviatoric components of the gradient of the loading surface for cyclic loading with respect to corresponding stress components.

The definition of the instantaneous plastic modulus is given as

$$h(\delta) = h_o - (h_o - h_A)(\delta/\delta_o)^{\gamma} \tag{9}$$

where δ is a measure of the relative configuration of the loading surface with respect to the equivalent plastic modulus surface defined as $\delta = |\tilde{s} - \tilde{s}_o|/p'$ where $\tilde{s}$ is the current stress deviator and $\tilde{s}_o$ is the deviatoric stress at the onset of reversal plastic loading. $\delta_o = |\tilde{s}_A - \tilde{s}_o|/p'$, where $\tilde{s}_A$ denotes the conjugate stress point on the equivalent plastic modulus plane given as the connecting point of the line parallel to $(\tilde{s} - \tilde{\alpha})$ at the reversal plastic loading with the origin of $\tilde{\alpha}$ to the surface. Fig.2 shows the two dimensional representation of the stress reversal mentioned above in the deviatoric plane. When the direction of the stress rate deviates from a radial loading, δ and δ_o were assumed to vary by adopting new values of $\tilde{s}_o$ and $\tilde{s}_A$.

The evolution of the back stress tensor is assumed to be given as

$$\dot{\alpha}_{ij} = c(\lambda_s)(s_{ij} - \alpha_{ij})/p' \tag{10}$$

Thus the movement of the back stress tensor is assumed to take place in the direction parallel to the vector connecting from the back stress to the current stress point. The quantity $c(\lambda_s)$ is determined as a consequence of the consistency condition, namely

$$\dot{f} = 0\,, \quad \dot{f} = (\partial f/\partial\sigma_{ij})\dot{\sigma}_{ij} + (\partial f/\partial\alpha_{ij})\dot{\alpha}_{ij} + (\partial f/\partial\lambda_s)\dot{\lambda}_s \tag{11}$$

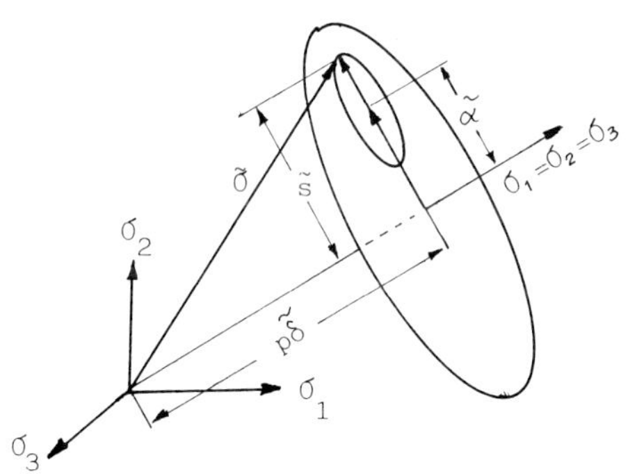

Fig.1 Schematic Representation of the Loading Surface

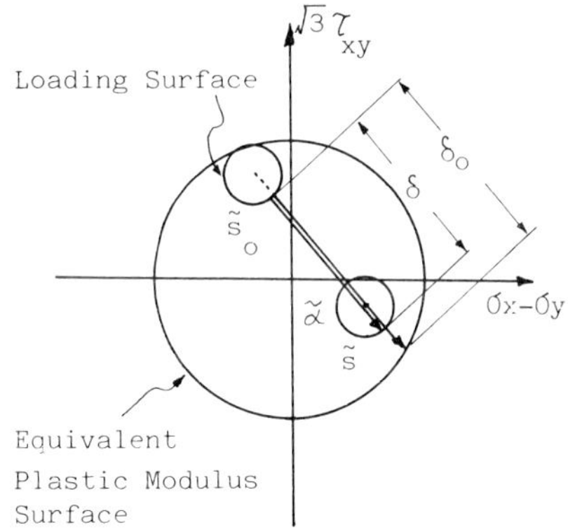

Fig.2 Concept of Loading Surface and Equivalent Plastic Modulus Plane

The dilatancy is assumed to be given as

$$\varepsilon^{p}_{kk} = \omega(\lambda_s)\dot{\lambda}_s \qquad (12)$$

where $\omega(\lambda_s)$ is a dilatancy factor depending on the stress path as will be exemplified in the next section.

It seems necessary to note the difference between the role of λ_t and λ_s. λ_t is a monotonic increasing function which operates for the virgin loading only, namely $\dot{\lambda}_t > 0$ when F=0 and $\hat{F} > 0$, and $\dot{\lambda}_t = 0$ for others. On the other hand λ_s operates in both virgin loading and cyclic loading within the equivalent plastic modulus surface. λ_s will be set equal to zero when the stress reversal occurs in the stress state beyond the phase transformation lines, which implies the shrinkage of the loading surface to a point; namely vanishment of the elastic domain. Thus the parameter λ_t has a role of memory regarding the previous maximum shear in terms of the stress ratio and the parameter λ_s plays roles in defining the gradual increase of shear resistance when sands are subjected to repeating low shear stress and the drastic change of volumetric deformation behavior during the stress reversal beyond the phase transformation lines. Fig.3 shows the evolution of the parameters λ_t and λ_s under typical loading events.

NUMERICAL RESULTS ON THE RESPONSE OF A SAND ELEMENT

After rewriting the constitutive model by a convenient stress and strain quantity for the triaxial and plane strain conditions, the programs were developed in order to simulate the cyclic behavior of sand in these conditions, which have been well established experimentally. Here in the triaxial condition R(θ) was set to be 1.0 for triaxial compression and

R(θ) = (3 − sinϕ)/(s + sinϕ) = K

for the triaxial extension. For the plane strain condition R(θ) was set to be 1.0.

Fig.4(a) shows the cyclic behavior of a sand element subjected to relatively low stress ratio modified by R(θ) in a drained condition. (Note that the stress ratio in the figures is a quantity modified by R(θ).) Fig.4(b) shows the corresponding dilatancy be-

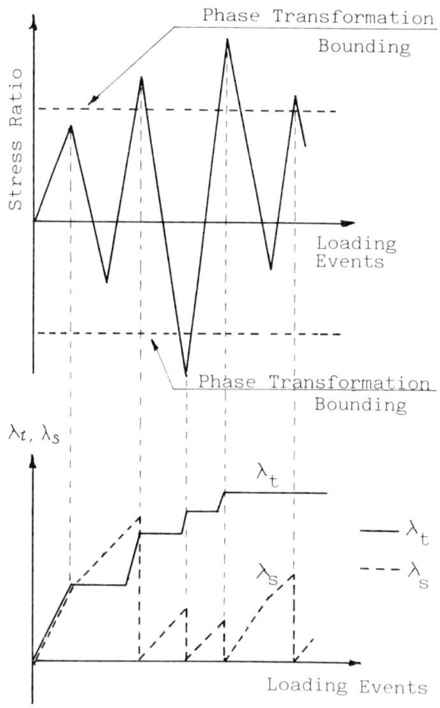

Fig.3 Evolution of λ_t and λ_s for Cyclic Loading

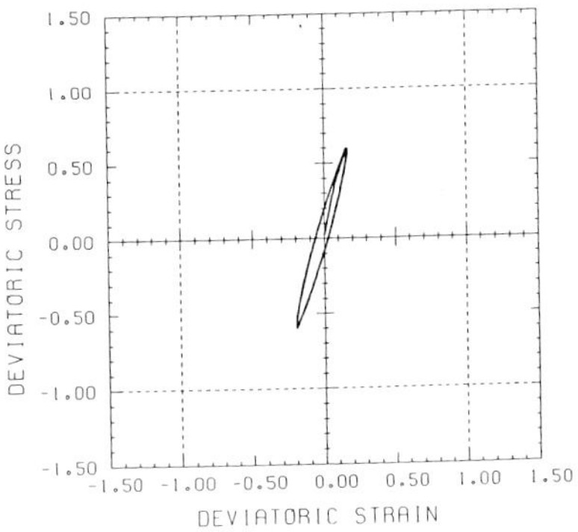

(a) Deviatoric Strain

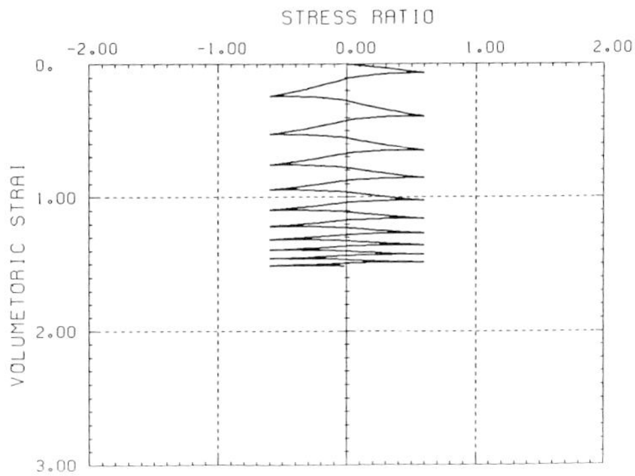

(b) Volumetric Strain

Fig.4 Cyclic Loading by Low Stress Ratio

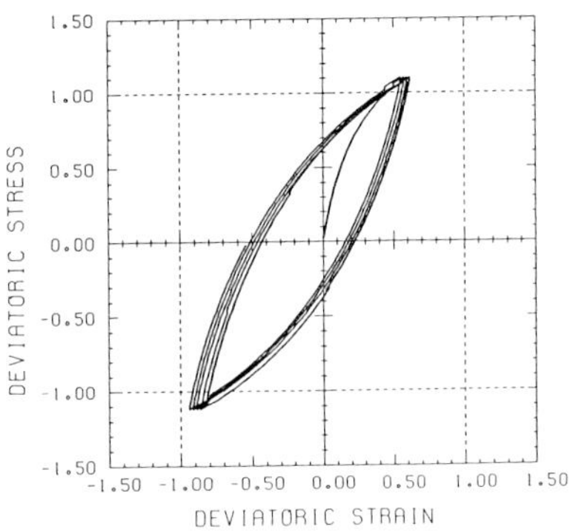

(a) Deviatoric Strain

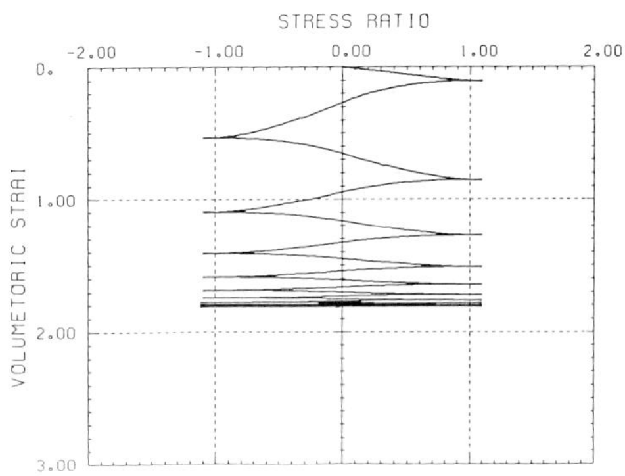

(b) Volumetric Strain

Fig.5 Cyclic Loading by Relatively High Stress Ratio

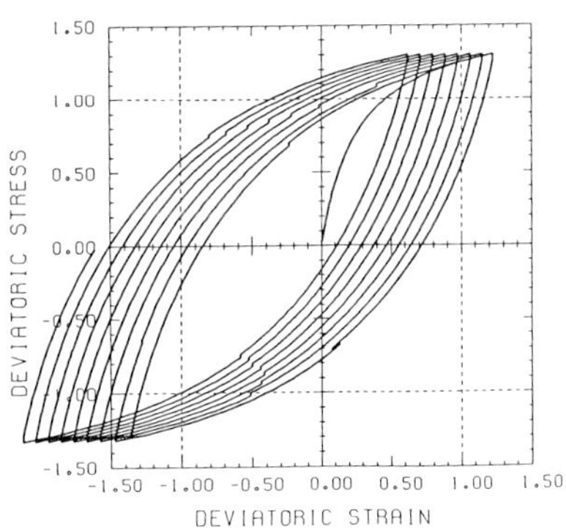

(a) Deviatoric Strain

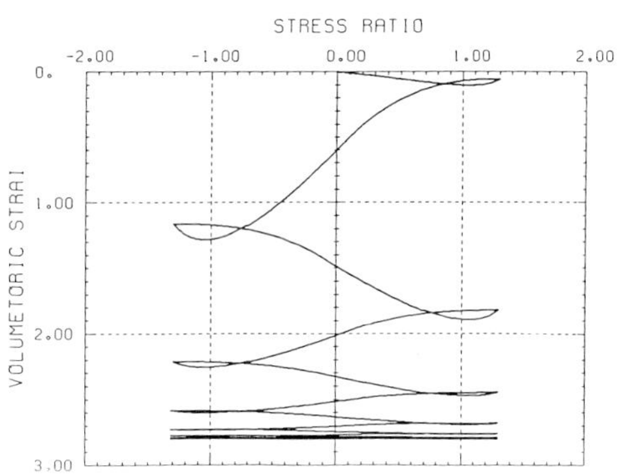

(b) Volumetric Strain

Fig.6 Cyclic Loading by High Stress Ratio

havior in cyclic loading. In this case volumetric strain increases as the number of cycles increases, although deviatoric strain does not accumulate in the process. Fig.5(a) and (b) indicate also the deviatoric strain and volumetric strain of the element under repeated load by stress ratio as high as the phase transformation line. Fig.6 indicates the response to the loading beyond the phase transformation line. In this case volumetric strain shows a characteristic feature of dilatancy behavior during stress reversal around the line. In the cases of high stress ratio cumulative strain was seen both in deviatoric and volumetric strains. It should be noted that the elastic strains are not included in the figures.

In the above calculations, the following values were used: $(1/a) = 2.0\times10^5$, $(1/b) = 1.82$ (in Eq.(5)); $a_o = -420$, $b_o = 0.4$ (in Eq.(7)) and $\gamma=0.6$ (in Eq(9)) The dilatancy factor was assumed as follows: for virgin loading $\Omega = M - \eta$, where $\eta = \sqrt{3J_2'}/R(\theta)p'$ and M is a material parameter defining the phase transformation line. For cyclic loading we assumed $\omega(\lambda_s) = c_o\exp(-c_1\lambda_s)(M - \eta)$ in loading phase and $\omega(\lambda_s) = c_o\exp(-c_1\lambda_s)(\eta_t - \eta)$ in unloading phase, where η_t denotes the stress state at the stress reversal point. c_o is given by the equation; $c_o = 0.6((\eta_f - \eta_t)/(\eta_f - \eta))^2$

By use of a parameter in the stress deviator $\sqrt{(\sigma_y - \sigma_x)^2 + 3\tau_{xy}^2/\beta}$ instead of $\sqrt{3J_2'}$, the effect of inherent anisotropy can be taken into account. Fig.7 shows instantaneous plastic deformation taking place during the rotation of principal stress axes at the stress point from which the magnitude and direction of the strain rate is presented by arrows. Fig.8 indicates experimental results of the second cycle of cyclic rotation test of principal axes by hollow cylindrical specimen. The tendency of direction and magnitude of strain increment resembles to calculated ones. Thus the model can predict the plastic deforma-

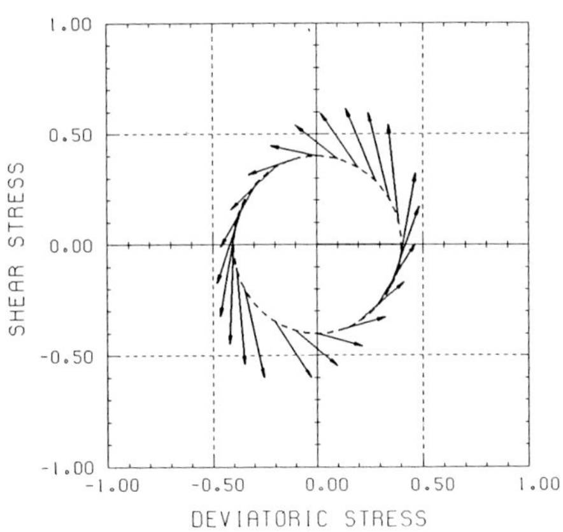

Fig.7 Strain Increment by Principal Stress Axes Rotation

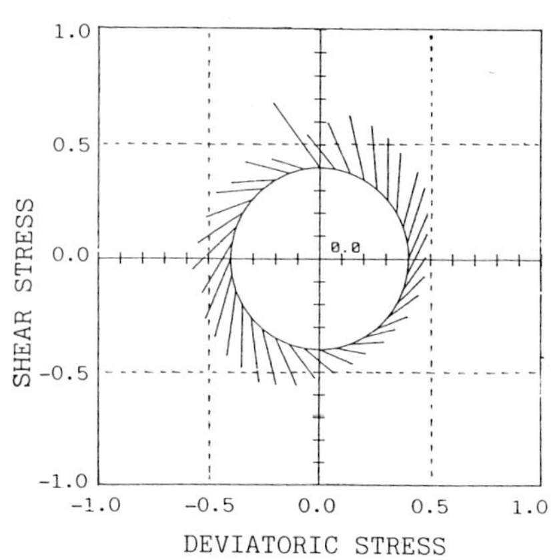

Fig.8 Experimental Result of Rotation of Principal Stress Axes

tion in the stress path along the isotropic loading surface for the virgin loading in a qulitative manner.

Examining closely the numerical results as well as the mathematical formaulation, one may notice that the proposed model must be modified in order to simulate more closely the deformation characteristics of sands under complex loading paths. Some of them can be resolved by adopting more elaborated functional forms, e.g., the shape of hysteresis curves and dilatant behaviors. The other will require fundamental discussions from both mathematical and physical point of view.

CONSIDERATION ON THE PROPOSED MODEL

From the proposed model fairly good approximation can be obtained for rather simple loading conditions but the model predicts sometimes unrealistic behavior under some complex stress paths. Consider the case that the current stress point lies on both the equivalent plastic modulus surface F and the loading surface f. The model allows the gradient of F and f to have different values; namely, non-intersection condition (ref.7) is not satisfied in the model, which leads to plastic deformation in the stress path along the equivalent plastic modulus surface. If the stress rate comes inside the surface with an acute angle with respect to grad $\hat{f}$, it produces a high amount of plastic deformation. While the stress rate goes outside of the surface we used grad$\tilde{F}\cdot\dot{\tilde{\sigma}}$ to define the plastic deformation by the assumption of the virgin loading Eq.(1). Thus it is possible that the stress rate coming into the surface produces higher amount of plastic deformation than that due to the stress rate going out from the surface. This is an apparent unrealistic prediction.

To circumvent this unrealistic prediction we must pose the non-intersection condition to the movement of F and f, which assures the continuity condition regarding the strain rate generated in any stress paths. However the use of the non-intersection condition fails to predict plastic deformation in the stress path along the isotropic loading function as far as the equivalent plastic modulus surface is assumed to be given by the isotropic loading function F. Thus the anisotropic hardening model must be developed in order to predict the plastic deformation during the rotation of the principal stress axes as far as the non-intersection condition is used. The introduction of the anisotropic loading function may not still lead to the reasonable prediction of the plastic deformation in the stress path associated with the rotation of the principal stress axes. The shape change of the bounding surface is expected to be of necessity. Thus the elasto-plastic models for comprehensive behavior of soils under dynamic loading conditions will be associated with considerable mathematical complexity.

CONCLUSIONS

A simple anisotropic hardening model was proposed to predict the cyclic behavior of soils. Some illustrative examples of numerical analysis were shown in order to see some features of the proposed model in triaxial stress state and plastic deformation by cyclic loading and the rotation of principal stress axes were predicted. Fairly good approximation of drained behavior of sands was obtained by this model. Further appropriate refinement will result in the precise description of the behaviors. One may find that the model with micromechanical sources of inelastic deformation is more suitable for the description of the behavior of geomaterials, at least from a physical point of view. It is expected that inelastic sources such as slips and fabric changes can account for the comprehensive behavior of soils with more mathematical complexity than the elasto-plastic constitutive model. For the purpose of dynamic analysis simplicity would be rather important.

REFERENCES

1) Desai and Siriwardane (1984): Constitutive Laws for Engineering materials with Special Emphasis on Geologic Materials, Prentice Hall

2) Pande and Zienkiewicz (1982): Soil Mechanics - Transient and Cyclic Loads: Constitutive Relations and Numerical Treatment, John Wiley & Sons

3) Gaboussi and Momen (1982): Modelling and Analysis of Cyclic Behavior of Sands Soil Mechanics - Transient and Cyclic Loads, pp313-342

4) Mroz and Norris (1982): Elastoplastic and Viscoplastic Constitutive Model for Soils with Application to Cyclic Loading, Soil Mechanics - Transient and Cyclic Loads, pp173-217

5) Pietruszczak and Poorooshasb (1986): On Modelling of Cyclic Behavior of Soils Developement in Soil Mechanics and Foundation Engineering - 2, pp139-184

6) Mroz, Norris and Zienkiewicz (1978): An Anisotropic Hardening Model for Soils and its Application to Cyclic Loading, Int. J. Numer. Anal. Methods Geomech., Vol2, pp203-221

7) Hashiguchi (1985): A Mathematical Refinement of The Two Surface Model Plastisity, J. JSCE No.362, I-4 pp259-265

8) Tobita, Yanagisawa and Kato (1986): A Formulation of Constitutive Equation for Granular Materials Like Sand Based on Slip and Its Application Models, J. JSCE, No.370,3-5 pp57-66

9) Tobita and Yanagisawa (1986): The Applicability and Limitation of the Anisotropic Hardening Model to Cyclic Deformation Behavior of Soils, Proc. JEES pp601-606

Micromechanics of Granular Materials, edited by M. Satake and J.T. Jenkins
Elsevier Science Publishers B.V., Amsterdam, 1988 — Printed in The Netherlands

MODELING OF CYCLIC BEHAVIOR OF SAND IN MULTI-DIRECTIONAL SHEAR

Y. YAMADA

Institute of Engineering Mechanics, University of Tsukuba, Tsukuba (Japan)

ABSTRACT

A model for cyclic deformation behavior of sand under three-dimensional stress conditions was developed using experimental data obtained from true triaxial tests. The proposed model is based on a postulate that the deformation of sand comprises of strain components associated with a set of imaginary two-dimensional slippings. The model can well simulate the measured cyclic behavior of sand both in drained and undrained tests, taking account of inherent anisotropy of sand specimens as well as anisotropic yielding in multi-directional shear.

INTRODUCTION

Using a true triaxial apparatus, Yamada and Ishihara (ref. 1,2) carried out monotonic loading tests and showed that the deformation of sand was affected not only by the relative magnitude of three principal stresses but also by their orientation relative to the direction of deposition. Yamada and Ishihara (ref. 3) also performed cyclic loading tests and confirmed the anisotropic nature of yielding in the three-dimensional deformation of sand.

In modeling deformation behavior of sand, therefore, it is necessary to take the above two kinds of anisotropy into consideration. On the basis of a postulate proposed by Matsuoka (ref. 4) and Tatsuoka (ref. 5), Yamada and Ishihara (ref. 6) constructed a multi-yield surface model for drained deformation of sand and examined its adequacy by the true triaxial data. Yamada and Wickremesinghe (ref. 7) then applied the model to the prediction of undrained behavior of sand in multi-directional shear. The model, however, could not well simulate the deformation behavior of sand during unloading and reloading because of adopting linear stress-strain relationships except for virgin loadings. This paper aims at improving the model so as to simulate more adequately the deformation behavior of sand during unloading and reloading.

TRUE TRIAXIAL TESTS

Experimental data obtained from true triaxial tests on loose sand specimens (ref. 1-3) are used in this study to construct and evaluate the model. In these tests saturated sand was poured vertically into a mold filled with water to form a specimen.

A rectangular coordinate system, $\sigma_x - \sigma_y - \sigma_z$, in which σ_z axis is chosen vertically coincident with the pouring direction is adopted to specify the direction of sand deposition. The three-dimensional stress conditions can also be represented by effective mean principal stress, p', octahedral shear stress, τ_{oct}, and an angular parameter, θ, which are defined as follows:

$$p' = 1/3 \, (\sigma_x' + \sigma_y' + \sigma_z') \qquad (1)$$

$$\tau_{oct} = 1/3 \sqrt{(\sigma_x' - \sigma_y')^2 + (\sigma_y' - \sigma_z')^2 + (\sigma_z' - \sigma_x')^2} \qquad (2)$$

$$\tan\theta = 3 \, (\sigma_y' - \sigma_x') \, / \, (2\sigma_z' - \sigma_x' - \sigma_y') \qquad (3)$$

Similarly volumetric strain, v, and octahedral shear strain, γ_{oct}, are defined by the corresponding strains, ε_x, ε_y and ε_z, as follows:

$$v = \varepsilon_x + \varepsilon_y + \varepsilon_z \qquad (4)$$

$$\gamma_{oct} = 2/3 \sqrt{(\varepsilon_x - \varepsilon_y)^2 + (\varepsilon_y - \varepsilon_z)^2 + (\varepsilon_z - \varepsilon_x)^2} \qquad (5)$$

All the specimens were isotropically consolidated to a confining pressure of 98kPa, then subjected to shear stresses so as to keep mean principal stress constant. Drained and undrained monotonic loading tests were performed along radial stress paths shown in Fig. 1(a). Depending on b-value defined by:

$$b = (\sigma_2' - \sigma_3') \, / \, (\sigma_1' - \sigma_3') \qquad (6)$$

the radial shear stress paths can be classified into five groups: b=0 (ZC and YC corresponding to triaxial compression), b=0.268 (RS15 , RS105 and Rs135), b=0.5 (RS30 , RS90 and RS150), b=0.732 (RS45 , RS75 and RS165) and b=1 (XE

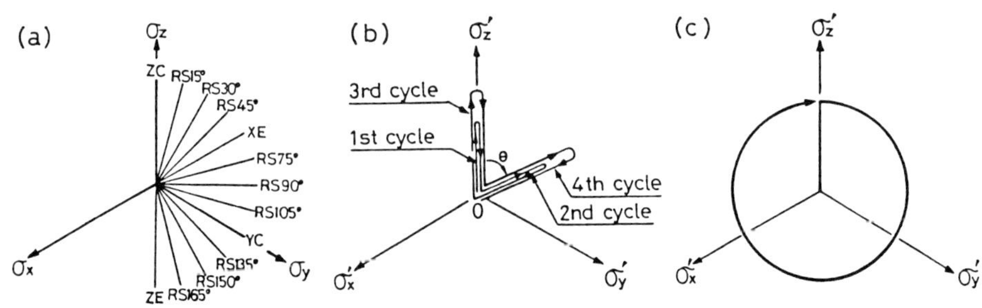

Fig. 1 Stress paths in the octahedral plane; (a) radial stress path, (b) multi-directional cyclic loading path, (c) circular stress path.

and ZE corresponding to triaxial extension). To examine the yield condition in the octahedral plane, drained and undrained cyclic loading tests were performed along a stress path shown in Fig. 1(b). Another kind of tests with a circular shape stress path illustrated in Fig. 1(c) was also carried out.

MODEL FOR DRAINED DEFORMATION

The present model for drained deformation of sand is basically the same as the previous ones (ref. 6,7) except for the stress-strain relationship during unloading and reloading.

The model was constructed based on the postulate proposed by Matsuoka (ref. 4) and Tatsuoka (ref. 5) that the deformation of sand is expressed as a summation of strains supposed to develop in the imaginary two-dimensional slip planes. In the stress condition $\sigma_i' > \sigma_j' > \sigma_k'$, three imaginary slippings are envisaged for two-dimensional stress systems, $\sigma_i' > \sigma_j'$, $\sigma_i' > \sigma_k'$ and $\sigma_j' > \sigma_k'$, and each principal strain is decomposed into two components as follows:

$$\varepsilon_i = f_{ij} + f_{ik} \tag{7}$$

$$\varepsilon_j = f_{jk} + g_{ij} \tag{8}$$

$$\varepsilon_k = g_{ik} + g_{jk} \tag{9}$$

where f_{ij} means compressive (positive) strain in i-direction and g_{ij} means expansive (negative) strain in j-direction caused by $\sigma_i' > \sigma_j'$ stress system. The strains generated by each two-dimensional slipping are given as a function of stress ratio, r_{ij}, defined by:

$$r_{ij} = (\sigma_i' + p') / (\sigma_j' + p') \tag{10}$$

Since the sand specimens exhibited anisotropic characteristics between vertical and horizontal directions, six stress-strain curves were specified for virgin loadings based on the drained monotonic loading tests (ref. 6). These curves are designated as f_{vh}, f_{hh}, f_{hv}, g_{vh}, g_{hh} and g_{hv}, where suffixes "v" and "h" mean vertical and horizontal directions, respectively, and approximated by power function as follows:

$$f_{ij} \text{ or } g_{ij} = m \; (r_{ij} - 1)^n \tag{11}$$

where m and n are material constants taking different values as listed in Table 1 depending on the stress systems.

TABLE 1
Values of parameters used for the stress-strain curves in virgin loadings.

	f_{vh}	f_{hh}	f_{hv}	g_{vh}	g_{hh}	g_{hv}
m	0.0131	0.0160	0.0230	-0.0110	-0.0126	-0.0185
n	1.85	1.85	1.74	2.20	2.20	2.02

In the original model a unique linear elastic modulus was used for all strain components in unloading and reloading, which necessarily resulted in deformation without volume change (ref. 6). In the actual tests, however, volume contraction was observed during unloading and reloading even at small stress ratios. To simulate this tendency the second model adopted slightly different linear moduli between positive and negative strain increments, yet it predicted less volume contraction during unloading process near failure compared with the measured behavior. This is because the moduli were determined only on the data obtained at relatively lower stress levels.

Hence the stress-strain curves were reexamined and a pair of power functions were assigned for the compressive and tensile strain increments produced during unloading and reloading. The following equations give the stress-strain relationship during unloading and reloading:

$$\Delta\varepsilon_c = 0.00300 \left|\Delta r_{ij}\right| \tag{12}$$

$$\Delta\varepsilon_t = -0.00205 \left|\Delta r_{ij}\right| \tag{13}$$

where $\Delta\varepsilon_c$, $\Delta\varepsilon_e$ mean increments of compressive (positive) and tensile (negative) strains, and Δr_{ij} means stress ratio increment.

Figure 2(a) illustrates the comparison between measured and predicted volumetric strains for ZC-YC test following a multi-directional cyclic loading path shown in Fig. 1(b). Similarly Fig. 2(b) plots the measured and predicted volumetric strains for a circular loading test whose stress path is shown in Fig. 1(c). In the latter figure stress ratio is calculated by:

$$\tau_{oct}^{*} = \cos\theta \; \tau_{oct} \tag{14}$$

which means the projection of octahedral shear stress on σ_z axis. As shown in these figure, the present model well simulates the volume contraction tendency during the second and third cycles which is not realized by the original one.

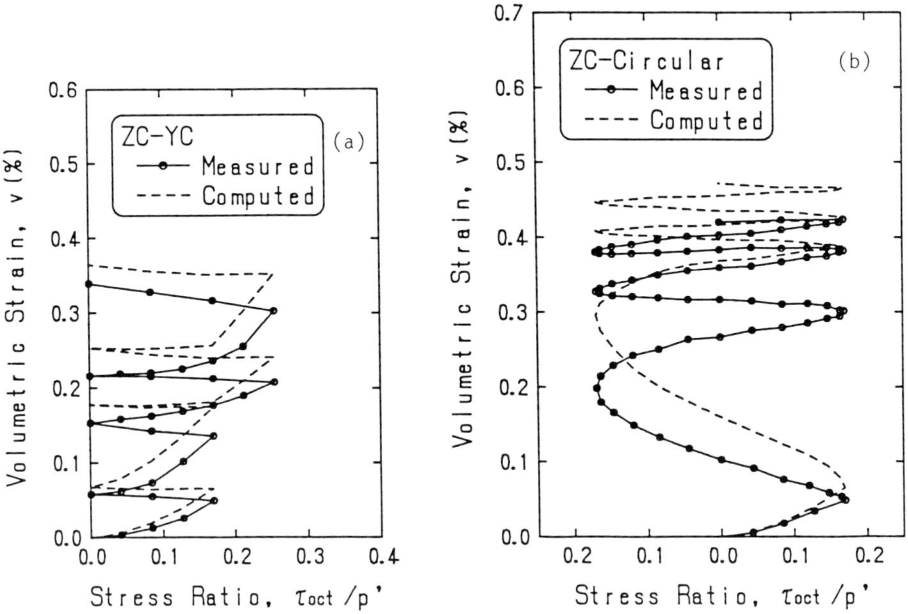

Fig. 2 Comparison between measured and predicted volumetric strain for drained test; (a) ZC-YC test, (b) Circular shear test.

MODEL FOR UNDRAINED DEFORMATION

The buildup of pore water pressure during undrained tests was predicted from the volumetric strain supposed to develop under drained conditions. The volumetric strain itself was calculated using the drained deformation model described above.

First the concept proposed by Martin et al. (ref. 8) was applied for the prediction. Their model employed the rebound modulus of one-dimensional consolidation curve to convert the volumetric strain into the undrained pore water pressure. In this study, however, the pore pressure calculated using the rebound modulus was found to be much larger than the measured value.

Lewin et al. (ref. 9) showed that the drained volume strain and the undrained stress path were closely related under three-dimensional stress conditions. Hence, an attempt to correlate directly the volumetric strain increment, Δv, in drained tests and pore water pressure increment, Δu, in undrained tests was made and the following relationship was obtained (ref. 7):

$$\Delta u = K_r \, \Delta v \tag{15}$$

$$K_r = 16.7 - 48.0 \ln(1 - p'/p_o') \quad \text{MPa} \tag{16}$$

This relationship was finally employed to predict the buildup of pore water pressure in undrained tests. For the initial stage of shearing, however, K_r-value calculated at $p'/p_o' = 0.9995$ was used.

Figure 3 shows the comparison between measured and predicted effective stress paths for monotonic loadings with b-value of 0.5 and 1.0. It is to be noted that the model simulates well the measured stress paths during loading, taking account of the stress state expressed by b-value and the effects of inherent anisotropy of sand specimens. In the case of the previous models, the predicted effective stress paths for unloading differed markedly from the actual behavior, because the stress-strain relationships were determined without consideration for the unloading carried out near failure. As can be seen in Fig. 3 the model predicts considerable increase in pore pressure during un-

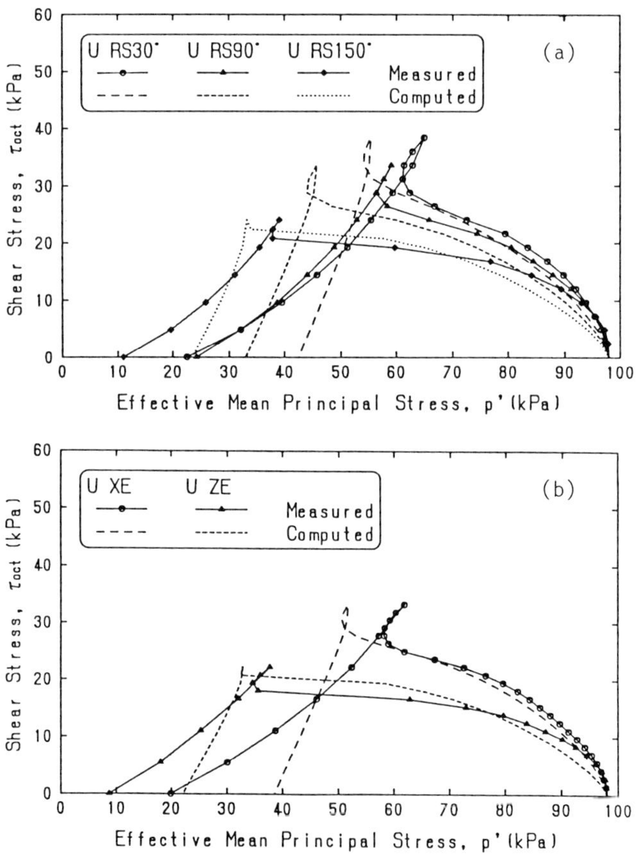

Fig. 3 Comparison between measured and predicted effective stress paths for undrained monotonic loadings; (a) b = 0.5, (b) b = 1.0.

loading, although it still underestimates the amount of pore pressure buildup.

Figure 4 compares the measured and predicted effective stress paths for undrained multi-directional cyclic loadings whose stress path is schematically shown in Fig. 1(b). As shown in these figures, the model can well simulate

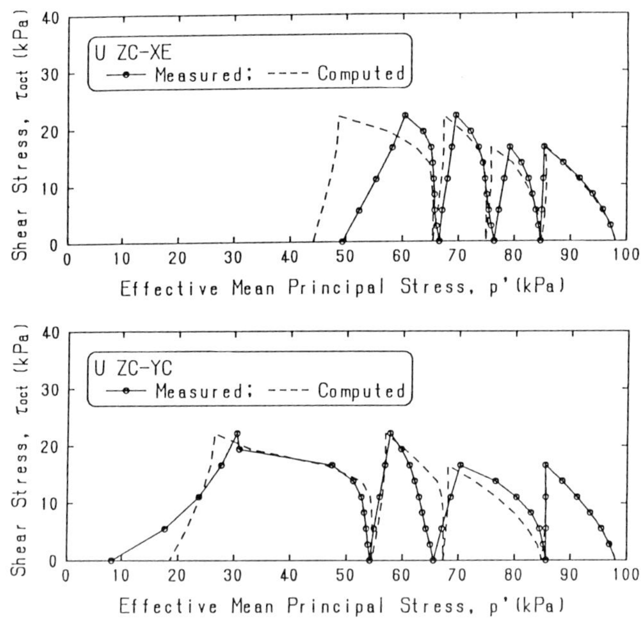

Fig. 4 Comparison between measured and predicted effective stress paths for undrained multi-directional loadings.

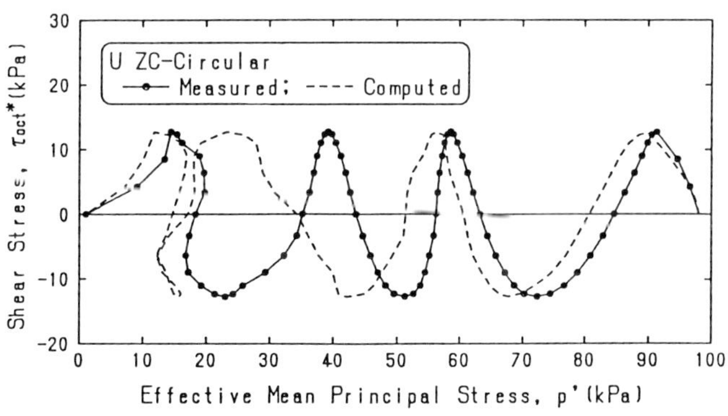

Fig. 5 Comparison between measured and predicted effective stress paths for an undrained circular loading.

the tendency of pore water pressure buildup during cyclic loadings including changes in shearing direction. Figure 5 plots p'- τ_{oct}^* stress paths for the circular shear test with a stress path shown in Fig. 1(c). It is shown that the model tends to overestimate the development of excess pore water pressure during earlier cycles. The present model, however, well simulates the effective stress path after the stress ratio reaches the value corresponding to the angle of phase transformation (ref. 10). These figures confirm that the model can take account of the effects of inherent anisotropy as well as the anisotropic yielding in multi-directional shear.

CONCLUSIONS

The multi-yield surface model (ref. 5,7) was improved so as to predict more reasonably the cyclic behavior of sand under three-dimensional stress conditions. The model can well simulate the measured cyclic behavior of sand both in drained and undrained tests, taking account of the inherent anisotropy of sand specimen as well as the anisotropic yielding in multi-directional shear.

REFERENCES

1 Y. Yamada and K. Ishihara, Anisotropic deformation characteristics of sand under three-dimensional stress conditions, Soils and Foundations, 19(2) (1979) 79-94.

2 Y. Yamada and K. Ishihara, Undrained deformation characteristics of loose sand under three-dimensional stress conditions, Soils and Foundations, 21(1) (1981) 97-107.

3 Y. Yamada and K. Ishihara, Yielding of loose sand in three-dimensional stress conditions, Soils and Foundations, 22(3) (1982) 15-21.

4 H. Matsuoka, Stress-strain relationship of sand based on the mobilized plane, Soils and Foundations, 14(2) (1974) 47-61.

5 F. Tatsuoka, Stress-strain behavior of an idealized anisotropic material, Soils and Foundations, 20(3) (1980) 75-90.

6 Y. Yamada and K. Ishihara, Modeling of drained deformation of sand under three-dimensional stress conditions, in: P.A. Vermeer and H.J. Luger (Ed.), Proc. Symp. on Deformation and Failure of Granular Material, Balkema, Rotterdam, (1982) 331-342.

7 Y. Yamada and D.S. Wickremesinghe, Multi-yield surface model for undrained behavior of sand, Proc. 8th Asian Reg. Conf. on Soil Mechanics and Foundation Engineering, Vol. 1, Kyoto, (1987) 129-132.

8 G.R. Martin, W.D.L. Finn and H.B. Seed, Fundamentals of liquefaction under cyclic loading, ASCE, J. Geotech. Div. 101(5) (1975) 423-438.

9 P.I. Lewin, Y. Yamada and K. Ishihara, Correlating drained and undrained 3D tests on loose sand, in: P.A. Vermeer and H.J. Luger (Ed.), Proc. Symp. on Deformation and Failure of Granular Material, Balkema, Rotterdam, (1982) 419-429.

10 K. Ishihara, F. Tatsuoka and S. Yasuda, Undrained deformation and liquefaction of sand under cyclic stresses, Soils and Foundations, 15(1) (1975) 29-44.

Micromechanics of Granular Materials, edited by M. Satake and J.T. Jenkins
Elsevier Science Publishers B.V., Amsterdam, 1988 — Printed in The Netherlands

PLASTICITY APPROACH TO SAND BEHAVIOUR UNDER PRINCIPAL STRESS AXES ROTATION

I.Towhata[1], D.E.Pradel[2], and K.Ishihara[3]
[1]Assoc. Prof. of Civil Eng., University of Tokyo, Tokyo, JAPAN.
[2]Former Graduate Student, University of Tokyo, Tokyo, JAPAN.
[3]Professor of Civil Eng., University of Tokyo, Tokyo, JAPAN.

SUMMARY

The yielding behaviour of hollow cylindrical samples of Toyoura sand was investigated by applying triaxial and torsion shear stresses simultaneously and thereby introducing a rotation of principal stress axes. The shape of the yield loci was clearly indicated by the test results. Moreover, the directions of plastic strain increments were found out not to be unique.

INTRODUCTION.

The effects of stress axes rotation on deformation of sand have been attracting concerns for the past merely seven years. However, following the pioneering works by Arthur et al. (ref.1), several researches, both experimental and numerical, have been conducted.

Ishihara and Towhata (ref.2) conducted undrained cyclic shear tests on hollow cylindrical samples using a triaxial torsion shear apparatus, and revealed that the stress axes rotation can generate the excess pore water pressures. Symes et al. (ref.3) drew similar conclusions from a monotonic rotation of stress axes. They also showed that shear deformation of sand develops when stress axes rotate under undrained conditions. Miura et al. (ref.4) and Ishihara and Momenzadeh (ref.5) carried out drained tests and indicated shear distortions of sand generated by rotation of principal stress axes. The test results obtained so far suggest that the effects of stress axes rotation is a mirror image of the anisotropic behaviour of sand; the inherently anisotropic nature of sand results in an development of irreversible deformation under stress axes rotation, which in turn creates additional anisotropy, induced anisotropy, in sand.

Attempts have been made also to develop numerical models of soil which can predict the deformation of sand undergoing rotation of principal stress axes (refs. 6, 7, and 8). It is quite interesting that the three literatures mentioned above did not follow the conventional elastoplastic approach. No wonder that the elastoplasticity is not the only one possibility. However, examination of experimental results from plasticity view point, such as yielding and direction of strain increment vectors, is still beneficial to further development of modelling.

TESTING PROCEDURES.

Laboratory testings were carried out by using a triaxial torsion shear apparatus, which employed a hollow cylindrical soil specimen, measuring 10 and 6 cm in its outer and inner diameters, respectively, and 19.3 cm in height. This apparatus could apply both triaxial and torsional shear stresses simultaneously on soil samples, generating rotation of principal stress axes. Stresses were loaded in stress-controlled manners.

Fig. 1 indicates the definitions of stress components. The normal stresses in axial, circumferential, and radial directions are denoted by σ_z, σ_θ, and σ_r, respectively, while the shear stress due to torsion by $\sigma_{z\theta}$. Since the cell pressures inside and outside the hollow cylinder were held identical, the circumferential and radial stresses were equal to each other. Note that the apparatus could develop two components of shear stresses simultaneously; namely, the torsional stress, $\sigma_{z\theta}$, and the stress difference, $(\sigma_z - \sigma_\theta)/2$. Strain components, ε_z, ε_θ, ε_r, and $\varepsilon_{z\theta}$, were defined in accordance with their stress counterparts. Note that the torsional shear strain, $\varepsilon_{z\theta}$, represents a tensor strain.

Using the stress and strain components thus defined, the increment of strain energy, dW, is given by

$$dW = \sigma_z \; d\varepsilon_z + \sigma_r + d\varepsilon_r + \sigma_\theta \; d\varepsilon_\theta + \sigma_{z\theta} \cdot 2d\,\varepsilon_{z\theta}$$

Considering $\sigma_r = \sigma_\theta$ and introducing $d\varepsilon_m = (d\varepsilon_r + d\varepsilon_\theta)/2$

$$dW = \frac{\sigma_z + \sigma_r + \sigma_\theta}{3}(d\varepsilon_z + d\varepsilon_r + d\varepsilon_\theta) + \frac{2}{3}(\sigma_z - \sigma_\theta)(d\varepsilon_z - d\varepsilon_m) + 2\sigma_{z\theta}\, d\varepsilon_{z\theta} \qquad (1)$$

in which the second and third terms of the righthand side are contributions from shear distortion.

The Toyoura sand was used for testing. The specific gravity being 2.65, it has the maximum and minimum void ratios of 0.977 and 0.597, respectively.

An air-dry Toyoura sand was pluviated down into a sample preparation mould. The height of fall was adjusted to achieve either loose or dense samples of sand. After an isotropic consolidation under 98.1 kN/m^2, the relative density,

Dr, of loose samples ranged from 26 to 39% while that of dense samples from 69 to 73%. All shear testings were carried out in drained manners.

TYPES OF SHEAR STRESS PATHS.

The loading schemes of shear stresses, which were applied to sand samples after isotropic consolidation, are explained with the aid of diagrams in Fig. 2. In this figure, the vertical and horizontal axes represent the torsional shear stress, $\sigma_{z\theta}$, and the stress difference, $(\sigma_z - \sigma_\theta)/2$, respectively, which are normalized by the effective mean principal stress, P';

$$P' = \frac{\sigma_z' + \sigma_r' + \sigma_\theta'}{3} = \frac{\sigma_z' + 2\sigma_\theta'}{3} \tag{2}$$

This normalization does not make data interpretation difficult, because the P' value was maintained constant during the period of shear. The direction of loading in Fig. 2 is denoted by α_σ .

One type of loading scheme, denoted as Y or L in Fig. 2 depending on sand density, has the stress state translate between α_σ and α_σ+30° directions with cyclic loading and unloading of shear stress components. This variation of loading direction means that the orientation of major principal stress varies periodically between $\alpha_\sigma/2$ and $\alpha_\sigma/2$+15° directions which are measured from the vertical direction. This loading was employed to study the yielding behavior of sand when stress axes rotate.

On the other hand, the M test in Fig. 2 is a monotonic loading with orientation of stress axes fixed. In what follows, the tests are identified by a combination of type of loading and its direction, α_σ ; e.g., Y60 is a testing on dense sand in which the direction of stress state varies between 60 and 90 degrees in Fig. 2.

Furthermore, the plastic components of strain increment were studied by running stress probe tests on dense samples. Soil specimens were first sheared to a stress state of α_σ equal to either 25 or 120 degrees, followed by loading and unloading of small stress increments (see Fig. 3). This type of tests is denoted by a symbol "F", α_σ angle, and the direction of stress increment; e.g., F25-45 has the initial stress state in the 25-degree direction and the stress increment in the 45-degree direction.

MEASURED STRESS-STRAIN RELATIONSHIP.

In the present loading schemes, two shearing phenomena occur simultaneously; one due to the torsional shear and the other due to the stress difference. Hence, the following discussions are made of each of them separately.

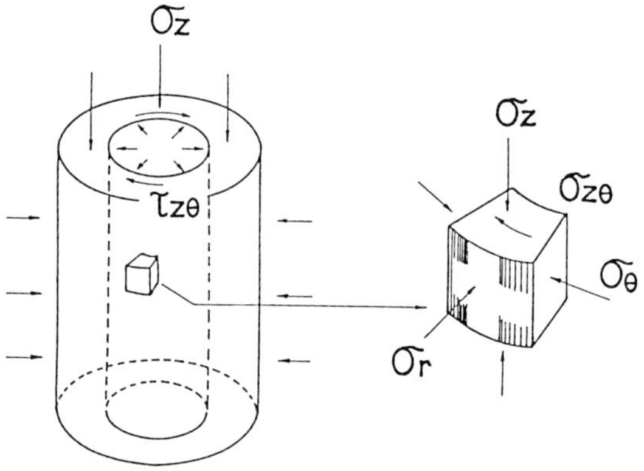

Fig.1 Definitions of stress components.

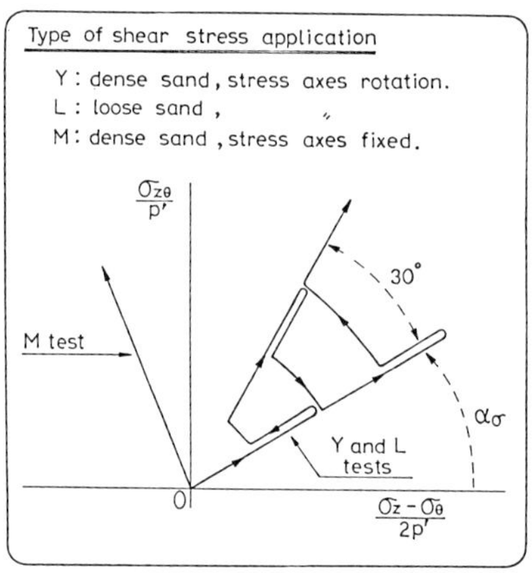

Fig.2 Loading paths for study of stress axes rotation.

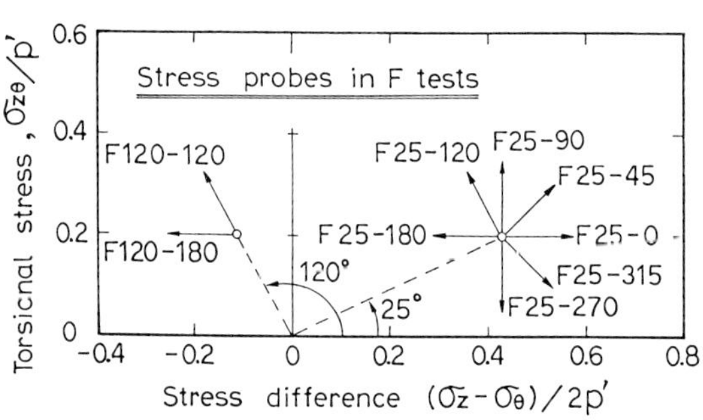

Fig.3 Loading schemes in stress-probe tests.

Fig. 4 shows the deformation of sand in the stress-difference manners (see Eq. 1) which was observed in Y30 test on a dense sand. Results of a monotonic loading test, M30, is indicated as well. It can be seen that, when the direction of principal stress rotates, the yielding does not necessarily occur at stress levels equal to the previous maximum value. Fig. 4 also indicates that deformation behaviors in Y30 test are similar to those in M30 test when the direction of loading, α_σ , of Y30 test is 30 degrees.

Fig. 5 indicates the torsional deformations observed in Y30 test. It is seen again that the yielding stress is not always equal to the previous maximum stress level; being greater or less, dependent on the direction of loading.

The stress-difference behaviour in Y60 test is shown in Fig. 6. Noteworthy is that the deformation developed in the negative direction when the stress was maintained constant and the torsional stress increased toward the pure torsion. This observation is related to the discussion on non-coaxiality of principal stresses and strain increments.

Similar observations were mode of loose sand as well. Fig. 7 indicates the stress-difference behaviour in L30 test. Due to the rotation of stress axes, yielding does not necessarily occur at the previous maximum stress. The stress-strain curve in the torsional manner of L90 test (Fig. 8) reveals a yielding at a stress level greater than the previous maximum value.

In summary, the experimental data presented in this section shows that the yielding of sand occurs at different stress levels dependent on the direction of loading or orientation of principal stress axes. Since the combination of stress-difference and torsional shear stresses plays an important role in yielding, the idea of separating the deformation of sand into two independent mechanism, one due to stress difference, and the other due to torsion, does not seem reasonable.

YIELD LOCI OF SAND.

The yielding behaviour of sand, both dense and loose, was studied by, first, detecting the yield stresses in the observed stress-strain curves and then plotting them in the stress plane. Although possible to be done otherwise, the yield stress in this study was detected following the idea shown in ref.9; the intersection of the extensions of the linear reloading stress-strain curve and the nonlinear portion after yielding. Apparently, this method is applicable to both stress-difference and torsional stress-strain diagrams. The extension of the nonlinear portion is made by superposing the curve in monotonic loadings.

Fig. 9 indicates the detected yield points of dense sand, and the shape of the yield loci is obtained when those points are connected with one another. Two types of loci, derived from stress-difference and torsional deformations, look similar, although not exactly identical. It is interesting to note that

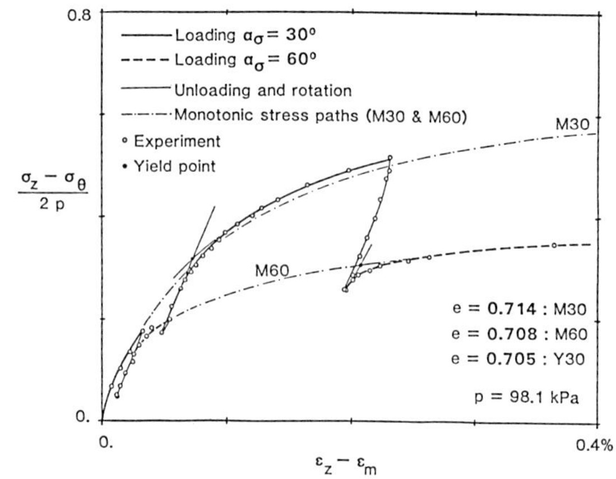

Fig.4 Deformation of dense sand in stress-difference manners in Y30 test.

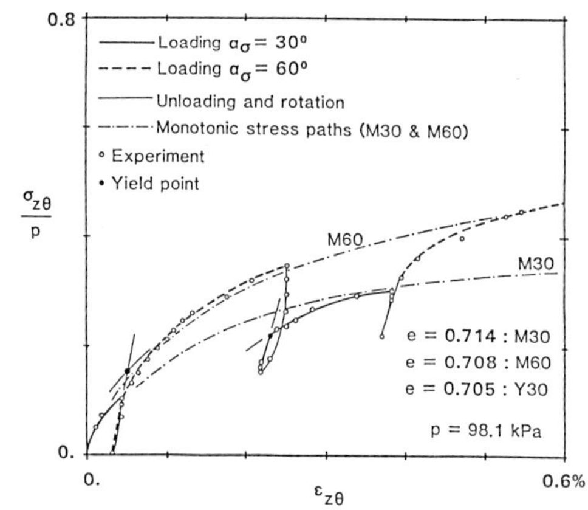

Fig.5 Deformation of dense sand in torsional manners in Y30 test.

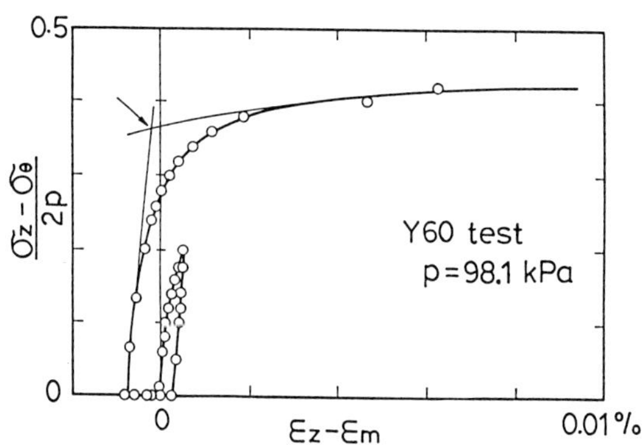

Fig.6 Deformation of dense sand in stress-difference manners in Y60 test.

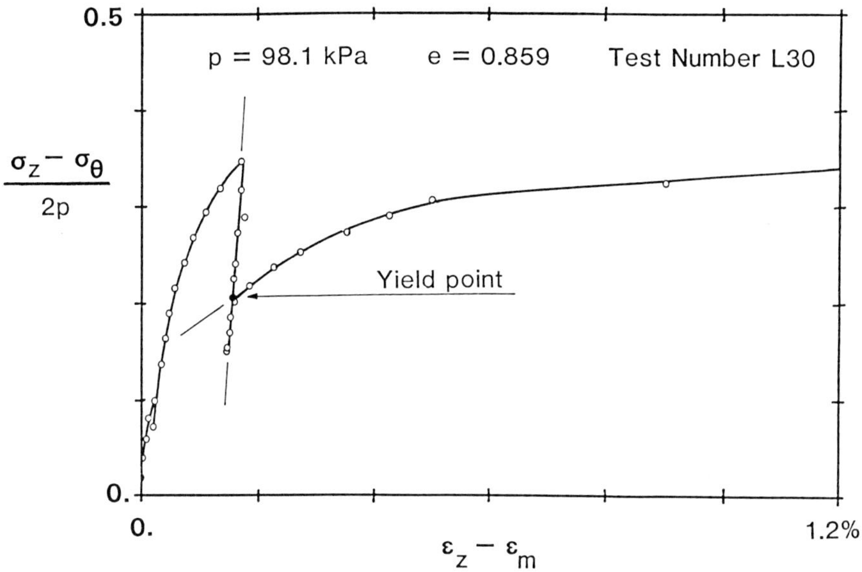

Fig.7 Deformation of loose sand in stress-difference manners in L30 test.

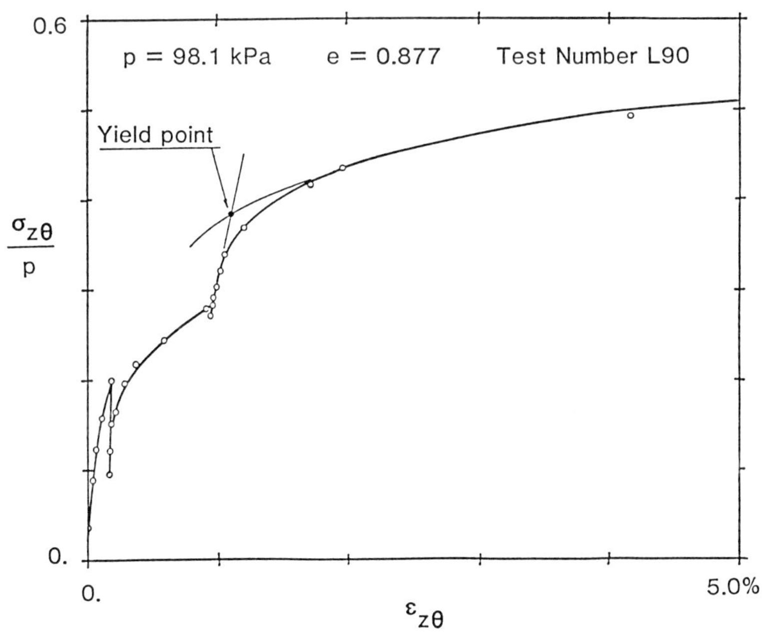

Fig.8 Deformation of loose sand in torsional manners in L90 test.

the yield locus at a low stress level is very nearly circular, while that at a higher stress level is distorted, larger in size toward the right of the figure. A similar distortion is seen clearly for loose sand as well (Fig. 10). Most probably the distortion of the yield locus at a higher stress level is due to the shape of the failure criterion. When the Coulomb's criteria;

$$\sigma_1' - \sigma_3' = (\sigma_1' + \sigma_3')\sin\phi \tag{3}$$

was plotted with the friction angle ϕ = 35 and 42 degrees, two curves in Fig. 11 were obtained. Note that the effective mean principal stress, P', was maintained constant, as was the case in the experiments. The distorted shape of the failure criteria is apparent.

DIRECTION OF PLASTIC STRAIN INCREMENTS.

Stress probe tests as illustrated in Fig. 3 were carried out on dense sand to examine the directions of plastic strain increments. The plastic strain increments are defined as those which remain unrecovered after a cycle of loading and unloading of a small stress increment. Referring to the strain increments in Eq. 1, the plastic strain increments of $\dot{\varepsilon}_z^p - \dot{\varepsilon}_m^p$ and $2\dot{\varepsilon}_{z\theta}$ were taken as horizontal and vertical coordinates, respectively.

Fig. 12 shows plastic strain increments detected in F25 tests. Apparently, the strain increment vectors are not parallel in direction. Note that the Drucker's discussion on normality (ref.10) fails when the directions of plastic strain increments are not unique. The variation of the direction of plastic strain increments are seen in Fig. 13 for F120 tests as well, although only two stress probes were employed (Fig. 3). Consequently, it may be reasonable to say that the direction of plastic strain increment depends at least partly on the stress increment, and that a use of single plastic potential function in an elastoplastic formulation would not be satisfactory.

CONCLUDING REMARKS.

A series of laboratory tests on dense as well as loose Toyoura sand was carried out, using a triaxial torsion shear apparatus with a hollow cylindrical specimen. Applying a stress difference and a torsional shear stress simultaneously in drained manners, the effects of stress axes rotation on yielding of sand were investigated. The conclusions drawn are as what follows;

(1) The intensity of stress components at yielding is not necessarily equal to

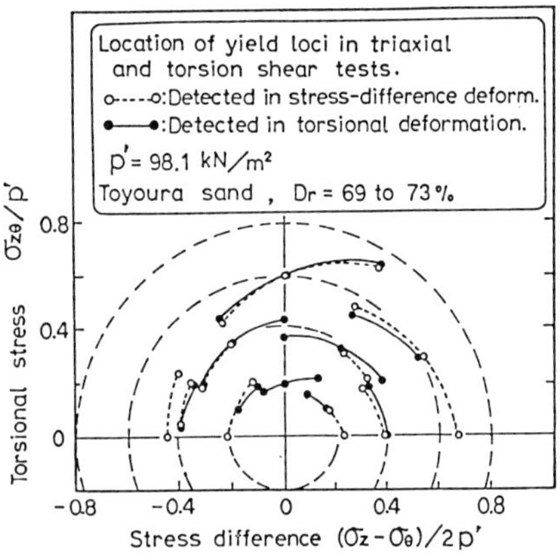

Fig.9 Experimental yield loci of dense sand.

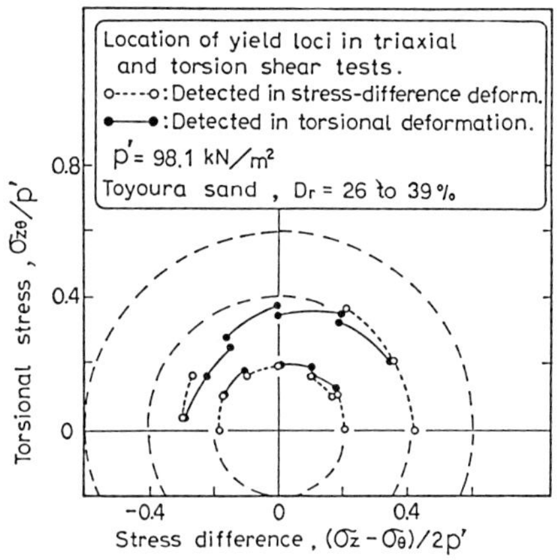

Fig.10 Experimental yield loci of loose sand.

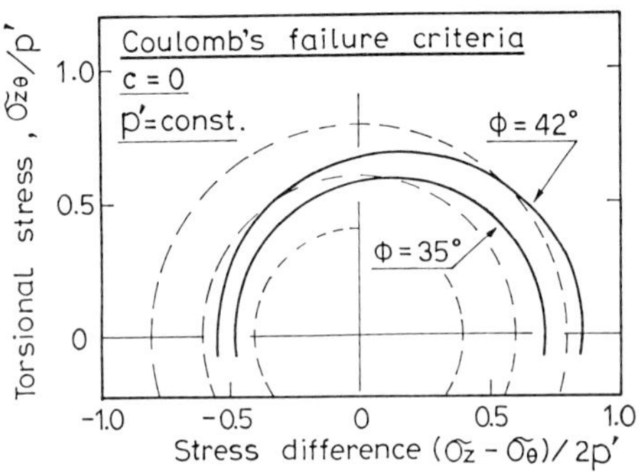

Fig.11 Obliquity in Coulomb's failure criteria with P'= const.

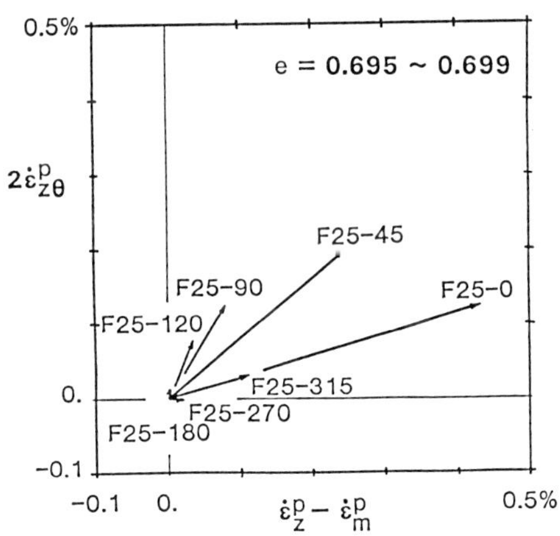

Fig.12 Plastic strain increments observed in F25 tests.

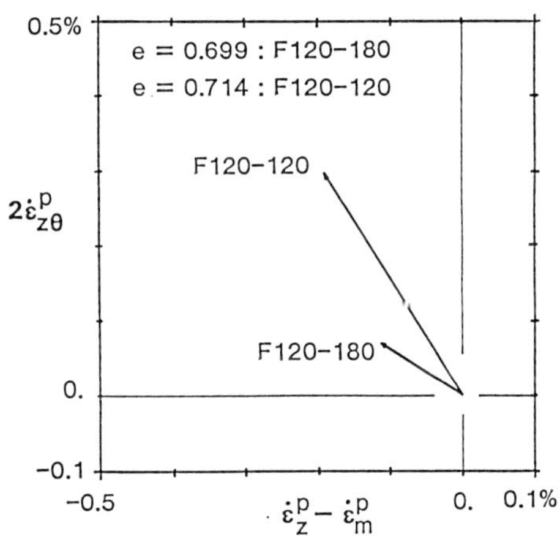

Fig.13 Plastic strain increments observed in F120 tests.

the previous maximum value. It increases or decreases in accordance with the rotation of stress axes.

(2) By studying the shear distortions of soil samples in stress-difference as well as in torsional manners, two sets of yield loci were obtained. They look nearly identical, and are not circular in shape at higher shear stress levels.

(3) The direction of plastic strain increment vectors is not unique. Thus, the use of a single plastic potential function in an elastoplastic formulation is not reasonable.

LIST OF REFERENCES.

1 J.R.F.Arthur, K.S.Chua, T.Dunstan, and J.I.Rodriguez del C. (1980) Principal stress rotation : a missing parameter, Proc. ASCE, 106 (GT4), 1980, 419-433.

2. K.Ishihara and I.Towhata (1983) Sand response to cyclic rotation of principal stress directions as induced by wave loads, Soils and Foundations, 23(4), 11-26.

3. M.J.R.P.Symes, A.Gens, and D.W.Hight (1984) Undrained anisotropy and principal stress rotation in saturated sand, Geotechnique, 34(1), 11-27.

4. K.Miura, S.Miura, and S.Toki (1986) Deformation behaviour of anisotropic dense sand under principal stress axes rotation, Soils and Foundations, 26(1), 36-52.

5. K.Ishihara and M.Momenzadeh (1986) Influence of rotation of principal stress directions on the cyclic behaviour of sand, Proc. 2nd Int. Symp. Numer. Models Geomech., Ghent, 827-835.

6. I.Towhata and K.Ishihara (1985) Modelling soil behaviour under principal stress axes rotation, Proc. 5th Int. Conf. Numer. Methods Geomech., Nagoya, 1, 523-530.

7. H.Matsuoka, Y.Iwata, and K.Sakakibara (1986) A constitutive model of sands and clays for evaluating the influence of rotation of the principal stress axes, Proc. 2nd Int. Symp. Numer. Models Geomech., Ghent, 67-78.

8. K.Miura, S.Toki, and S.Miura (1986) Deformation prediction for anisotropic sand during the rotation of principal stress axes, Soils and Foundations, 26(3), 42-56.

9. F.Tatsuoka and F.Molenkamp (1983) Discussion on yield loci for sands, in J.T.Jenkins and M.Satake (Eds.), Mechanics of Granular Materials : New Models and Constitutive Relations, Elsevier, Amsterdam, pp.75-87.

10. D.C.Drucker (1951) A more fundamental approach to plastic stress-strain

Micromechanics of Granular Materials, edited by M. Satake and J.T. Jenkins
Elsevier Science Publishers B.V., Amsterdam, 1988 — Printed in The Netherlands

EFFECT OF FABRIC ON LIQUEFACTION AND DENSIFICATION OF SATURATED SOIL: EXPERIMENTS AND THEORY

S. Nemat-Nasser

Department of Applied Mechanics and Engineering Sciences, University of California, San Diego, La Jolla, California 92093

SUMMARY

It has been known that the inherent and induced anisotropy or fabric has considerable influence on the response and failure modes of granular masses. Some definitive recent experiments which clearly demonstrate this phenomenon in relation to the densification and liquefaction potential of saturated sands, are briefly reviewed, together with associated micromechanically based theoretical observations.

INTRODUCTION

Cohesionless granular materials support the applied forces through contact friction. Therefore, the nature and distribution of the contacts are expected to have considerable influence on the overall mechanical response of the granular mass. The term "fabric" has been used to characterize this kind of microstructure. It relates to the distribution of the particles, their sizes and orientations, and, in particular, to the distribution of the contact normals and contact areas (which reflect the magnitude of the contact forces), as well as other geometrical entities. In this summary we shall not deal directly with the characterization of fabric, since the matter has been fully examined in other articles within this volume; see, e.g. the paper by Mehrabadi *et al.* (ref. 1). What is of concern here is to show how seemingly minute variations in a given loading history can change the fabric of the granular mass to such an extent that the load-bearing capacity of the sample is changed by orders of magnitude.

In particular, we shall examine the effect of pre-straining on the liquefaction potential of saturated sands, and point out the results of recent experiments by Nemat-Nasser and Tobita (ref. 2) and Nemat-Nasser and Takahashi (ref. 3), which clearly have brought into focus the effect of induced fabric on the potential to failure by liquefaction of undrained samples.

RESULTS AND DISCUSSION

It has been shown by Finn *et al.* (ref. 4) and confirmed by Ishihara *et al.* (ref. 5), and other investigators, that, once a saturated undrained sample of sand is liquefied in simple shearing, it will have essentially no resistance to reliquefaction in subsequent tests. These tests were all performed under a *load-controlled* condition, so that, with the confining pressure held fixed, the tests were terminated after liquefaction initiation, by reducing the shear *stress* to zero.

Faced with the above experimental facts, in the late seventies the present writer became very interested in this phenomenon, and sought to: (1) develop a micromechanical model which might clarify the phenomenon; and (2) seek to check the experimental results independently and in light of the micromechanics of the process. The work on the micromechanics of this phenomenon was reported in Nemat-Nasser (1980, ref. 6), and the experimental effort was done in collaboration with Dr. Tobita and Mr. Takahashi; see (ref. 2 and ref. 3).

Micromechanical Model:

The model considers two-dimensional shearing of a layer of a granular mass under uniform applied normal compressive stress, σ, and uniform shear stress, τ, which is viewed positive when clockwise. The macroscopic shear flow is the result of the microscopic motion (rolling and sliding) of grains relative to each other at *active* contact points. Fig. 1 shows this.

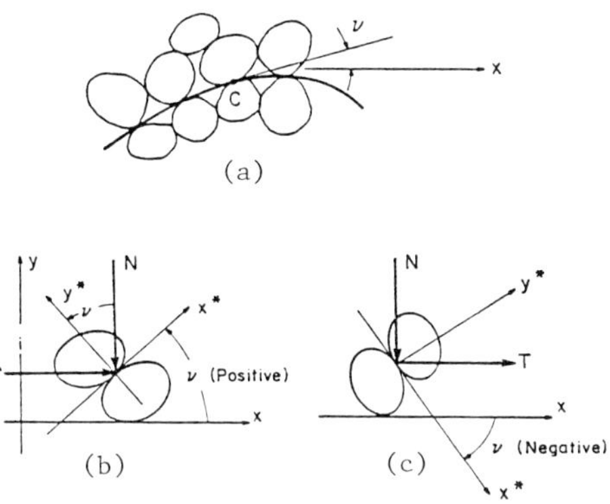

(a)

(b) (c)

Figure 1

The angle ν is called the (micro) dilatancy angle. It varies from *active* contact to *active* contact, and when there is a very large number of such contacts, one may assume a continuous change and introduce a distribution-density function $p(\nu)$ in such a manner that $p(\nu_0)\,d\nu$ represents the volume fraction of active granules whose dilatancy angles are between ν_0 and $\nu_0 + d\nu$. *ν is positive when it produces volume expansion.* By considering the balance of forces transmitted across each active contact, and by equating the rate of frictional dissipation at active contacts within a unit volume to the rate of the overall stress-work, Nemat-Nasser (ref. 6) obtains the following dilatancy equation:

$$\frac{1}{V}\frac{dV}{d\gamma} = \frac{1}{\cos\phi_\mu}\int_{\nu_0^-}^{\nu_0^+} p(\nu)\cos(\phi_\mu + \nu)\sin\nu d\nu \tag{1}$$

in which ϕ_μ = "grain-to-grain" friction angle; and the Mohr-Coulomb failure criterion is used to characterize the *local* (at the micro-level) flow process.

Nemat-Nasser (ref. 6) points out that upon shearing and under normal pressure, granules with negative dilatancy angles are activated first, leading to a distribution-density function, $p(\nu)$, which initially is biased toward negative dilatancy angles. Intuitively, this follows from the physical observation that the local normal force N hinders the motion of an *active* granule with positive dilatancy angle ν, whereas it assists when ν is negative (Figs. 1*(b)* and*(c)*). Hence, upon shearing under confinement, an initial densification is expected and is invariably observed.

As shearing proceeds, the distribution-density function $p(\nu)$ tends to become biased toward positive dilatancy angles, and this leads to subsequent dilation. During this stage, a greater number of active contacts has positive dilatancy angles, and this number increases with increasing shear strain amplitude, up to the strain corresponding to the peak stress. Now, if, after a microstate of this kind is attained, unloading begins, then some of the granules with suitably large dilatancy angles may actually start a *downward* motion under the action of the normal force N, which leads to some densification. However, it is reasonable to expect, and our experiments on photoelastic granules confirm (see ref. 7), that at zero shear stress there still exists a strong bias toward positive dilatancy angles for clockwise shearing. Now, upon load reversal (i.e., upon shearing in the counterclockwise direction), the previously positive dilatancy angles behave now as negative ones. Thus, a strong tendency toward densification would be expected during load reversal. Hence, prestraining to suitably large shear strain amplitudes in drained conditions should lead to immediate liquefaction when load reversal occurs under undrained conditions.

Experimental Results:

To test the liquefaction potential of a preliquefied sample in light of the micromechanical model, Nemat-Nasser and Tobita (ref. 2) compared the reliquefaction potential of two similar samples which were liquefied by cyclic shearing under constant confining pressure. The difference between the two tests was the state at which the initial liquefaction was terminated during the final cycle: (1) In one test the final cycle in the initial liquefaction was terminated at *zero shear stress*. (2) In the second, otherwise similar, sample, the final cycle was terminated at *zero shear strain*. It was found that, during reliquefaction, the first sample (which was stopped at zero shear *stress)* showed essentially no resistance to reliquefaction, whereas the second sample (which was stopped at zero shear *strain)* was essentially as resistant to liquefaction as a virgin sample. These and related results were reconfirmed in a series of carefully planned experiments by Nemat-Nasser and Takahashi (ref. 3).

Another related phenomenon is prestraining under drained conditions and its effect on the liquefaction potential of the saturated undrained sample. Guided by the micromechanical model, Nemat-Nasser and Tobita (ref. 2) showed that, if prestraining at a suitably large shear strain is terminated at *zero shear stress*, then the sample will have strong induced anisotropy and hence, in subsequent undrained cyclic shearing, will liquefy during the first cycle. On the other hand, if the prestraining is terminated at *zero shear strain*, the sample will be left with essentially no induced anisotropy, and will be at least as resistant to liquefaction as the corresponding virgin sample.

In an effort to further verify these facts and, in addition, to examine the influence of the sample preparation on the mechanical response of cohesionless sands in cyclic shearing, Nemat-Nasser and Takahashi (ref. 3) have made a series of *strain-controlled* tests on Monterey No. 0 sand samples. The same apparatus as the one used by Nemat-Nasser and Tobita (ref. 2) is employed, except that the horizontal shearing device is modified in such a manner as to control the horizontal stroke, and to measure the corresponding resulting horizontal force. Two sample preparation techniques are used: moist tamping and pluviating dry sand through air. The basic conclusions of these experimental results are as follows:

1. In cyclic simple shearing, the resistance to reliquefaction (undrained) or densification (drained) of a preliquefied sample actually increases, because of the concomitant densification, if the preliquefaction is terminated at zero residual shear *strain*, but this resistance becomes very small if the preliquefaction is terminated at zero residual shear *stress*.
2. The inherent anisotropy associated with sample preparation techniques affects both the densification and liquefaction potential of the sample.

3. Within each cycle of simple shearing, the *induced anisotropy is essentially wiped out in the neighborhood* of zero shear *strain*, and the anisotropy that exists at this state is basically due to the sample preparation techniques (i.e., it is the *inherent* anisotropy), provided that the sample is not very loose and the strain amplitude is not very large. (Note that the state of *zero shear strain* is essentially the same as that of the state of *zero dilatancy* which is, indeed, the basic underlying factor.)
4. For simple shearing, the distribution of the dilatancy angles characterizing the fabric may be related to the shear strain and, in this manner, the densification pattern may be estimated.

Since the simple shearing experiments are performed on small circular cylinders, the state of stress in the sample is quite complicated. In order to verify the validity of the results, a new series of tests has now been initiated at the author's laboratory. These tests are being performed on a large hollow cylindrical sample under carefully controlled conditions so as to ensure a uniform state of stress and deformation in the sample. The results of these experiments will be reported soon after they are completed.

ACKNOWLEDGMENTS

This work has been supported by Air Force Office of Scientific Research under Grant No. AFOSR-87-0079 to the University of California, San Diego.

REFERENCES

1. Mehrabadi, M. M., Nemat-Nasser, S., Shodja, H. M., and Subhash, G., "Some Basic Theoretical and Experimental Results on Micromechanics of Granular Flow."
2. Nemat-Nasser, S., and Tobita, Y., "Influence of Fabric on Liquefaction and Densification Potential of Cohesionless Sand," *Mechanics of Materials*, Vol. 1, 1982, pp. 43-62.
3. Nemat-Nasser, S., and Takahashi, K., "Liquefaction and Fabric of Sand," *Journal of Geotechnical Engineering Division*, ASCE, Vol. 110, No. 9, 1984, pp. 1291-1306.
4. Finn, W. D. L., Bransby, P. L., and Pickering, D. J., "Effects of Strain History on Liquefaction of Sand," *Journal of the Soil Mechanics and Foundations Division*, ASCE, Vol. 96, No. SM6, 1970, pp. 1917-1934.
5. Ishihara, K., Tatsuoka, F., and Yasuda, S., "Undrained Deformation and Liquefaction of Sand under Cyclic Stress," *Soils and Foundations*, Vol. 15, 1975, pp. 29-44.
6. Nemat-Nasser, S., "On Behavior of Granular Materials in Simple Shear," *Soils and Foundations*, Vol. 20, 1980, pp. 59-73.
7. Nemat-Nasser, S., Mehrabadi, M. M., Shodja, H., Subhash, G., "A Microscopic Study of the Granular Material Behavior in Pure Shear," in preparation.

Micromechanics of Granular Materials, edited by M. Satake and J.T. Jenkins
Elsevier Science Publishers B.V., Amsterdam, 1988 — Printed in The Netherlands

THE VALIDITY OF THE EFFECTIVE STRESS CONCEPT IN SOIL MECHANICS

F.Oka
Department of Civil Engng., Gifu University, Gifu 501-11, Japan

SUMMARY

The validity and the limit of the effective stress concept in soil mechanics is shown based on the theory of two-phase mixture. It has been emphasized that the behavior of saturated soil with water depends on the balance of force and the compressibility of constituents.

INTRODUCTION

The effective stress concept proposed by Terzaghi(1923,1943) plays an important role in soil mechanics and is quite useful for analyzing geotechnical problems. It is still an open problem, however, why effective stress is effective. The aim of this paper is to give an interpretation of effective stress within the frame of continuum mechanics, and to show the validity, limits, and region of this concept.

Terzaghi introduced the effective stress concept in order to describe the deformational behavior of saturated soil with water, based on experimental results on the strength and deformation of soil. Terzaghi's effective stress is defined as follows:

$$\sigma' = \sigma - u \tag{1}$$

where σ' is the effective stress, σ is the total stress, and u denotes the pore water pressure.

Based on the pore water pressure data measured by Rendlic(1937), this concept was thereafter examined and expanded upon by Bishop(1960) and Skempton(1969). In later years, Nur and Byerlee(1971), Grag and Nur(1973),Verruijt(1982),Zienkiewicz(1982) and Carroll and Katsube(1983) developed this concept even further. The rational basis of this concept,however, still remains unclear.

Using this concept, Terzaghi succeeded in developing the one-dimensional consolidation theory. And recently, other important problems such as liquefaction and three-dimensional consolidation have been analyzed by the two-phase mixture theory first proposed by Biot(1962). In these analyses, the effective stress concept is introduced as a principle.

At present, it is essential for saturated soil to be treated as a two-phase mixture. Therefore, the author will attempt to give a physical interpretation of the effective stress concept from a theory of mixture.

Up to now, interpretations of effective stress have mainly been based on the concept of balance of force, which considers an interparticle or intergranular contact area. The author has emphasized,however, that the behavior of saturated soil depends on both the balance of force and the compressibility of constituents. It is therefore reasonable to consider that pore water constrains the volumetric motion of soil as a whole, since the compressibility of water is much less than that of the soil skeleton.

CONSTITUTIVE EQUATIONS OF ELASTIC SOLID-FLUID MIXTURE FOR VOLUMETRIC DEFORMATION AND EFFECTIVE STRESS

In this section, we will consider the constitutive equation of a two phase-elastic mixture for volumetric deformation, because pore water only constraints the volumetric deformation of saturated soil. The linear stress-strain relations for volumetric deformation are written as:

$$\sigma_{kk}^{(s)} = \gamma_1 \varepsilon_{kk}^{s} + \gamma_2 \varepsilon_{kk}^{f} \tag{2}$$

$$\sigma_{kk}^{(f)} = \gamma_3 \varepsilon_{kk}^{f} + \gamma_4 \varepsilon_{kk}^{s} \tag{3}$$

where σ_{ij} is the stress tensor defined at the surface of the mixture, ε_{ij} is the strain tensor, and γ_i are material constants. The indices f and s denote a fluid phase and a solid phase, respectively.

Equations (2) and (3) show that stress in the solid phase is dependent on the volumetric deformation in fluid phase, except for the case of a fully drained condition. Under the fully drained condition, the strain of pore water is zero. Eqs.(2) and (3) correspond to Biot's linear theory of porous elastic solid saturated with water. As for the meaning of the material constants in Biot's equation, Ishihara(1968) has already clarified the relation between material constants in the linear theory of mixture and the conventional soil parameters. We have written this paper in accordance with the relations and symbols in Ishihara's work.

Following Ishihara(1968), Eqs.(2) and (3) are rewritten as:

$$\varepsilon_{kk}^{s} = (k_c\sigma^{(s)} - \alpha_c\sigma^{(f)})/(\alpha_b k_c - \alpha_b^{\,2}) \tag{4}$$

$$\varepsilon_{kk}^{f} = (\alpha_b\sigma^{(f)} - \alpha_c\sigma^{(s)})/(\alpha_b k_c - \alpha_c^{\,2}) \tag{5}$$

where $\sigma^{(s)}=\sigma_{kk}^{(s)}/3$, $\sigma^{(f)}=\sigma_{kk}^{(f)}/3$ and α_b, α_c and k_c are material constants.

$\varepsilon_{kk}^{(s)}$ and $\varepsilon_{kk}^{(f)}$ can be described by e_p(ratio of the volume change of the pore space to the pore volume prior to deformation), e_s (volumetric strain of soil particles), e_w(volumetric strain of pore water) and the porosity n as follows:

$$\varepsilon_{kk}^{(s)}=ne_p+(1-n)e_s \qquad (6) \qquad \varepsilon_{kk}^{(f)}=e_w+(1-n)(e_s-e_p) \qquad (7)$$

In an unjaeketed test where the specimen is not covered with a membrane and the pore water pressure is equal to the cell pressure, we have the following relations.

$$\varepsilon_{kk}^{(s)}=e_p=e_s \qquad (8) \qquad e_s=C_s u \quad (9) \qquad e_w=C_1 u \qquad (10)$$

$$C_s=-\frac{(n-1)k_c+n\alpha_c}{\alpha_b k_c-\alpha_c^2} \qquad (11) \qquad C_1=\frac{(n-1)\alpha_c+n\alpha_b}{\alpha_b k_c-\alpha_c^2} \qquad (12)$$

where C_s is the compressibility of soil particles and C_1 is the compressibility of pore water.

In the drained test, $\sigma_{kk}^{(f)}=0$. Therefore, the following relations are obtained from Eqs.(4) and (5).

$$\varepsilon_{kk}^{s}=\frac{k_c}{\alpha_c k_c-\alpha_c^2}\sigma^{(s)}=C_b\sigma^{(s)} \qquad (13) \qquad \varepsilon_{kk}^{f}=\frac{-\alpha_c}{\alpha_b k_c-\alpha_c^2}\sigma^{(s)} \qquad (14)$$

where C_b is the bulk compressibility of soil.

PARTIAL STRESS UNDER UNDRAINED CONDITIONS

We will discuss partial stress acting on the solid and fluid phases under the undrained condition, where the specimen is only subjected to the cell pressure and the fluid flow is constrained. Under the undrained condition, $\varepsilon_{kk}^{(s)}=\varepsilon_{kk}^{(f)}$. Therefore, we obtained the following relation from Eqs.(4) and (5).

Using the relation between total stress and partial stress($\sigma=\sigma^{(s)}+\sigma^{(f)}$), the partial stress for the fluid phase is given as:

$$\sigma^{(f)} = \frac{\alpha_c + k_c}{k_c + 2\alpha_c + \alpha_b}\sigma \tag{15}$$

Combining Eqs.(11),(12) and (15) the partial stress is obtained by:

$$\sigma^{(f)} = \frac{C_b - C_s}{C_b/n-(1+1/n)C_s C_1}\sigma \quad (16) \qquad \sigma^{(s)} = \frac{(1/n-1)C_b - C_s/n + C_1}{C_b/n-(1+1/n)C_s + C_1}\sigma \quad (17)$$

From these equations, we get the following relation only when $C_s=0$ and $C_1/C_b=0$.

$$\sigma^{(f)} = n\sigma \quad (18) \qquad \sigma^{(s)} = (1-n)\sigma \quad (19)$$

Equations (18) and (19) denote the fact that when the soil particles and pore fluid are incompressible, the stress distribution is only dependent on the area ratio of soil constituents. However, stress distribution generally depends on the compressibilities of soil particles and pore fluid.

EFFECTIVE STRESS

Supposing an undrained condition where the drainage is prevented by an impermeable membrane, let's obtain the surrounding applied total stress when the measured pore pressure is u'.

Partial stress for fluid phase $\sigma^{(f)}$ is given as:

$$\sigma^{(f)} = nu' \tag{20}$$

From Eqs.(16),(17), and (20) partial stress for solid phase $\sigma^{(s)}$ is given by

$$\sigma^{(s)} = \frac{(1/n-1)C_b - C_s/n + C_1}{C_b - C_s} nu' \tag{21}$$

Total stress σ^{i}, which is balanced with this partial stress is therefore obtained by

$$\sigma = mu' \tag{22}$$

$$m = \frac{C_b - C_s + n(C_1 - C_s)}{C_b - C_s} \tag{23}$$

It is worthy to note here that the inverse of m is equal to Bishop's pore water pressure coefficient B. The isotropic deformation caused by the measured pore water pressure can be neglected due to the low compressibility of water. The general stress system can be divided into two parts. One is a stress component which corresponds to σ^i and the other is one which corresponds to the residual.

The difference between the general stress components and σ^i can be defined by effective stress σ^e as:

$$\sigma_{ij}^e = \sigma_{ij} - \sigma^i_{ij} \tag{24}$$

where σ_{ij}^e is an effective stress.

Next, we will examine the value of m. Table I shows the data of compressibility for various geological materials. The relation between m and C_l/C_b given by Eq.(22) is shown in Fig.1 for several values of porosity n.

From this figure, one can see m is almost equal to unity in the soil and soft rock cases. The value of m increases, however, with a decrease in the value of C_b. It is evident that Terzaghi's effective stress equation is applicable in cases the ratio of the compressibility of pore fluid to that of the soil skeleton is small, and the volumetric strain in the undrained condition is negligible. However, Terzaghi's equation is meaningless when the compressibility of pore fluid is high. This condition corresponds to that of dry soil with air.

Using Eq.(15), when C_l is large under the undrained condition, $\sigma^{(f)}$ becomes zero and $\sigma^{(s)}$ becomes equal to the total stress at the limit of C_l. In this case, the total stress analysis is useful for describing the behavior of sand. This case corresponds to that of dry soil or unsaturated soil with low saturation.

Firstly, let us consider saturated porous tuff which has recently been analyzed by effective stress. When n=0.37 , C_s=2.7x10^{-6} cm^2/kgf, and c_b=160 cm^2/kgf, m is 1.156. Under the undrained condition, the volumetric strain is 1.975x10^{-4} assuming that the measured pore pressure is about 10 kgf/cm^2. In other words, the isotropic applied pressure of 1.1u'(u':pore water pressure) is necessary for soil to be balanced with the internal pore water pressure. In this case, the effective stress differs from the conventional one by 10 %.

Secondly, we will consider saturated dense sand. When n=0.5, C_s=2.7, C_l=48x10^{-6} cm^2/kgf, and C_b=1800 cm^2/kgf, m is 1.013 and the corresponding volumetric strain is 2.54x10^{-5} when the measured pore water pressure is 1 kgf/cm^2.

COMPARISON OF PROPOSED THEORY AND CLASSICAL INTERPRETATION OF TERZAGHI'S EFFECTIVE STRESS

Skempton(1960) generalized the concept of effective stress and challenged tp conclude that Terzaghi's effective stress equation was not true effective stress, but an excellent approximation. He proposed that the effective stress equation should be as follows:

(for shear strength) $\sigma^{e}=\sigma-(1-a_{c}\tan\psi/\tan\phi')u'$ (25)

(for volume change) $\sigma^{e}=\sigma-(1-C_{s}/C_{b})u'$ (26)

where a_c is a contact area, ϕ' is an effective stress angle of shearing resistance, and ψ is an angle of intrinsic friction. However, Skempton's equations have the following defects.

(1) The definition of effective stress for shear strength is different from that for compression.

(2) The empirical parameters, which are introduced in the deduction of effective stress in order to connect the intergranular force with the external force, have not the physical basis.

(3) Skempton's approach is dependent on the assumption that Coulomb's strength equation is valid a priori.

(4) In the derivation of Eq.(26), the compressibility of pore fluid is not considered.

Mitchell(1961) has already pointed out the third shortcomings. Almost all consideration given to the effective stress concept by many researchers(e.g. Lambe, Scott, Mitchell) follow Skempton's study, and rely only on the balance of force. Nur and Byerlee(1971) and Grag and Nur(1973) proposed an exact effective stress law for porous saturated rocks. They considered only unjaketed test conditions, and did not deal with the undrained condition which is important when we discuss about effective stress in soil mechanics. Carroll and Katsube(1983) showed that the volumetric strain due to change in porosity is determined by the effective stress based on the similar concept by Nur et al.(1971). Recently, Zienkiewicz(1981),Verruijt(1982) and Vardoulakis et al.(1985) have disccussed the effective stress concept. For instance, Zienkiewicz(1981) proposed a new effective stress equation in which the compressibility of soil particles and fluid are considered. However, since they have assumed effective stress as an intergranular or interparticle stress a priori, the expression is not rational.

EFFECTIVE STRESS FOR UNSATURATED SOIL

Bishop proposed the effective stress equation for unsaturated soil as follows:

$$\sigma^{e}=\sigma-u_{a}-\chi(u'-u_{a}) \qquad (27)$$

where u_a is the pore-air pressure and χ is a parameter related to saturation.

The explanation of this expression is usually based on the balance of force around the interparticle contact area. This type of expression is meaningless because it depends only on the balance of force. Equation (26) expresses effective stress when dry soil $\chi=0$ and the effective stress equation becomes

$$\sigma^e = \sigma - u_a \tag{28}$$

As is well known, the effective stress expressed by Eq.(28) is not applicable to dry soil, and thus, is not widely used for such. Unfortunately, an effective stress equation similar to that for saturated soil with water does not exist for unsaturated soil.

CONCLUDING REMARKS

In this paper, the effective stress concept has been examined considering the balance of force and the compressibility of soil constitutents from a two-phase mixture point of view. It should be noted that the validity of Terzaghi's effective stress equation is exact for saturated soil only when the soil particles and the pore fluid are incompressible, and is approximate for soil in which the ratio of the compressibility of soil particles to that of the soil skeleton, and the ratio of fluid compressibility to that of the soil skeleton are very small.

REFERENCES

1 Terzaghi,K.1943. Theoretical Soil Mechanics.John Wiley & Sons,New York:11-15.
2 Bishop, A.W. 1960. The principle of effective stress. Norwegian Geotechnical Institute, No.32:1-5.
3 Skempton,A.W. 1960. Effective stress in soils, concrete and rocks. Conf. Pore pressure and suction in soils: 4-16.
4 Rendulic, L. 1937. Ein Grundgesetz der Tonmechanics und sein experimenteller Beweis,Bauingenieur, 18:459-467.
5 Nur,A. and J.D. Byerlee 1971. An exact effective stress law for elastic deformation of rock with fluids, J. Geophys. Res., 76:6414-6419.
6 Grag,S.K. and A.Nur 1973. Effective stress laws for fluid-saturated porous rocks, J.Geophys. Res., 78:5911-5921.
7 Carroll,M.M. and N.Katsube 1983:The role of Terzaghi's effective stress in linearly elastic deformation,J. Energy Resources Technology,ASCE,105:509-511
8 Verruijt, A. 1982. Some remarks on the principle of effective stress. Proc. IUTAM Conf. on Deformation and Failure of Granular Materials, Delft:167-170.
9 Zienkiewicz, O.C. 1982.Basic formulation of static and dynamic behaviour of soil and other media. Numerical Methods in Geomechanics, ed. by J.B. Martins, D. Reidel Pub. Com.:39-55.
10 Biot,M.A. 1962.Mechanics of deformation and acoustic propagation in porous media. J. Appl. Phys., Vol.33:1482-1498.
11 Ishihara,K.1968. Propagtion of compressional waves in a saturated soil. Proc. Int. Symp. of Wave Propagation and Dynamic Properties of Earth Materials:195-206.
12 Skempton,A.W. 1960.Significance of Terzaghi's concept of effective stress, From Theory to Practice in Soil Mechanics, John Wiley & Sons:42-53.
13 Akai,K.,T.Adachi and Nishi 1978. Elasto-plastic behavior of sedimentary soft rocks, Proc. JSCE.,Vol.271:83-95(in Japanese).
14 Mitchell,J.K. 1976. Fundamentals of soil behavior. John Wiley & Sons:186-196.
15 Lambe,T.W.and R.V.Whitman 1969.Soil Mechanics.John Wiley and Sons, 241-250.
16 Scott,R.F.1963.Principle of Soil Mechanics. Addison Wisely Pub.Com.Ltd.,163-166.
17 Vardoulakis,I. and D.E. Beskos 1986. Dynamic behavior of nearly saturated porous medis. Mechanics of Materials, Vol.5:87-108.
18 Bishop,A.W. and G.E.Blight 1963. Some aspects of effective stress in saturated and partly saturated soils. Geotechnque, Vol.12,No.2:125-144.

Table 1 Compressibility of Materials

Material	Compressibility $(cm^2/kgf) \times 10^{-6}$ C_b	C_s
Quartzinic sand stone	5.8	2.7
Quincy granite(100ft deep)	7.5	1.9
Vermont marble	17.5	1.9
Concrete(approx. value)	20.0	2.5
Sedimentary Rock*(porous tuff)	110-680	2.3-2.7
Dense sand	1800	2.7
Loose sand	9000	2.7
London clay(over consolidation)	7500	2.0
Gasport Clay(normally consolidation)	60000	2.0

$C_l=48\times10^{-6}$ cm^2/kgf(compressibility of water)
(after Skempton(1960): at pressure of $1kgf/cm^2$,
* after Akai et al. 1978)

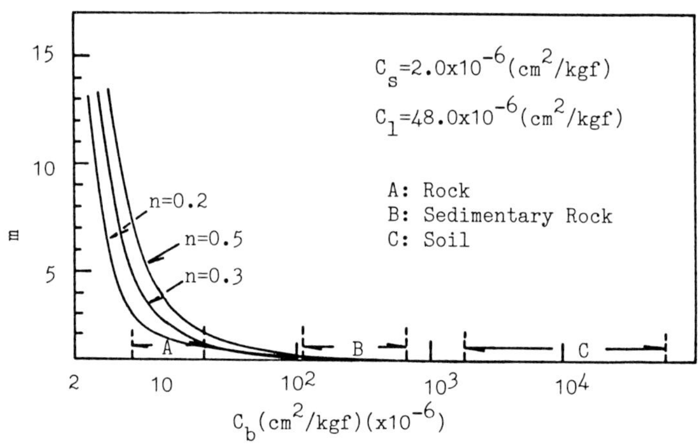

Fig.1 Relationship between m and C_b

Micromechanics of Granular Materials, edited by M. Satake and J.T. Jenkins
Elsevier Science Publishers B.V., Amsterdam, 1988 — Printed in The Netherlands

STRESS ANALYSIS ON THE COMPACTION OF GRANULAR MATERIALS BY FEM

T. NAGAO, Y. HATAMURA, H. MATSUI and T. MIKAMI

Dept. Mechanical Eng. for Production, Faculty of Eng., University of Tokyo, Hongo 7, Bunkyo-ku, Tokyo (Japan)

ABSTRACT

The triaxial compression tests for ceramic powder has been carried out to obtain the empirical stress-strain relationship. By using this relationship, the process of the compaction of ceramic powder has been simulated by newly developed simplified FEM. The result by FEM has been compared with the result of the experiments and they are in good agreement.

INTRODUCTION

For more than 20 years we have been trying to establish the mechanics of granular materials. The fundamental equations of granular materials are composed of stress equilibrium, moment equilibrium, strain-displacement relations, compatibility equations, stress-strain relations, continuity equation, geometrical relation and stress-force relations.[1] Especially we have

- Fundamental Equations
  - Stress Equilibrium
  - Moment Equilibrium
  - Strain-Displacement Relations
  - Compatibility Equations
  - Stress-Strain Relations
  - Continuity Equation
  - Geometrical Relation
  - Stress-Force Relations

Fig.1 Mechanics of Granular Materials

found that if the force distribution around a grain is assumed to be ellipsoidal, its principal values are given by $P_J=3\sigma_J/(2r\Lambda)$, where P_J denotes the principal value, σ_J the principal stress, r the mean radius of grains, and Λ the number of contacts in unit volume.[2] And it has been also found that the displacement between the adjacent grains w is proportional to 1/m th power of P, the contact force between them, where m=1,2 or 3 depending on the plastic deformation of the grain surface in contact.[3][4] Then the theoretical constitutive equations have been derived[2][5][6], and FEM stress analysis has been tried using these equations.[1] This has been much complicated and took so long time to calculate. So we have tried this time a simplified FEM using the stress-strain relaitons from triaxial tests, which has been applied to the compaction of ceramic powder in a cylinder, and compared with the experimental data.

EXTENSION OF FEM INTO GRANULAR MATERIALS[7]

Different from an elastic body, granular materials shows two kinds of non-linearity, the non-linearity introduced through the material properties and the one introduced through large deformations and geometrical changes in the structure. In order to apply FEM to this non-linear problem, incremental procedure is used here. The explanation of the incremental procedure for the axi-symmetric analysis is as follows. The unknown parameters are displacements u_r and u_z, stresses $\boldsymbol{\sigma}=(\sigma_r, \sigma_z, \sigma_\theta, \tau_{rz})^T$ and strains $\boldsymbol{\varepsilon}=(\varepsilon_r, \varepsilon_z, \varepsilon_\theta, \gamma_{rz})^T$. The fundamental equations are written as follows.

equilibrium equations :
$$\frac{\sigma_r}{r} + \frac{d\sigma_r}{dr} + \frac{d\tau_{rz}}{dz} = 0 \qquad (1)$$
$$\frac{d\sigma_z}{dz} + \frac{\tau_{rz}}{r} + \frac{d\tau_{rz}}{dr} = 0 \qquad (2)$$

stress-strain relations : $d\boldsymbol{\sigma}=Dd\boldsymbol{\varepsilon}$ (3)

strain-displacement relations :
$$d\varepsilon_r = \frac{\partial \delta u_r}{\partial r} \qquad (4)$$
$$d\varepsilon_z = \frac{\partial \delta u_z}{\partial z} \qquad (5)$$
$$d\varepsilon_\theta = \frac{\delta r}{r} \qquad (6)$$
$$d\gamma_{rz} = \frac{\partial \delta u_r}{\partial z} + \frac{\partial \delta u_z}{\partial r} \qquad (7)$$

Because of the impossibility to treat differential equations directly by FEM, the principle of virtual displacement is used in place of the equilibrium equations, which is written

$$\int (\sigma_r \, d\varepsilon_r + \sigma_z \, d\varepsilon_z + \sigma_\theta \, d\varepsilon_\theta + \tau_{rz} \, d\gamma_{rz}) \, dv = \int (f_r \, du_r + f_z \, du_z) \, ds \tag{8}$$

where du_r and du_z are virtual displacements, $d\varepsilon_r$, $d\varepsilon_z$, $d\varepsilon_\theta$ and $d\gamma_{rz}$ are virtual strain increments, f_r and f_z are forces per unit boundary surface area, v is the region of analysis, and s is the boundary surface.

The region is divided into triangular elements and each node and each element is given a number. An arbitrary finite element, n, is defined by nodes, j,k,l. Let the displacement increments at any point within each element be defined as follows:

$$\delta u_r = \alpha_1 + \alpha_2 r + \alpha_3 z \tag{9}$$

$$\delta u_z = \alpha_4 + \alpha_5 r + \alpha_6 z \tag{10}$$

where α_1, α_2, α_3, α_4, α_5, α_6 are constant. Then the stress increment and the strain increment are constant within each element. So the following vectors can be defined for the element n.

$$u_n = (u_{rj}, u_{zj}, u_{rk}, u_{zk}, u_{rl}, u_{zl})^T \tag{11}$$

$$\sigma_n = (\sigma_r, \sigma_z, \sigma_\theta, \tau_{rz})^T \tag{12}$$

$$\varepsilon_n = (\varepsilon_r, \varepsilon_z, \varepsilon_\theta, \gamma_{rz})^T \tag{13}$$

where u_{rj} and u_{zj} represent the displacement of node j. On this condition, the strain-displacement relation is given by

$$\Delta\varepsilon_n = N_n \, \Delta u_n \tag{14}$$

where

$$N_n = \frac{-1}{A} \begin{bmatrix} z_k - z_l & 0 & z_l - z_j & 0 & z_j - z_k & 0 \\ 0 & r_l - r_k & 0 & r_j - r_l & 0 & r_k - r_j \\ N_{kl} & 0 & N_{lj} & 0 & N_{jk} & 0 \\ z_l - r_k & z_k - z_l & r_j - r_l & z_l - z_j & r_k - r_j & z_j - z_k \end{bmatrix} \tag{15}$$

$$A = \begin{bmatrix} 1 & r_j & z_j \\ 1 & r_k & z_k \\ 1 & r_l & z_l \end{bmatrix} \tag{16}$$

$$N_{JK} = (r_J z_K - r_K z_J)/r^* + (z_J - z_K) + z^*(x_K - x_J)/r^* \tag{17}$$

$$((J,K)=(j,k),(k,l),(l,j))$$

$$r^* = (r_j + r_k + r_l)/3 \tag{18}$$

$$z^* = (z_j + z_k + z_l)/3 \tag{19}$$

The relationship between the stress before the small change of boundary condition, σ_n^*, and the stress after it, σ_n, in the element n is written as follows:

$$\begin{aligned} \sigma_n &= \sigma_n^* + \Delta\sigma_n \\ &= \sigma_n^* + D \, \Delta\varepsilon_n \\ &= \sigma_n^* + D \, (N_n \Delta u_n) \end{aligned} \tag{20}$$

where $\Delta\sigma_n$ and $\Delta\varepsilon_n$ are the stress increment and the strain increment, respectively. Let dE_1 and dE_2 represent the left side and the right side of Eq.(8) , respectively. Then the contribution of the element n to dE_1, $(dE_1)_n$, is written as follows:

$$\begin{aligned}(dE_1)_n &= d\varepsilon_n{}^T \sigma_n v_n \\ &= (du_n{}^T N_n{}^T)\ \sigma_n v_n \\ &= (du_n{}^T N_n{}^T)\ \{\sigma_n^{\cdot} + D\ (N_n \varDelta u_n)\}\ v_n \\ &= du_n{}^T\ (B_n + K_n \varDelta u_n)\end{aligned} \tag{21}$$

where v_n is the volume of the element n,

$$B_n = v_n N_n{}^T \sigma_n^{\cdot} \tag{22}$$

$$K_n = v_n N_n{}^T D N_n \tag{23}$$

Then let u_n, B_n, and K_n, which are defined for the element n, be extended as follows.

$$u = (u_{r1}, u_{z1}, u_{r2}, u_{z2}, \qquad , u_{rm}, u_{zm})^T \tag{24}$$

$$B_n^{\cdot} = (0, \quad ,0, \underset{(2j-1)}{B_{n1}}, B_{n2}, 0, \quad 0, \underset{(2k-1)}{B_{n3}}, B_{n4}, 0, \quad 0, \underset{(2l-1)}{B_{n5}}, B_{n6}, 0 \quad ,0)^T \tag{25}$$

$$K_n^{\cdot} = \begin{bmatrix} 0 & 0 & 0 & 0 & \\ & K_{n11}\ K_{n12} & K_{n13}\ K_{n14} & K_{n15}\ K_{n16} & (2j-1) \\ & K_{n21}\ K_{n22} & K_{n23}\ K_{n24} & K_{n25}\ K_{n26} & \\ 0 & 0 & 0 & 0 & \\ & K_{n31}\ K_{n32} & K_{n33}\ K_{n34} & K_{n35}\ K_{n36} & (2k-1) \\ & K_{n41}\ K_{n42} & K_{n43}\ K_{n44} & K_{n45}\ K_{n46} & \\ 0 & 0 & 0 & 0 & \\ & K_{n51}\ K_{n52} & K_{n53}\ K_{n54} & K_{n55}\ K_{n56} & (2l-1) \\ & K_{n61}\ K_{n62} & K_{n63}\ K_{n64} & K_{n65}\ K_{n66} & \\ 0 & 0 & 0 & 0 & \\ & (2j-1) & (2k-1) & (2l-1) & \end{bmatrix} \tag{26}$$

where B_{nj} is the j component of B_n and K_{njk} is the (j,k) component of K_n.

Then (dE1)n is given by

$$(dE_1)n = du^T\ (B_n^{\cdot} + K_n^{\cdot} \varDelta u) \tag{27}$$

So the (dE1) is given by

$$dE_1 = \Sigma (dE_1)n \tag{28}$$

to obtain

$$dE_1 = du^T\ \{\Sigma B_n^{\cdot} + (\Sigma K_n^{\cdot})\ \varDelta u\} \tag{29}$$

which reduces to

$$dE_1 = du^T\ (B + K \varDelta u) \tag{30}$$

where

$$K = \Sigma K_n^{\cdot} \tag{31}$$

$$B = \Sigma B_n^{\cdot} \tag{32}$$

Assuming the external force acts only on the nodes, the external force can be written

$$F = (f_{r1}, f_{z1}, f_{r2}, f_{z2}, \qquad f_{rm}, f_{zm}) \tag{33}$$

Then dE_2 is given by

$$dE_2 = du^T F \tag{34}$$

Because dE_1 is equal to dE_2 at any virtual displacement increments,du, the following equation can be derived.

$$B + K\varDelta u = F \tag{35}$$

Because the stress equilibrates before the change of the boundary condition,

$$\Sigma\ (d\varepsilon_n{}^T \sigma_n{}^{\bullet} v_n) = du^T F^{\bullet} \tag{36}$$

where $F^{\bullet}$ is external force before the change of boundary condition, to obtain

$$\Sigma\ (du_n{}^T v_n N_n \sigma_n{}^{\bullet}) = du^T F^{\bullet} \tag{37}$$

which reduces to

$$du^T B = du^T F^{\bullet} \tag{38}$$

Because of Eq.(38) at any virtual displacement increments, the following equation can be obtained.

$$B = F^{\bullet} \tag{39}$$

From Eqs.(38) and (39),

$$K\varDelta u = \varDelta F \tag{40}$$

can be derived.

By solving this linear equation, the unknown $\varDelta u$ and $\varDelta F$ can be obtained.This process is repeated until the state reaches a certain final condition.

The stress-strain relations should be derived from the analysis of the mutual interactions between each particle. But in this paper they are derived from the results of the triaxial test.

First, the relationship between the principal stress increment and the strain increment is assumed to be written in the following form.

$$d\varepsilon_K = \{d\sigma_K - \nu\ (d\sigma_L + d\sigma_J)\}\ /F \tag{41}$$

$$d\varepsilon_J = \{d\sigma_J - \nu\ (d\sigma_K + d\sigma_L)\}\ /F \tag{42}$$

$$d\varepsilon_L = \{d\sigma_L - \nu\ (d\sigma_J + d\sigma_K)\}\ /F \tag{43}$$

$$d\gamma_{JK} = d\tau_{JK}/G \tag{44}$$

where J,K,L indicate the directions of the principal stresses . F,ν and G are functions of the state of stress.

When the relationship between the deviator stress and the axial strain for the constant confinig pressure as shown in Fig.2 and the relationship between the hydrostatic pressure and the volumetric strain as shown in Fig.3 are given, the way of the determination of the value of F, ν and G, is as follows.

(1) The way of the determination of the value of F

As derived from Eqs.(41),(42) and (43), F is the ratio of the stress increment to the strain increment when only one principal stress is increased. So F is the gradient of the curves shown in Fig.2. It is a problem how to determine the confining pressure and the deviator stress. In this paper, σ_{min} and

$(\sigma_{max}-\sigma_{min})$ are used, respectively.

(2) The way of the determination of the value of

Adding up the three equations (41),(42) and (43), the following equation is derived:

$$d(\varepsilon_J+\varepsilon_K+\varepsilon_L)=3\ (1-2\nu)\ d(\sigma_J+\sigma_K+\sigma_L)/F \tag{45}$$

to obtain

$$d\varepsilon_v=(1-2\nu)d\sigma_w/F \tag{46}$$

where

$$d\varepsilon_v=d(\varepsilon_J+\varepsilon_K+\varepsilon_L) \tag{47}$$

$$d\sigma_w=d(\sigma_J+\sigma_K+\sigma_L)/3 \tag{48}$$

Solving Eq.(46) for ν, the following equation can be obtained.

$$\nu=1/2-F\ (d\varepsilon_v/d\sigma_w)\ /6 \tag{49}$$

where $d\varepsilon_v/d\sigma_w$ can be derived from the gradient of the curve shown in Fig.3, $(\sigma_J+\sigma_K+\sigma_L)/3$ being used for .

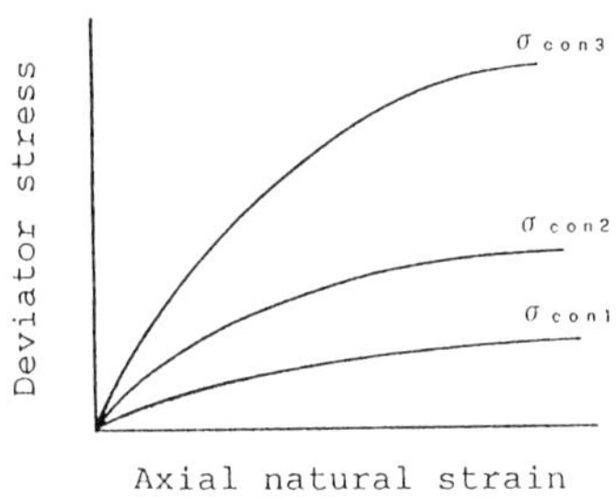

Fig.2 Relationship between the axial strain and the deviator stress.

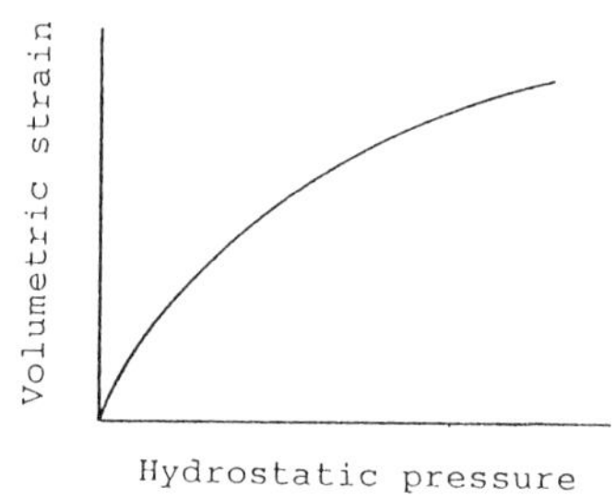

Fig.3 Relationship between the hydrostatic pressure and the volumetric strain.

(3) The way of the determination of the value of G

G is derived by the same relation of an elastic body which is written as follows:

$$G=F/\{2\ (1+\nu)\} \tag{50}$$

RESULT OF THE TRIAXIAL TEST

The ceramic powder used in this test is Al_2O_3 , which average particle size is 0.5μm in diameter.

The relationships between the stress difference and the axial strain at some confining pressures are shown in Fig.4. The relationship between the

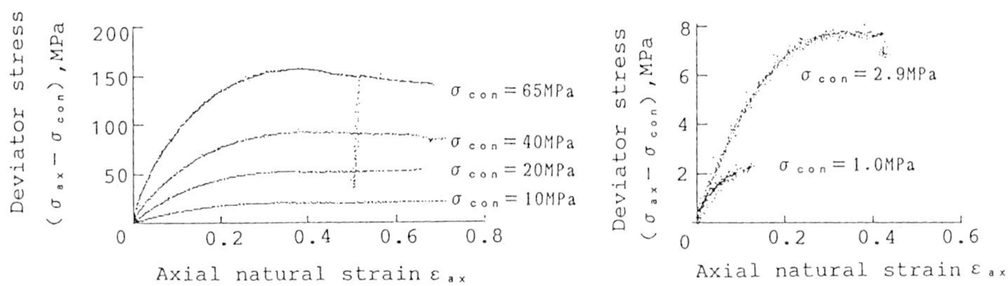

Fig.4 Relationship between the axial strain and the deviator stress by triaxial tests.

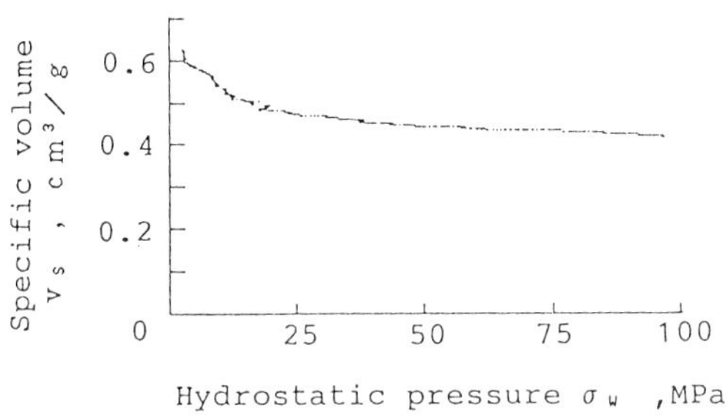

Fig.5 Relationship between the hydrostatic pressure and the specific volume by triaxial tests.

hydrostatic pressure and the specific volume is shown in Fig.5. The Mohr envelope is shown in Fig.6. As a result the internal friction angle of this ceramic powder is determined as 32.7 degrees.

The relation between the axial strain and the dimensionless deviator stress, which is the deviator stress divided by the deviator stress at the failure point is shown in Fig.7. Because of the agreement of the data at the various confinig pressures as shown in Fig.7, the relation between the axial strain and the dimensionless deviator stress before the failure is approximated by

$$1-\sigma^{*}=a(\varepsilon_{y}-\varepsilon_{ax})^{b} \tag{51}$$

where ε_y is the axial strain at the failure point, a and b are constant. In this case, a and b were determined as 11.4 and 2.21, respectively.

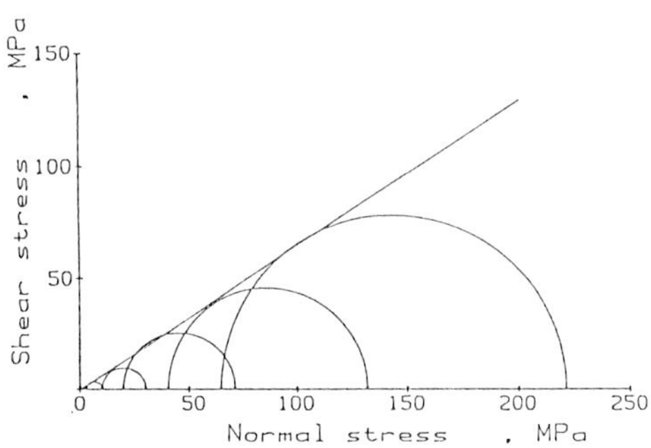

Fig.6 Mohr envelopeby triaxial tests.

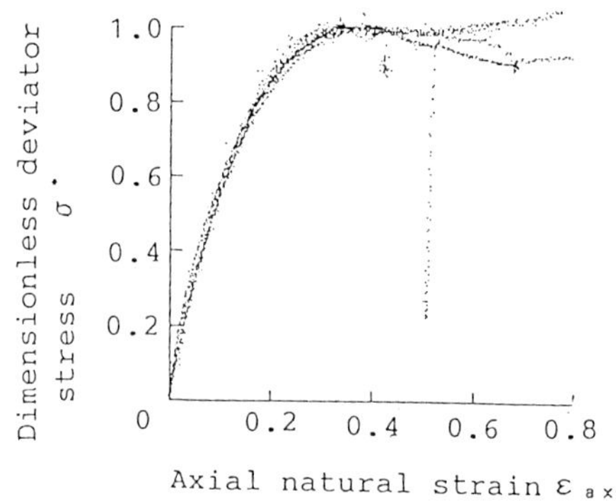

Fig.7 Relationship between the axial strain and the dimensionless deviator stress.

Assuming the form of the relation between the hydrostatic pressure and the specific volume is given by

$$\sigma_w = c(v_0 - v)^d \qquad (52)$$

where v_0 is 0.94 cm^3/g , the specific volume when the hydrostatic pressure is zero, c and d are constant, c and d are determined as follows: $c = 1.90 \times 10^3$, $d = 5.63$ (σ_w<10MPa); $c = 2.43 \times 10^5$,$d = 10.7$ (10MPa<σ_w<30MPa); $c = 2.95 \times 10^7$, $d = 18.6$ (σ_w>30MPa).

In this paper,the stress-strain matrix is derived from these two approximate relations.

SIMULATION OF THE COMPACTION PROCESS BY FEM COMPARED WITH THE EXPERIMENTAL RESULT

The compaction process in a cylindrical mold is examined, and the result of the simulation by FEM is compared with the experimental result.

The experiment is carried out as follows. First the ceramic powder(17.0g) is put in the cylindrical mold as shown in Fig.8. Then the ceramic powder is compacted with the punch until the mean pressure on the punch reaches 500MPa. In this experiment the weight on the punch, the weight on the base, the height of the powder and the distribution of the stress on the lateral wall are detected.

As for the boundary condition at the simulation by FEM, the powder is supposed to be fixed to the surface of the punch and the base and it is supposed to slide on the lateral wall with the friction. And the friction coefficient is assumed to be constant all over the lateral wall, 0.23.

The relation between the mean pressure on the punch and the height of the powder, both calculated and experimental, is shown in Fig.9. The distribution

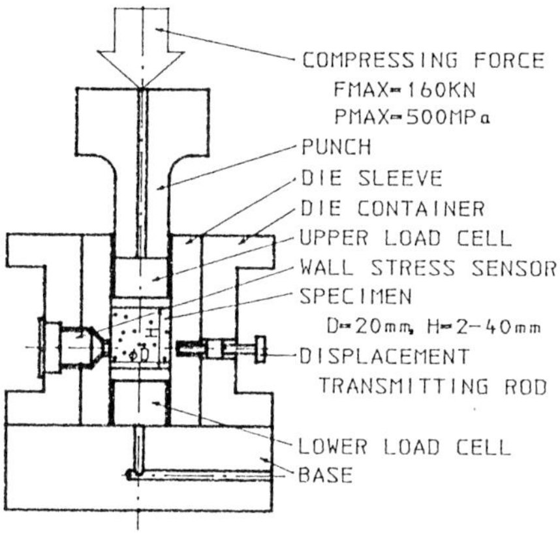

Fig.8 The mold for the compaction of granular materials.

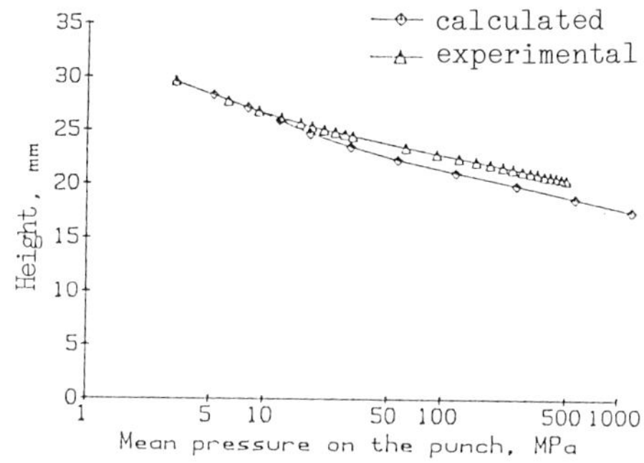

Fig.9 Relationship between the mean pressure on the punch and the height of the powder.

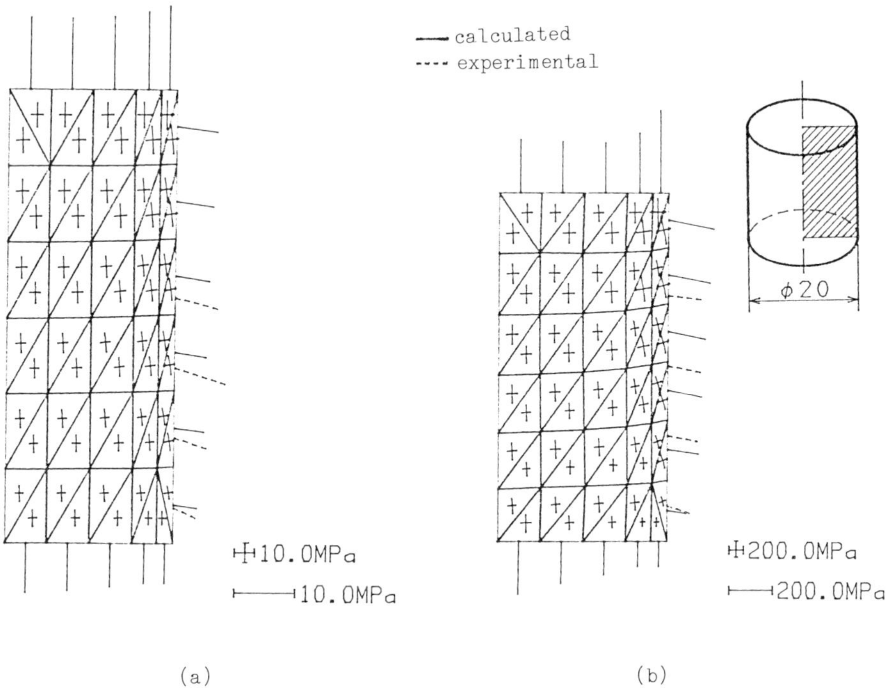

Fig.10 Calculated stress distributions on the surface of the mold and calculated principal stress distributions in the mold with experimental stress distributions on the lateral wall. (a)Mean pressure on the punch is 12MPa. (b)Mean pressure on the punch is 260MPa.

of the stress on the wall, both calculated and experimental, are shown in Fig.10. The distribution of the principal stress in the mold and the state of the deformation calculated by FEM are also shown in Fig.10. The calculated distribution of the stress on the wall and the calculated relation between the mean pressure on the punch and the height of the powder agree with the experimental results well, although there is some difference between the calculated height and the experimental height when the pressure on the punch is over 15 MPa. This difference is supposed to be due to the linearization at the incremental procedure of FEM and the limit of the confinig pressure at the triaxial test which is under 65MPa. Fig.10 shows well that the distributions of the stress and the strain in the mold are affected by the friction on the lateral wall.

CONCLUSION

(1) A simplified FEM has been developed for the axi-symmetric case of compaction of granular materials by utilizing the stress-strain relationship obtained from triaxial tests.

(2) This method has been applied to the compaction of aluminum oxide powder in a cylinder. The experimental values and the calculated ones of the wall stress and the height of powder have been in good agreement. Moreover the distribution of internal stress and the resultant stress on punch surfase and base have been calculated.

REFERENCES

1 T. Nagao and S. Katayama, A Stress Analysis on the Compressive Behavior of Granular Materials by the Two-dimensional Finite Element Method, Bull. JSME, 23 (1980) 1763-1770.

2 T. Nagao, The Stress-Strain Relations of Granular Materials (1st Rep.), Bull. JSME, 21 (1978) 1077-1084.

3 T. Nagao, A Study of the Statics of Granular Materials (1st Rep.) Bull. JSME, 10 (1967) 775-785.

4 T. Nagao, A Study of the Statics of Granular Materials (2nd Rep.), Bull. JSME, 11 (1968) 866-874.

5 T. Nagao, The Stress-Strain Relations of Granular Materials (2nd Rep.), Bull. JSME, 22 (1979) 148-155.

6 T. Nagao, The Stress-Strain Relations of Granular Materials (3rd Rep.), Bull. JSME, 22 (1979) 636-641.

7 O.C. Zienkiewicz, The Finite Element Method in Engineering Science, McGraw Hill (1975).

Micromechanics of Granular Materials, edited by M. Satake and J.T. Jenkins
Elsevier Science Publishers B.V., Amsterdam, 1988 — Printed in The Netherlands

APPLICATION OF MARKOV MODEL TO Ko-COMPRESSION AND CONSOLIDATION OF PARTICULATE MATERIAL

R. KITAMURA[1] and M. SHINCHI[2]

[1]Dept. of Ocean Civil Engineering, Kagoshima University, 1-21-40 Kourimoto Kagoshima 890 (Japan)

[2]Technical Clerk of Kagoshima Prefectural Government (Former Graduate Student of Kagoshima University)

SUMMARY

In this paper the Markov model which has been proposed for the particulate material by Kitamura (ref. 1) is applied to the Ko-compression and the consolidation behaviors. The procedures of the numerical experiment are explained for the Ko-compression and the consolidation. The modelling of voids is also proposed for the permeability because it is necessary when the Markov model is applied to the consolidation. The stress-strain relation, and the excess pore water pressure and deformation-time relations obtained by the numerical experiment are discussed. It is found out that the Markov model can express the mechanical behaviors of the Ko-compression and the consolidation.

INTRODUCTION

Kitamura (ref. 1) proposed the mechanical model of particulate material which is called the Markov model and applied it to the shear deformation. In the Markov model the motion of particles in the particulate material is regarded as the Markov process in the deformation process. The contact angle is adopted as a random variable. The contact angle is defined as the angle between the normal line to the tangential plane formed at the contact point of the adjacent particles and the reference axis. The basic equation of the Markov model is derived from the n-dimensional basic equation of the Markov process as follows.

$$\frac{\partial}{\partial s}\omega(\eta, s) = -\sum_{i=1}^{2}\frac{\partial}{\partial \beta_i}\{A_i(\eta, s)\cdot\omega(\eta, s)\} + \sum_{i,j=1}^{2}\frac{\partial^2}{\partial \beta_i\cdot\partial \beta_j}\{B_{ij}(\eta, s)\cdot\omega(\eta, s)\} \quad (1)$$

$$A_i(\eta, s) = \lim_{\Delta\to 0}\frac{1}{\Delta}\int_0^{2\pi}\int_0^{\pi/2}(\beta_{i,s+\Delta} - \beta_{i,s})\cdot P(\eta_s, \eta_{s+\Delta})\cdot d\beta_1 d\beta_2 \quad (2)$$

$$Bij(\eta, s) = \lim_{\Delta \to 0} \frac{1}{2\Delta} \int_0^{2\pi} \int_0^{\pi/2} (\beta i,_{s+\Delta} - \beta i,_s) \cdot (\beta j,_{s+\Delta} - \beta j,_s) \cdot P(\eta_s, \eta_{s+\Delta}) \cdot d\beta_1 d\beta_2 \qquad (3)$$

where, $\eta = (\beta_1, \beta_2)$:contact angle,

$\omega(\eta, s)$:probability density function of contact angle η at the stress state s,

$Ai(\eta, s)$:drift vector,

$Bij(\eta, s)$:diffusion tensor,

$P(\eta_s, \eta_{s+\Delta})$:transition probability of contact angle η when the stress state changes from s to $s+\Delta$.

The coefficients in the basic equation, Eqn.(1), which are called the drift vector and the diffusion tensor are determined by introducing the new concepts of the potential slip plane and the potential barriers. The diffusion tensor B_{ij} is zero when $i \neq j$, because the change of contact angles, β_1 and β_2 can be assumed to be independent. The idea of the discontinuous motion of particles is also introduced where the discontinuous motion means that the contact point of particles disappears and/or appears. This discontinuous motion is similar to the dislocation of the crystal solid. The basic equation of the Markov model is numerically solved by the finite difference method. The energy transferred into the particulate material plays the key role when the basic equation is solved. Then the constitutive relation is derived.

In the previous research the Markov model was applied to the shear deformation of particulate material with the anisotropic fabric (ref. 2), under the drained and the undrained conditions where three principal stresses are different (refs. 2, 3), under the constant stress ratios' condition (ref. 3) and under the Ko-compression condition (ref. 4). It was found out that the Markov model could express these mechanical behaviors and this model was a comprehensive and promising one that was much different from the ones based on the continuum mechanics.

In this paper the Markov model is applied to the Ko-compression with the dense and the loose specimens, and the consolidation. Then it is made clear that the Markov model can express the constitutive relation of particulate material under the Ko-compression and the consolidation.

APPLICATION TO Ko-COMPRESSION

The Ko-compression is defined as the compression where the lateral

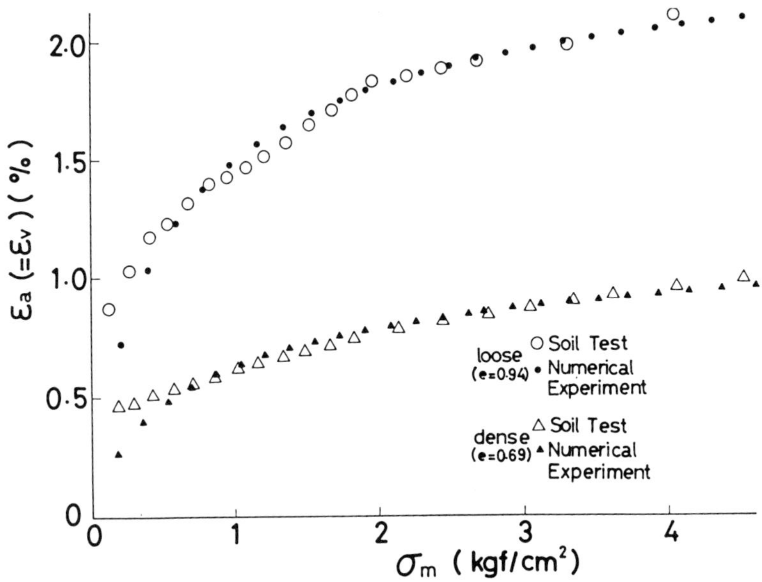

Fig. 1. Stress-strain relation under Ko-compression

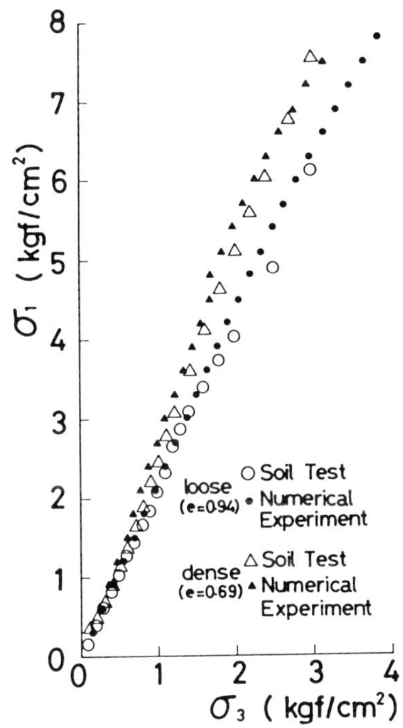

Fig. 2. Stress path under Ko-compression

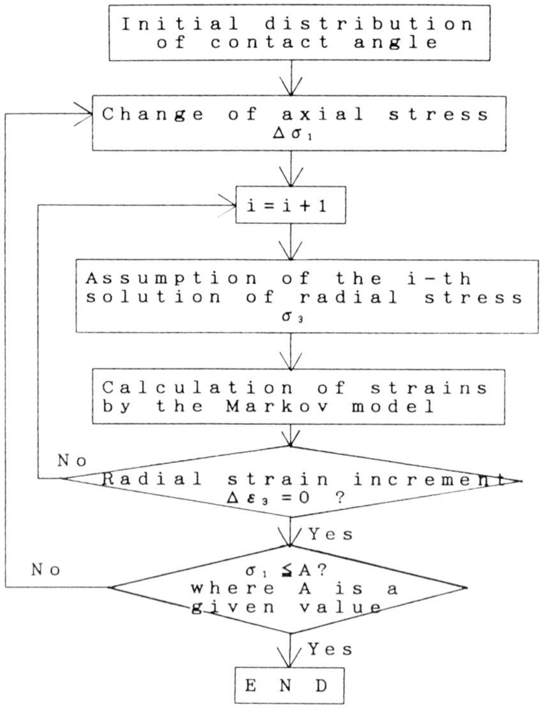

Fig. 3. Calculation procedure for Ko-compression

deformation is constraint, i.e., $\Delta\varepsilon_r=0$. The Ko-value means the stress ratio σ_3/σ_1 when the lateral strain increment $\Delta\varepsilon_r$ is zero in the deformation process, where σ_1 and σ_3 are the maximum and the minium principal stresses respectively.

The soil test on Toyoura sand is carried out by using the double cell type's triaxial testing apparatus. The Ko-condition is generated by keeping the volume of the water in the inner cell constant. The marks, o and Δ, in Fig. 1 designate the stress-strain relations obtained from the soil test with two kinds of void ratios, i.e., the loose and the dense states, where ε_a and ε_v are the axial and the volumetric strains respectively and σ_m is the mean effective principal stress. Figure 2 shows the stress paths of the loose and the dense specimens under the Ko-compression. It is found out from Fig. 2 that the Ko-value depends on the initial void ratio.

Figure 3 shows the calculation procedure of the numerical experiment for the Ko-compression by the computer. The strain is calculated by the same computer program as the one for the constant stress ratio's deformation process (ref. 2). The iteration method is used to obtain the stress state when the lateral strain increment $\Delta\varepsilon_r$ is zero in the compression process. In Figs. 1 and 2 the results of the numerical experiment is shown by the marks ● and ▲. The input data are

TABLE 1 Used values for Ko-compression

initial void ratio	total energy		recovarable energy	disapparance X_1		disapparance X_2	
e_0	ac	bc	Cs	λ_1	λ_2	λ_3	λ_4
0.69	0.35	1.062	0.041	340	15	645	-45
0.94	0.70	0.967	0.041	315	20	628	-35

listed in TABLE 1. The initial void ratios are same as the ones of soil test on Toyoura sand. The model parameters a_c and b_c are used to estimate the total energy transferred into the specimen in the deformation process. The model parameter C_s is used to estimate the recoverable (elastic) energy and called the swelling index in the soil mechanics. The model parameter λ_i (i=1-4) is introduced to estimate the discontinuous motion of particles. It is found out from Figs. 1 and 2 that the Markov model can express the deformation of Toyoura sand under the Ko-condition.

APPLICATION TO CONSOLIDATION

The consolidation behavior is divided into two parts, i.e., the primary and the secondary consolidation. The primary consolidation is known as a time dependent behavior as the water among voids is squeezed out and the soil mass is compressed whereas the secondary consolidation is considered to be the visco-plastic behavior of soil skeleton.

In this paper the Markov model is applied to the primary consolidation behavior, i.e., the deformation due to the change in the excess pore water pressure is investigated. To evaluate the phenomenon that the pore water is squeezed out of voids, it is necessary to model the voids in the particulate material because the primary consolidation behavior closely relates to the permeability behavior. Figure 4 shows the model of voids and particles in the particulate material, i.e., the rectangular parallelepiped which is called the element volume in this paper is composed of the pipe of which the diameter is D and the inclination angle is θ, and the other solid part. The pipe and the other solid part correspond to the voids and the soil particles respectively. If the diameter D and the inclination angle θ are adopted as the random variables, they have the probability density functions which are presented by $P_d(D)$ and $P_c(\theta)$. The permeability coefficient and the void ratio can be obtained by using these probability density functions as follows. Deriving Eqns. (4) and (5), the Hagen-Poisseuille's law for the laminar flow through the pipe is applied.

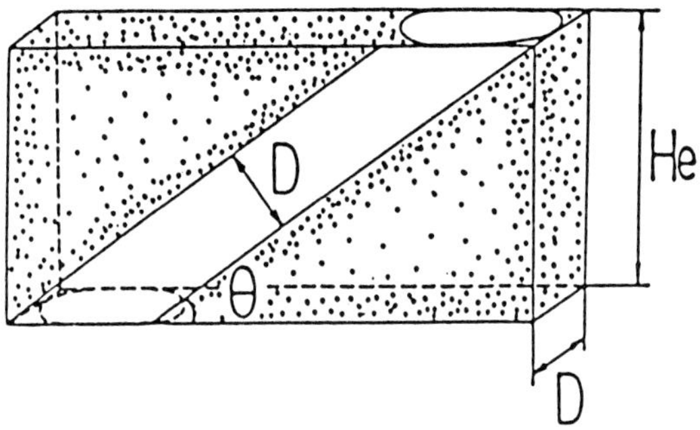

Fig. 4. Modelling of voids

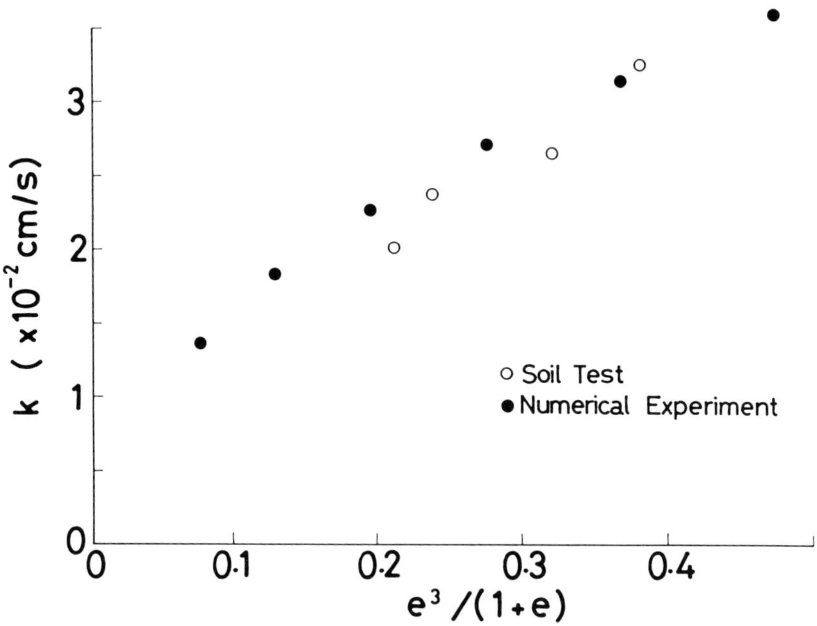

Fig. 5. Relation between permeability coefficient and void ratio

TABLE 2 Used values for permeability

Height of element volume	Mean diameter of pipe	Standad deviation of D	Coefficient of viscousity	distribution of inclination angle
He (cm)	Dm (cm)	Ds (cm)	μ_w (gf·s/cm^{-2})	ζ
0.0327	0.9×He	0.1×He	1.161×10^{-5}	0.159155

$$k = \int_0^{\infty} \int_{-\pi/2}^{\pi/2} \frac{D^2 \cdot \gamma_w \cdot \sin\theta}{32 \cdot \mu} \; Pc(\theta) \cdot Pd(D) \; d\theta \, dD \qquad (4)$$

$$e = \int_0^{\infty} \int_{-\pi/2}^{\pi/2} \frac{D \cdot \pi}{\sin\theta} \Big/ \left(\frac{D}{\sin\theta} + \frac{He}{\tan\theta} - \frac{D \cdot \pi}{4 \cdot \sin\theta} \right) \cdot Pc(\theta) \cdot Pd(D) \; d\theta \, dD \qquad (5)$$

where, k : permeaility coefficient,

e : void ratio,

μ : coefficient of viscousity,

γw : unit weight of fluid

Figure 5 shows the relation between the permeability coefficient and the void ratio which are obtained by the constant head permeability test on Toyoura sand and the numerical experiment. The marks ○ and ● in Fig. 5 denote the results of the soil test and the numerical experiment respectively. In the numerical experiment the log normal distribution is adopted as $P_d(D)$ and the opposite pentagonal distribution to the one of contact angle is adopted as $P_c(\theta)$ (refs. 2, 3, 5). TABLE 2 shows the values used in the numerical experiment for the permeability behavior. In TABLE 2 the height of the element volume is the same order as the mean diameter of particles. The mean diameter and the standard deviation of D are tentatively 0.9 and 0.1 times as large as the height of element volume, H_e, to fit the numerical result to the soil test one. The value of ζ which prescribes the probability density function of θ is same as the one of the probability density function of contact angle. The coefficient of viscousity is same as the constant head permeability test. It is found out from Fig. 5 and TABLE 2 that the modelling of voids as shown in Fig. 4 can express the permeability behavior of particulate material by using the reasonable model parameters.

The application of the Markov model to the consolidation becomes possible by combining the procedure for the deformation under the constant stress ratio (ref. 4) with the one for the permeability. Figure 6 is the flow chart which shows the procedure to apply the Markov model to the consolidation. The numerical experiment is carried out by supposing the isotropic compression test on Toyoura sand. TABLE 3 shows the input data for the consolidation, where the values concerning the permeability are same as the ones in TABLE 2 and the values concerning the deformation are determined by the same way as the one used for Toyoura sand in the previous paper (refs. 2-4). Figure 7 shows the result of the numerical experiment. The isotropic consolidation test on Toyoura sand was carried out by the conventional triaxial testing apparatus. The pore water

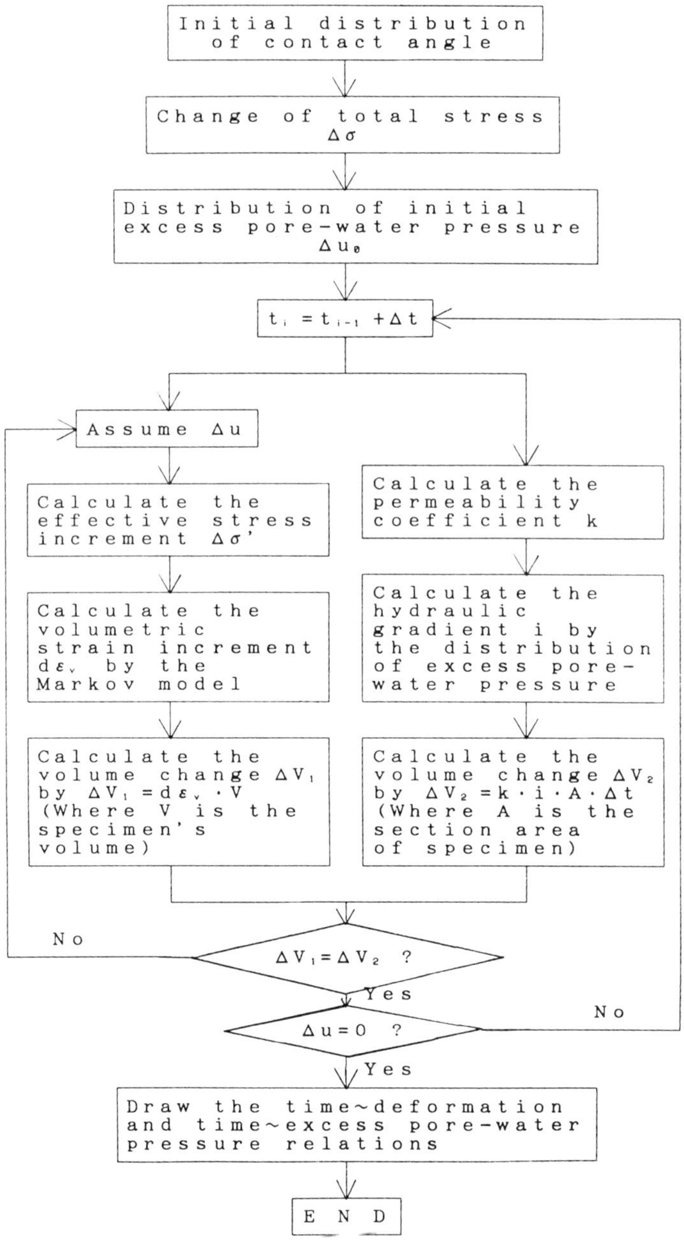

Fig. 6. Calculation procedure for consolidation

TABLE 3 Used values for consolidation

initial void ratio	total energy		recovarable energy	disapparance X_1		disapparance X_2	
e_0	ac	bc	Cs	λ_1	λ_2	λ_3	λ_4
0.88	0.47	1.24	0.041	380	2	450	-12

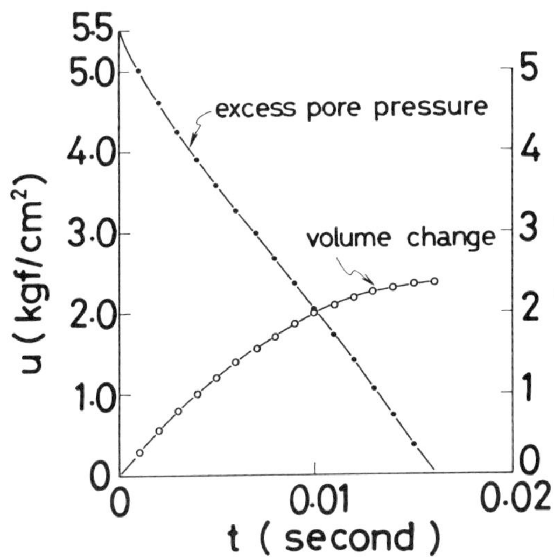

Fig. 7. Excess pore pressure, deformation-time relation

pressure was measured at the bottom pedestal and the pore water was drained from the upper cap, but the pore water pressure couldn't be measured because the time for the drainage was too short to generate the excess pore water pressure in the specimen and to measure it. The similar result is obtained by the numerical experiment as shown in Fig. 7, i.e., the excess pore water pressure becomes zero after about 0.016 seconds.

CONCLUSIONS

The Markov model is applied to the Ko-compression and the consolidation behaviors of particulate material. The modelling of voids is conducted in order to investigate the permeability behavior of particulate material and to derive the permeability coefficient and the void ratio. The Markov model is applied to the consolidation by combining the procedure for the deformation under the constant stress ratio with the one for the permeability behavior. The calculation procedure by the computer are represented for the Ko-compression and the consolidation. The results of the numerical experiment are compared with the soil test results although the time is too short to measure for the consolidation test on Toyoura sand by the triaxial testing apparatus. The results of the numerical experiment make clear that the Markov model can express not only the shear deformation but also the compression and the consolidation behaviors of particulate material. Then the Markov model is a comprehensive and promising one that can follow the various mechanical behaviors of particulate material.

ACKNOWLEDGEMENT

This research is sponsored by the grant-in-aid of the Ministory of Education of Japanese Government. The authors wish their gratitude to the organization.

REFERENCES

1 R. Kitamura, Deformation of granular soil as a Markov process, Proc. of the U.S.-Japan Seminar on Continuum-Mechanical and Statistical Approaches in the Mechanics of Granular Materials. 1978, pp.155-161.
2 R. Kitamura, A Markov model of particulate material with anisotropic fabrics, Proc. of the Int. Conf. on Numerical Methods in Engineering; Theory and Application, Balkema, 1985, pp.455-464.
3 R. Kitamura, A Markov model for deformation of particulate material, Proc. of 5th Int. Conf. on Numerical Methods in Geomechanics, Balkema, 1985, pp.547-554.
4 R. Kitamura and M. Shinchi, The Markov model for mechanical behavior of particulate material, Proc. of Int. Conf. on Computational Mechanics, Vol.2, Springer-Verlag, 1986, pp.XI-161-166.
5 R. Kitamura and M. Shinchi, Modelling of mechanical behaviors of particulate material based on the Markov process, Int. Sympo. on Slurry Flows, 1986, pp.41-50.

Micromechanics of Granular Materials, edited by M. Satake and J.T. Jenkins
Elsevier Science Publishers B.V., Amsterdam, 1988 — Printed in The Netherlands

DISTRIBUTED DISTURBANCE AND CONSTRAINT IN SOILS

O.O. Ogunbekun, A. Dalili, J.R.F. Arthur and T. Dunstan
Civil Engineering Dept., University College, Gower St., London WC1E6BT (U.K.)

ABSTRACT

Experiments in which a variety of boundary conditions were imposed in plane strain drained shear tests on dense Leighton Buzzard sand are reported. The results show that transient distributed disturbances or extremely localised yet evenly distributed restraints on the boundaries of the elements can have large effects on the strength and stress-strain relationships without significantly altering internal strain distributions or dilation rates. The suggested mechanism is the influencing of the stability of the particles through changes in the contact forces.

INTRODUCTION

The particulate nature of soils leads to inhomogenieties in deformation; discontinuity is inevitable at the level of individual particles and higher levels involving groups of particles. It is also widely recognised that for particulate soils the continuum concept of stress is an idealisation of contact force distributions which are usually far from uniformly distributed over the existing contacts. These large departures from the continuum behaviour, which we readily accept, might be expected to lead to appreciable effects on stress-strain behaviour, but experimentalists have not reported any readily attributable behaviour. Element tests with a wide variety of imposed boundary conditions normally lead to a good approximation to uniform boundary strains and uniform boundary stresses at some macro scale. There appear to be mechanisms which suppress inhomogenieties in deformation to a quite surprising degree presumably by changes in the inter-particle forces.

The large and continually changing inhomogenieties in contact forces that are regularly observed in studies of photo-elastic granular materials and computer simulations of deforming granular materials support this assumption (refs.1,2,3). Here then individual contact forces often adjust to establish an approximation to continuum behaviour, and the question arises as to what other effects forces applied to individual particles might have? As apparatuses developed to perform element tests have grown more diverse opportunities to find an answer to this question were inevitable. A plane strain apparatus designed and developed at University College at the behest of The Christian Michelsen Institute of Norway has proved especially useful in this respect because it can be operated with subtly different boundary conditions. Results from this apparatus which has become known as the Biaxial Tester (BT) and the more established Directional Shear Cell (DSC) will be presented. The data will relate to dense dry sand only but similar behaviour has been observed in loose sand and remoulded kaolin tested undrained.

Large effects have been observed on both stress-strain relationships and shear strength. It appears that there may be three useful marker conditions corresponding to maximum distributed incremental destabilisation, normal average conditions and maximum distributed constraint. There are no corresponding variations in dilation for the drained tests; dilation rate appears fixed by the average geometry of the particle packing. The results may be something of a challenge to modellers and also have considerable bearing on predictions in civil and chemical engineering (geotechnics and bulk handling).

APPARATUSES

The BT and DSC have been fully described elsewhere (refs 4,5,6); both shear soil in plane strain or a close approximation to plane strain and in the case of the DSC the use of the apparatus was restricted to constant principal stress directions with no application of shear stress to sample boundaries. An account of the BT describing the different modes of operation is given to allow the tests to be understood and for the same reason the DSC is briefly outlined to end this section.

The central plane of strain for the BT is shown in Figure 1a. Contact sensors detect any change in position of the centres of the stretched rubber membranes through which the major and minor principal stresses are applied to the sample boundaries. The sample itself is cubical (100mm) and contained in a thin rubber membrane which can be sealed and the air removed from the voids to stiffen the sample whilst placing it in the apparatus: subsequently the voids are restored to atmospheric pressure for testing. One of the plane strain sample surfaces is subjected to a controlled intermediate principal stress applied through another membrane whilst the other can be viewed through a glass plate. Both these plane strain boundaries can be lubricated in the conventional way with three layers of special lubricating grease separated by two sheets of thin rubber, or alternatively a more sophisticated boundary condition can be applied. This latter condition will be described a little later.

The main novel feature of the BT is the changing dimensions of the membranes through which the stresses are applied. These changes are made incrementally to return the contact sensors to the pre-increment state. The process is illustrated in Figure 1 from (a) to (c). The effect is to adjust the boundary strains to follow those of the sample central axes and to apply the correct stresses to the whole of these boundaries. This is achieved without applying any net shear traction to the sample boundaries because the rubber membranes are stretched or released to their new dimensions by an external mechanism. Moreover as the surfaces of the sample and pressure membranes are degreased the normal pressure across them locks the two together so that the same external mechanism also changes the dimensions of the sample membrane. When desired a similar process of unrestrained dimension following can be applied to the plane strain ends. The stretched rubber plane strain end sheets which achieve this are shown in Figure 2; the telescopic frame which stretches these partially reinforced end sheets is attached by the corner pins to the corner guide blocks which are shown in Figure 1 and moved by the external mechanism as for the other elements. Notice that the action of the sheets is such that the unreinforced area always changes to match that of the straining sample; this is achieved by the combination of one way and two way reinforced sections shown in Figure 2.

The following are three different boundary conditions under which plane strain tests have been carried out in the BT, no changes being made in σ_2 :

BT1, with all boundary surfaces changed incrementally by external means to match the strain on the central axes of the sample;

BT2, with the boundary surfaces applying σ_1 and σ_3 changed incrementally to match the strain on the central axes of the sample and the plane strain ends conventionally lubricated (devices shown in Figure 2 not used),

BT3, with the pressure membranes applying σ_1 and σ_3 changed incrementally to match the strain on the central axes of the sample but the normal grease and thin rubber sheet lubrication separating the sample membrane from the outer surfaces on all six sample faces,

The thickness of the rubber membranes are as follows:

Pressure membranes 0.75mm

Plane strain and stretched membranes 0.50mm

normal sample membrane and lubrication membrane 0.25mm

The Young's modulus of these membranes is approximately 1.4×10^3 kPa. It will become clear that this stiffness and these thicknesses are significant because of the potential of the rubber sheets to provide distributed restraintor disturbance over a boundary surface. The significance is of course related to the stress level of the tests, decreasing as this level increases; in all the tests reported on here the minor principal stress was held constant at 14kPa, a very low level.

It was necessary to use the DSC in the role of a simple plane strain apparatus to complete this series of tests. Figure 3 shows a typical cross-section (note the absence of shear sheets). Unlike the BT there is no incremental adjustment of the pressure applying boundary surfaces here and thickened rubber sheet can be incorporated in the plane strain ends of the sample membrane

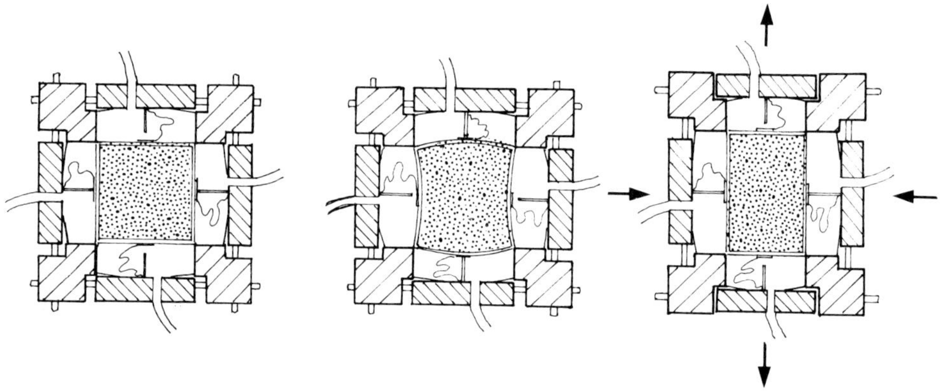

(a) Sample before shearing

(b) Sample deformation after a small stress increment

(c) Apparatus & sample after boundary adjustment

Fig 1 Principle of Biaxial Tester (exaggerated deformation)

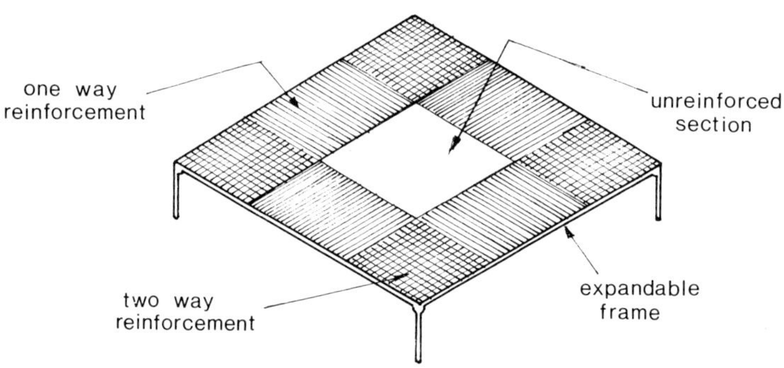

Fig 2 Specially designed plane strain membrane

without inevitable constraints related to the BT mechanism. Conventional boundary lubrication was employed on all six surfaces, but as well as carrying out tests with normal sample membranes others used sample membranes thickened on the plane strain ends to 1.6,2.6 or 3.6mm.

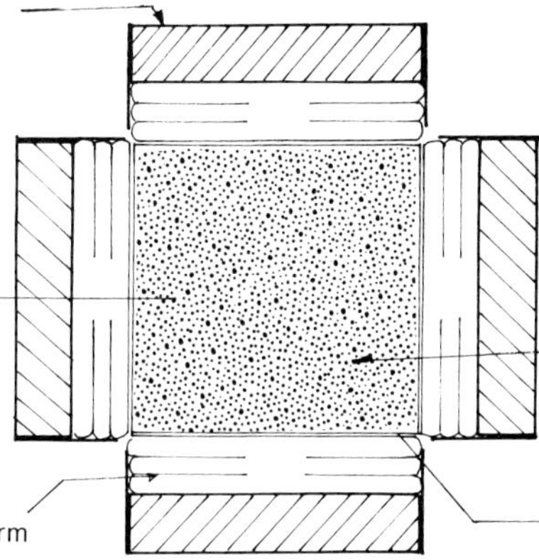

Fig 3 The simplified DSC plane strain apparatus

MATERIALS AND TEST CONDITIONS

Standard Leighton Buzzard sand was tested dry and in plane strain with full volume change permitted in all the experiments reported here. The sand is a sub-rounded quartz, closely graded (650-800μm) with a specific gravity of 2.66. All samples were formed by pluviation normal to the subsequent plane of strain to a voids ratio of 0.52. +/- 0.01 (90% relative density).

TESTING PROGRAMME

The programme explored the effects of the changes in boundary conditions one through three which are listed above for the BT and the further variation provided by the DSC which is also indicated above.

TEST DATA

Figure 4 presents the main stress-strain data for the central planes of strain of each of the representation tests. Data from two tests under each boundary condition are included to show the high degree of repeatability which is readily achieved with these apparatuses. There are three clearly differentiated groups of results which have been formed as a result of the differing boundary conditions. The shear strength of the central group agrees well with the accepted value of ϕ' measured in several other apparatuses at comparable stress levels (refs. 7,8,9.10). The other two groups show ϕ' either raised or lowered by 5° with corresponding differences in pre-failure stiffnesses. Figures 5 and 6 compare strains measured in the central plane of strain with those measured in planes of strain near the plane strain ends. These comparisons demonstrate rather good uniformity of strains, certainly as good as earlier comparisons of end and central plane strains for tests giving the accepted value of ϕ' (ref.11). It should also be noted from Figure 4 that none of the tests showed significant differences in dilation rate. All the samples failed by the formation of rupture layers.

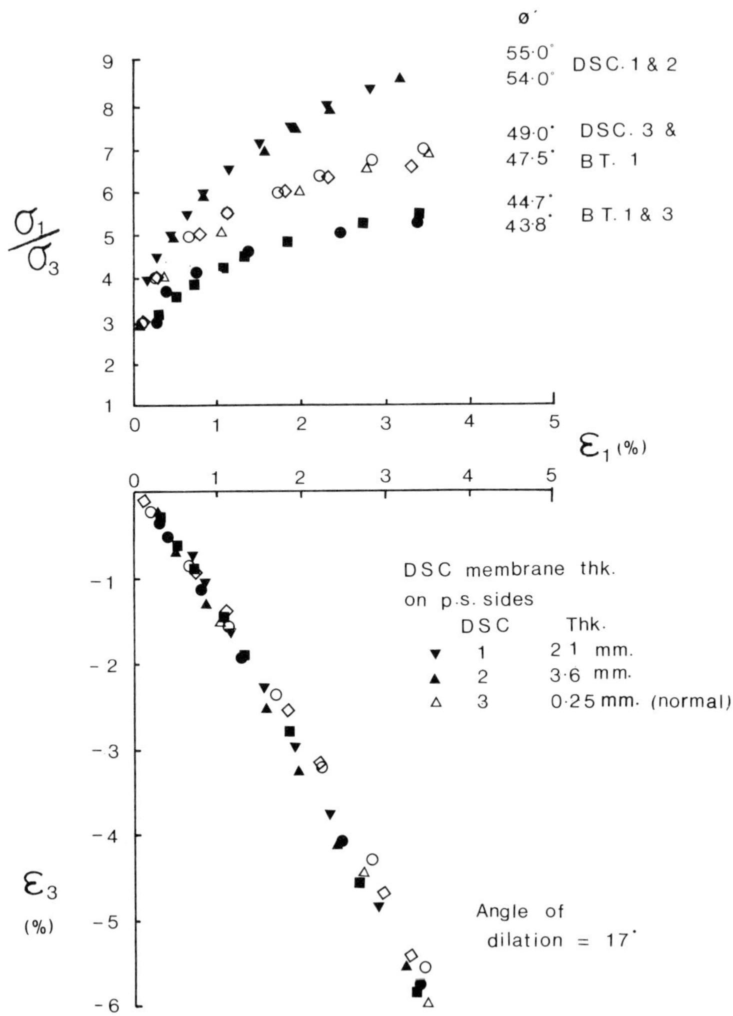

Fig 4 Effects of different boundary constraints on the stress–strain behaviour of dense Leighton Buzzard sand tested in different devices.

IMPLICATIONS OF THE DATA

As these tests were all made on initially homogeneous samples in flexible boundary apparatuses which applied uniform boundary stresses these strain distributions establish that all were true element tests in spite of the differences in the boundary conditions. It can be surmised that the differences in boundary conditions produced rather uniform yet different contact force distribution within the sample for the same geometric packing (equal dilation rates).

The results bring the present basis of membrane corrections into question.

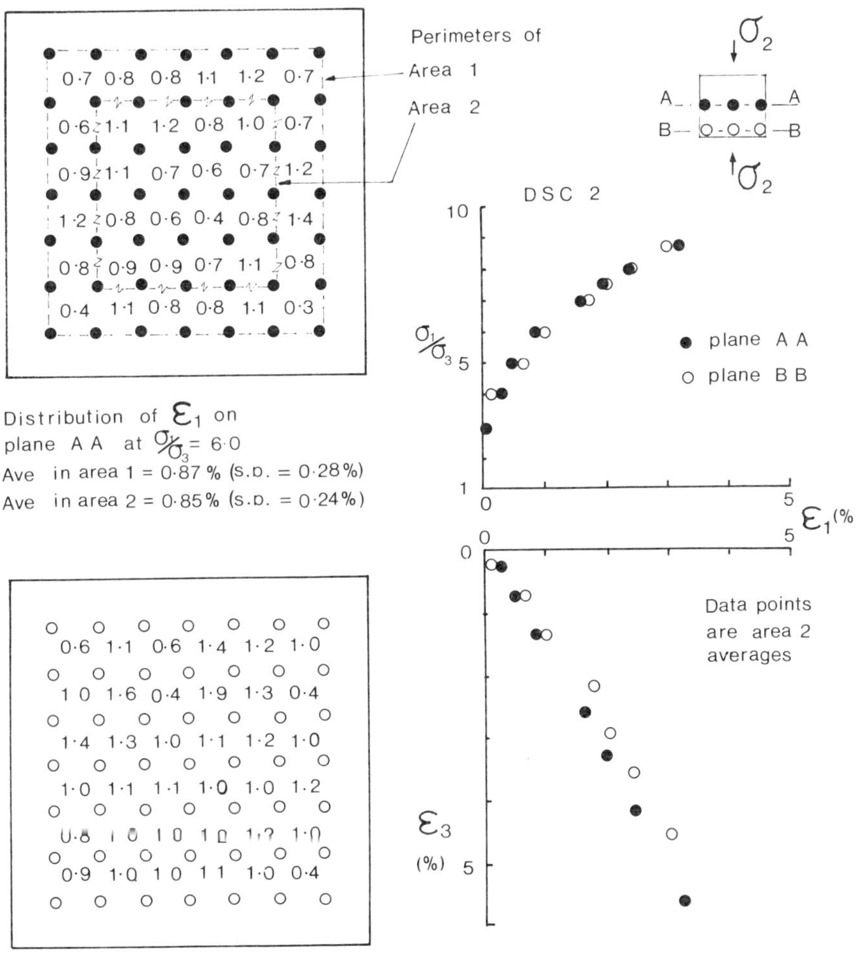

Distribution of ε_1 on plane A A at $\sigma_1/\sigma_3 = 6{\cdot}0$
Ave in area 1 = 0·87 % (S.D. = 0·28 %)
Ave in area 2 = 0·85 % (S.D. = 0·24 %)

Distribution of ε_1 on plane B B at $\sigma_1/\sigma_3 = 6{\cdot}0$
Ave in area 1 = 1·03 % (S.D = 0·37 %)
Ave in area 2 = 1·17 % (S.D = 0·33 %)

Fig 5 Comparison of strain distribution and resulting stress-strain data on plane A A and plane B B

The uniform stretching of the sample membrane could not produce more than 21% of the difference between the central and low group of results. In fact there is no difference at all between BT1 and BT3 results although in one case the membranes are strained externally and in the second the extra load has to be sustained by the sample. This suggests that the conceptual basis of current membrane corrections is wrong. It is probable that membranes stretch extremely non-uniformly applying in some cases surprisingly large lateral loads to individual grains. Clearly this can have profound effects on contact force

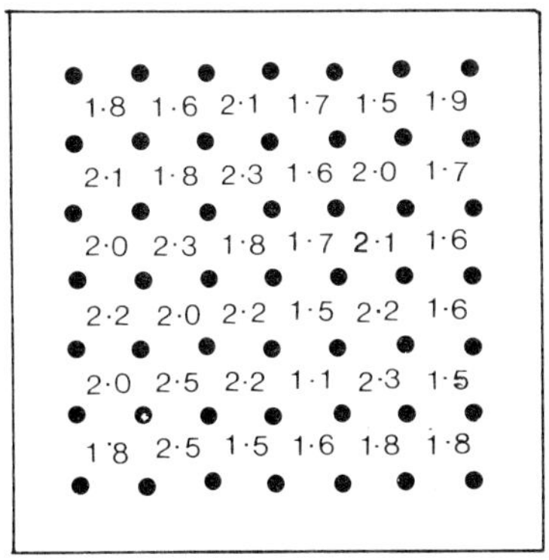

Distribution of ε_1 on plane A A at $\sigma_1/\sigma_3 = 4{\cdot}9$
Ave. in area 1 = 1·90 % (S.D. = 0·32%)
Ave. in area 2 = 1·98 % (S.D. = 0·36%)

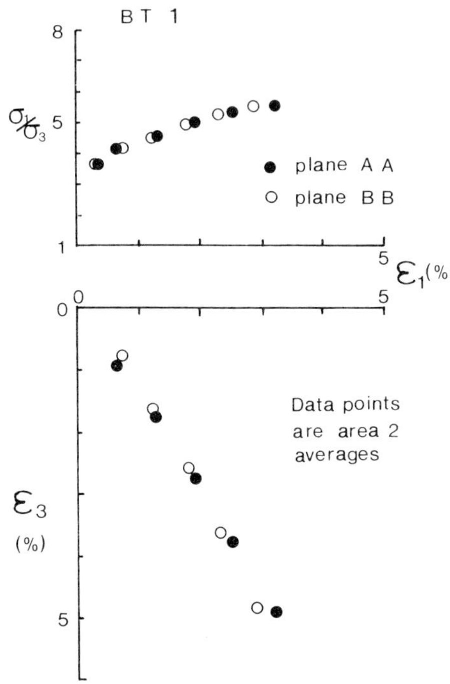

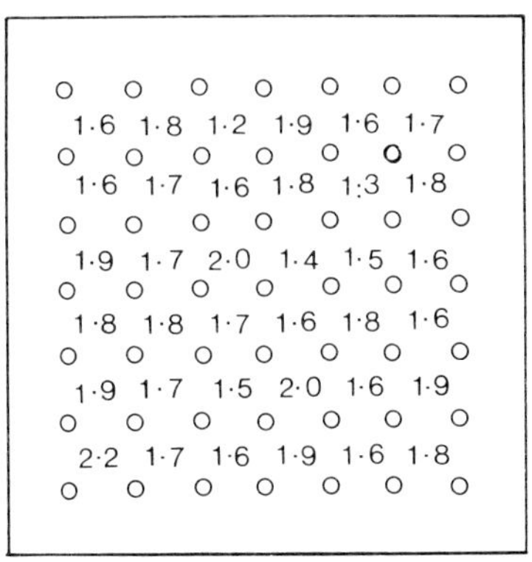

Distribution of ε_1 on plane B B at $\sigma_1/\sigma_3 = 4{\cdot}9$
Ave. in area 1 = 1·76 % (S.D. = 0·27%)
Ave. in area 2 = 1·72% (S.D. = 0·19%)

Fig 6 Comparison of strain distribution and resulting stress strain data on plane A A and plane B B distributions within the sample.

The cause of the low results for groups BT1 and BT3 appears to be the incremental nature of the BT's operation; as the sample reaches equilibrium under each stress increment a destabilising boundary movement is applied to get the boundary strains to match the central axes strains. A transient stress increment with different principal axes is applied in a distributed way through the boundaries during this stage; increased strain and apparently a less efficient contact force distribution result.

The replacement of the plane strain end strain matching boundaries (Figure 2) with conventional lubricated ends stabilises the sample sufficiently to attain the accepted normal strength and stress-strain behaviour (BT2). This demonstrates the ability of a granular material to transmit a distributed lateral restraint normal to the plane of strain. The incremental destabilisation still occurs but its effects are mitigated.

A strong response is obtained in the DSC in which the incremental destabilisation process is absent. The DSC achieves the accepted response when normal thin membrane plane strain ends are used (Fig.4 & ref.9), but the introduction of thicker plane strain ends of 1.6mm or greater produces the stiff strong response shown in Figure 4. Again the influence of distributed restraint laterally transmitted across the plane of strain is demonstrated. It brings the realisation that the long cherished belief of experimentalists that perfect lubrication is required on plane strain ends for the ideal element test is incorrect. We have by accident in our normal lubrication technique evolved a good approximation (at least at these low stress levels) to the distributed lateral restraint which the material itself generates internally as a result of the inherent inhomogenieties in deformation and contact force distribution. Correct treatment of the plane strain ends can be checked: samples with widely different dimensions in the intermediate principal stress direction should yield identical results.

The formation of rupture layers in samples with virtually identical packing geometry at such different stress ratios suggests that the controlling condition is reaching maximum dilation in continuum strain.

The different boundary conditions by which the extreme stiffness and strength results were obtained suggests that these may be true limits.

ACKNOWLEDGEMENTS

The continuing cooperation of the Christian Michelsen Institute in Bergen is gratefully acknowledge. The work is part of a programme under the general direction of the SERC Particulate Technology Group. We are grateful for the financial support of the SERC for part of the work reported here. Our thanks also go to Mr John Ford for making the equipment so well.

REFERENCES

1 Drescher, A. and G. de Josselin de Jong, J.Mech.Phys.Solids,20,1972,337-351.

2 Oda, M. and J.Konishi, Soils & Fdns,14,1974,4:25-38.

3 Cundall,P.A. and O.D.L.Strack,Geotechnique,29,1979,1:47-65.

4 Arthur,J.R.F.,Dunstan,T. and G.G.Enstad,Int.J.Bulk Strg in Silos,12,1985,7-10

5 Arthur,J.R.F.,Dunstan,T. and K.S.Chua,Geotechnique,27,1977,1:13-30

6 Arthur,J.R.F.,Bekenstein,S.,Germaine,J.T. and C.C.Ladd,ASTM,STP740,1981,516-540

7 Stroud,M.A., The behaviour of sand at low stress levels in the SSA, Ph.D.Th,1971, Cambridge

8 Arthur,J.R.F. and Assadi,A.Proc.9th I.Conf.SMFE,Tokyo,1,1977,19-22

9 Wong,R.K.S. and J.R.F.Arthur,Geotechnique,35,1985,4:471-481

10 Jewell,R.A. and C.P.Wroth,Geotechnique,37,1987,1:53-68

11 Wong,R.K.S. and J.R.F.Arthur,ASTM Geotech.Test J.,8,1985,3:101-110

Micromechanics of Granular Materials, edited by M. Satake and J.T. Jenkins
Elsevier Science Publishers B.V., Amsterdam, 1988 — Printed in The Netherlands

VOLUME CHANGE IN SMALL STRAIN AXISYMMETRIC DEFORMATIONS OF A GRANULAR MATERIAL

James T. Jenkins

Department of Theoretical and Applied Mechanics, Cornell University, Ithaca, New York 14853

SUMMARY

We consider a granular material consisiting of identical spheres, adopt a familiar model for the contact force, and employ a simple statistical characterization of the contact distribution. We suppose that the average strain is small and that the deformation at a contact is given in terms of it. Using the resulting stress–strain relation, we calculate the volume change when the material is subjected to axisymmetric deformations (triaxial compression or extension) at fixed pressure. The predictions agree with the results of experiments on glass spheres only when an initial depositional anisotropy in the contact distribution is included.

INTRODUCTION

This research is part of a collaboration involving theory, numerical simulation (refs. 1, 2), and experiments (refs. 3, 4, 5) devoted to the rate–independent behavior of idealized granular materials. The goal of the research is to understand what determines the state of the granular material and to develop macroscopic means for measuring this state. Knowledge of the state of the material permits the prediction of its incremental behavior.

Here we restrict attention to small strains and consider the most elementary stress paths, those corresponding to triaxial compression or triaxial extension carrried out at fixed pressure. We focus on the volume strains that accompany the shear strains as the deviatoric stress increases from zero. These are due to the development of anisotropy in the orientational distribution of the normal contact force and to the presence of any anisotropy in the initial orientational distribution of contacts. We compare the calculated volume strains to those observed in the corresponding experiments (ref. 3).

THEORY

We consider a granular material consisting of identical elastic spheres of diameter σ that interact through frictional Hertz–Mindlin contacts (ref. 6). There are assumed to be n such spheres per unit volume.

The orientational distribution of contacts $A(\underset{\sim}{\alpha})$ is defined so that $A(\underset{\sim}{\alpha})d\alpha$ is the probable number of contacting pairs with their line of centers in the element of solid angle $d\alpha$ centered on the unit vector $\underset{\sim}{\alpha}$.

The contact force for a contacting pair with orientation $\underset{\sim}{\alpha}$ has components normal and tangential to the plane of contact:

$$F_i = P\alpha_i - T_i \ , \tag{1}$$

with $\underset{\sim}{T} \cdot \underset{\sim}{\alpha} = 0$.

By considering the forces transmitted over an element of area by contacting pairs (ref. 7, 8), an expression for the components of the stress tensor $\underline{\underline{t}}$ may be obtained in terms of the contact distribution and the contact force.

$$t_{ik} = -\frac{1}{2} \sigma n \int A F_i \alpha_k \, d\alpha \ , \tag{2}$$

where the integration is over all solid angle. For these noncentral forces, the antisymmetric part of the stress is balanced by the divergence of a couple stress. However, from (2), it is easy to see that, at any stage of a deformation, the mean pressure, $3p \equiv -\mathrm{tr}\ \underline{\underline{t}}$, depends only upon the distribution of the normal component of the contact force.

The underlying kinematic assumption is that the displacement $\underset{\sim}{u}$ of a point of contact relative to the center of a sphere can be calculated from the average strain $\underline{\underline{e}}$ (refs. 9, 10):

$$u_i = (\sigma/2)\, e_{ik}\alpha_k \ . \tag{3}$$

This neglects slip at contacts and the relative rotation of contacting spheres; however it is expected to be valid over an initial, perhaps small, range of the axisymmetric experiments we wish to model.

In an axisymmetric experiment we take the principal strains positive in compression and call the axial strain E and the radial strain e. In terms of these, the volume strain Δ (positive for contraction) is

$$\Delta = E + 2e \tag{4}$$

and the shear strain γ (half the maximum shear strain) is

$$\gamma = (E - e)/2 \ . \tag{5}$$

Then the normal displacement δ of a contact point toward a center is, from (3),

$$\delta = (\sigma/6)(\Delta - 2\gamma + 6\gamma \cos^2\theta) \ , \tag{6}$$

where θ is the polar angle from the axis of symmetry. The corresponding tangential displacement $\underset{\sim}{s}$ is

$$\underset{\sim}{s} = \sigma\gamma \sin\theta \cos\theta \, \underset{\sim}{e}_\theta \,, \tag{7}$$

where $\underset{\sim}{e}_\theta$ is the unit vector in the direction of increasing θ.

In a uniaxial deformation, the solution for the normal component of the contact force is (ref. 6)

$$P = M(6\delta/\sigma)^{3/2} \,, \tag{8}$$

where M is given in terms of the shear modulus μ and Poisson's ratio ν of the material of the sphere by

$$M = \frac{2}{9\sqrt{3}} \frac{\mu\sigma^2}{1-\nu} \,. \tag{9}$$

The corresponding form for the tangential component is less certain (refs. 11, 12). Here, for axisymmetric loading, we write $\underset{\sim}{T} = T(\theta)\underset{\sim}{e}_\theta$ and $\underset{\sim}{s} = s(\theta)\underset{\sim}{e}_\theta$ and take

$$T = fP\left\{1 - \left[1 - \frac{16\mu a}{3(2-\nu)} \frac{s}{fP}\right]^{3/2}\right\} \tag{10}$$

for $0 \le \theta \le \pi/2$, where f is the friction coefficient, a is the radius of the circle of contact:

$$a = \left[\frac{3(1-\nu)\sigma}{16\ \mu} P\right]^{1/3} \,, \tag{11}$$

and s and P are given by (7) and (8), respectively. Equation (10) is the integral of an incremental form determined by Mindlin and Deresiewicz (ref. 11) that, in an axisymmmetric loading, applies over almost the entire range of θ. We note that the numerical simulations (ref. 2) indicate that the predicted behavior is not likely to be sensitive to the detailed form of the tangential force.

If an initial unstrained state is assumed to be isotropic, then $A = k/4\pi$, where k is the average number of contacts per sphere (the coordination number). For isotropic compression ($\gamma = 0$) from this state, (2), used with (8) and the appropriate simplification of (6), integrates to provide the relation between the pressure p_0 and the volume strain Δ_0 in the succession of isotropic states:

$$p_0 = \frac{kvM}{\pi\sigma^2} \Delta_0^{3/2} \,, \tag{12}$$

where $v \equiv \pi\sigma^3 n/6$ is the solid volume fraction.

If, next, the pressure is held fixed at the value given in (12) and the material is sheared, (2) becomes

$$p_0 = \frac{kvM}{2\pi\sigma^2}\int_0^{\pi} (b+c\cos^2\theta)^{3/2}\sin\theta\, d\theta \ , \tag{13}$$

where $b \equiv \Delta_0 + \epsilon_v - 2\gamma$, with ϵ_v the additional volume strain due to the shearing, and $c \equiv 6\gamma$.

It is a straightforward matter to carry out the integrations for axisymmetric compression ($c > 0$) and axisymmetric extension ($c < 0$). The results are, for $c>0$,

$$8\Delta_0^{3/2} = (5b + 2c)\sqrt{b+c} + \frac{3b^2}{\sqrt{c}} \ln\left[\frac{\sqrt{c} + \sqrt{b+c}}{\sqrt{|b|}}\right] \ ; \tag{14}$$

while, for $c<0$,

$$8\Delta_0^{3/2} = (5b + 2c)\sqrt{b+c} + \frac{3b^2}{\sqrt{-c}} \sin^{-1}\left[\sqrt{\frac{-c}{b}}\right] \ , \tag{15}$$

when $b \geq -c$, and

$$8\Delta_0^{3/2} = \frac{3\pi}{2}\frac{b^2}{\sqrt{-c}} \ , \tag{16}$$

when $b \leq -c$.

When expressed in terms of ϵ_v and γ, these provide implicit relations for the additional volume strain in axisymmetric compression and extension carried out at constant pressure. In Fig. 1 we plot this volume strain versus the maximum shear strain with both strains normalized by the initial volume strain. Values of the normalized shear strain at which the normal force at some contacts first drops to zero are $2\gamma/\Delta_0 = 0.85$ in compression and $2\gamma/\Delta_0 = -0.47$ in extension. Beyond these values, the kinematic assumptions should be regarded with some suspicion.

Note that the curve is nearly symmetric, indicating that, although the stress paths differ significantly, the volume change at corresponding values of the shear strain is about the same. The prediction is that the material will expand when subjected to either axisymmetric compression or extension. However, in the experiments (ref. 3), the material expands when compressed and contracts when extended; so the model is, at best, incomplete.

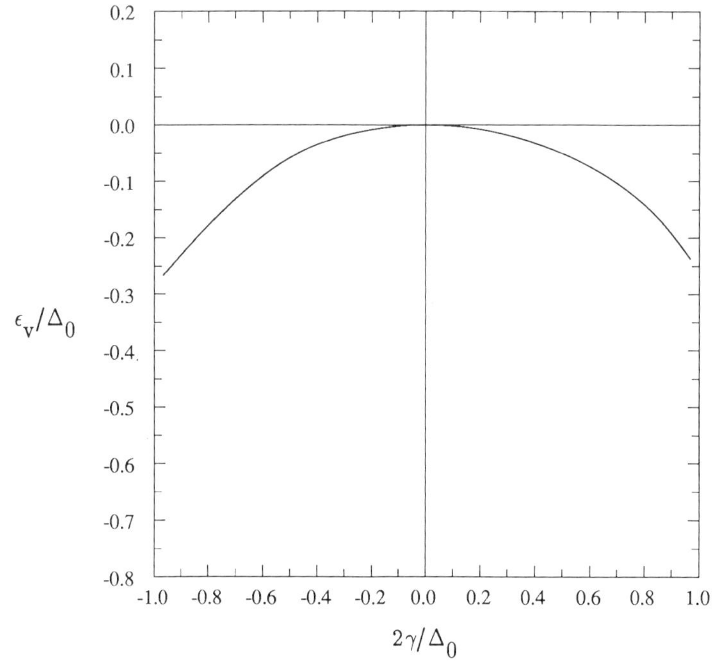

Fig. 1. Normalized volume strain versus normalized shear strain for an isotropic distribution of contacts.

A sample used in an experiment was prepared by raining glass spheres through air under gravity and then compacting mechanically to a volume fraction of 0.63. Clearly there was the possibility that, in this process, contacts were preferentially oriented toward the vertical. Consequently, we next suppose that the contact distribution is not isotropic but is given in terms of a symmetric traceless tensor $\underline{\underline{A}}$ (ref. 13):

$$A(\underset{\sim}{\alpha}) = \frac{k}{4\pi}(1 + A_{ik}\alpha_i\alpha_k) \ , \tag{17}$$

where $\underline{\underline{A}}$ is based on the unit vector $\underset{\sim}{h}$ along the vertical axis of symmetry,

$$A_{ik} = -\epsilon(\delta_{ik} - 3h_ih_k) \ , \tag{18}$$

and ϵ is the magnitude of the depositional anisotropy. Then

$$A(\underset{\sim}{\alpha}) = \frac{k}{4\pi}[(1-\epsilon) + 3\epsilon\cos^2\theta] \ . \tag{19}$$

With depositional anisotropy, the application of an isotropic pressure at zero deviatoric stress will result in shear strains in addition to volume strains. Using the forms given above for the contact force and the anisotropic contact distribtution in the expression for the stress (2), both strains could, in principle, be calculated. However, the experimental data indicates (ref. 14) that the shear strains are only several percent of the volume strains. Consequently, we can suppose that, at the initiation of shearing, the shear strain is zero and the volume strain is related to the pressure through (12).

In this event, using (19) in the contraction of (2) and carrying out the resulting integration in the expression corresponding to (13) gives, for $c > 0$,

$$8\Delta_0^{3/2} = 4\epsilon\frac{(b+c)^{5/2}}{c} + \left[1 - \epsilon\left(1 + \frac{b}{2c}\right)\right]\left\{(5b + 2c)\sqrt{b + c} + \frac{3b^2}{\sqrt{c}}\ln\left[\frac{\sqrt{c} + \sqrt{b + c}}{\sqrt{|b|}}\right]\right\}; \tag{20}$$

while, for $c > 0$,

$$8\Delta_0^{3/2} = 4\epsilon\frac{(b+c)^{5/2}}{c} + \left[1 - \epsilon\left(1 + \frac{b}{2c}\right)\right]\left\{(5b + 2c)\sqrt{b + c} + \frac{3b^2}{\sqrt{-c}}\sin^{-1}\left[\sqrt{\frac{-c}{b}}\right]\right\}; \tag{21}$$

when $b \geq -c$, and

$$8\Delta_0^{3/2} = \frac{3\pi}{2}\left[1 - \epsilon\left(1 + \frac{b}{2c}\right)\right]\frac{b^2}{\sqrt{-c}} \tag{22}$$

when $b \leq -c$.

In Fig. 2 we plot the normalized volume strain versus the normalized shear strain for $\epsilon = 0.20$ and 0.50. Near the origin, these curves exhibit the same qualitative behavior as that observed in the experiments. The experimental points shown in Fig. 2 are from separate compression and extension tests initiated at $\Delta_0 = 0.27\%$ and 0.39%, respectively (refs. 3, 14). The figure suggests that the compression test had an initial anisotropy with a strength near 0.20 while the strength of the initial anisotropy in the extension test was about 0.50.

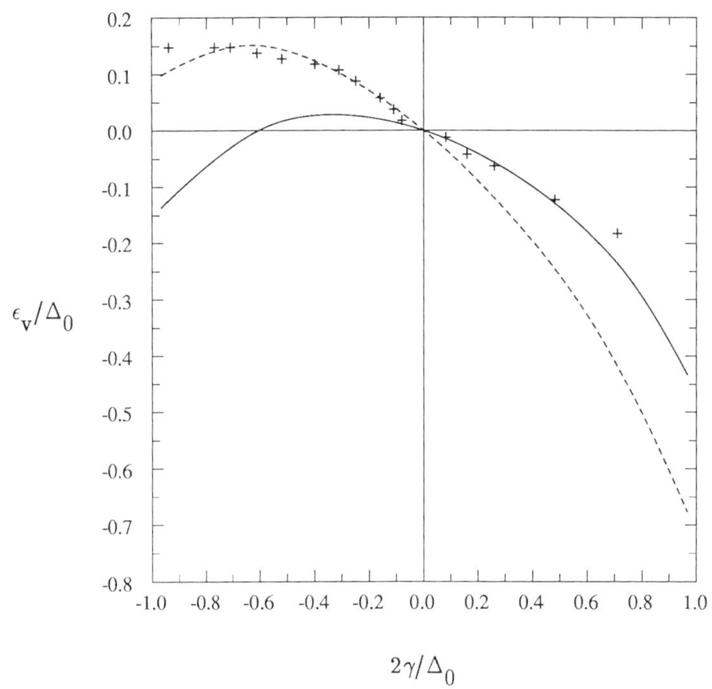

Fig. 2. Normalized volume strain versus normalized shear strain for two anisotropic contact distribtutions: $\epsilon = 0.20$(full line) and $\epsilon = 0.50$(dashed line). The points are experimental data.

CONCLUSION

It is encouraging that such a simple, straight–forward calculation of the volume change agrees so well with the experimental results. The calculation should make clear that, when small strains are developed in response to loading at a fixed pressure, it is the anisotropy of the material that is responsible for its volume change. In particular, the anisotropy in the orientational distribution of contacts strongly influences both the direction and magnitude of the volume change. This may have important implications for the study of liquefaction.

The apparent success in predicting the volume change motivates the more complicated calculations of the changes in shear stress and axial shear modulus with shear strain during small strain axisymmetric loading at fixed pressure. The former involves the tangential component of the contact force; the latter requires the normal and tangential contact stiffnesses and consideration of stress increments associated with rotations.

The simple, essentially elastic, model upon which the calculations for loading are based can also serve as the starting point for consideration of the inelastic behavior. For example, in determining the volume change upon unloading, contacts that have totally relaxed during loading can be deleated. The idea is that these contacts are prevented, by the slip of their neighbors, from regaining force upon unloading.

In the small strain regime, such simple modeling of the inelastic behavior is possible because the particle rearrangements are small. In any event, a successful model contains the characterization of state that we seek. Here, in the case of small strain volume change at fixed pressure, the state of the material is determined by its initial volume strain and by appropriate gross measures of the anisotropy in the orientiational distribution of contact area.

ACKNOWLEDGEMENT

This work was supported by a grant from the Geotechnical Engineering Division of the U.S. National Science Foundation. The author is grateful to P.A. Cundall and I.Ishibashi for many valuable discussions.

REFERENCES

1 P.A. Cundall, Distinct element model of rock and soil structure, in: E.T. Brown (Ed.), Analytical and Computational Methods in Engineering Rock Mechanics, Allen–Unwin, London, 1987, pp. 129–162.
2 P.A. Cundall Computer simulations of dense sphere assemblies, in this volume.
3 Y.C. Chen, I. Ishibashi, and J.T. Jenkins. Dynamic shear modulus and fabric. Part I. Depositional and induced fabric, Geotechnique 38 (1988) 25–32.
4 I.Ishibashi, Y.C. Chen, and J.T. Jenkins, Dynamic shear modulus and fabric. Part II. Stress reversal, Geotechnique 38 (1988) 33–37.
5 I.Ishibashi and Y.C. Chen, Dynamic shear moduli and their relationship to fabric of granular materials, in this volume.
6 R.D. Mindlin, Compliance of elastic bodies in contact, J. Appl. Mech., 16 (1949) 259–268.
7 A.E.H. Love, A Treatise on the Mathematical Theory of Elasticity (4th Ed.), University Press, Cambridge, 1927, pp. 616–619.
8 J.H. Irving and J.G. Kirkwood, The statistical mechanical theory of transport processes, IV: the equations of hydrodynamics, J. Chem. Phys., 18 (1950) 817–829.
9 P.J. Digby, The effective elastic moduli of porous granular rocks, J. Appl. Mech., 48 (1981) 803–808.
10 K. Walton, The effective elastic moduli of a random packing of spheres, J. Mech. Phys. Solids, 35 (1987) 213–226.
11 R.D. Mindlin and H. Deresiewicz, Elastic spheres in contact under varying oblique forces. J. Appl. Mech., 20 (1953) 327–344.
12 K. Walton, The oblique compression of two elastic spheres, J. Mech. Phys. Solids, 26 (1976), 139–150.
13 K. Kanatani, Distribution of directional data and fabric tensors, Int. J. Engng. Sci., 22 (1984) 149–164.
14 I.Ishibashi, personal communication.

Micromechanics of Granular Materials, edited by M. Satake and J.T. Jenkins
Elsevier Science Publishers B.V., Amsterdam, 1988 — Printed in The Netherlands

SOME BASIC THEORETICAL AND EXPERIMENTAL RESULTS ON MICROMECHANICS OF GRANULAR FLOW

M. M. Mehrabadi,[1] S. Nemat-Nasser,[2] H. M. Shodja,[3] G. Subhash[2]

[1]Department of Mechanical Engineering, Tulane University, New Orleans, Louisiana 70118

[2]Department of Applied Mechanics and Engineering Sciences, University of California, San Diego, La Jolla, California 92093

[3]Department of Civil Engineering, The Technological Institute, Northwestern University, Evanston, Illinois 60201

SUMMARY

In order to establish guidelines for modeling the macroscopic behavior of granular materials, an experimental study of the evolution of the microstructure of an assembly of granular materials under a uniform confining pressure and subjected to a pure shear was conducted. The granular material used in the study consisted of photoelastically sensitive rod-shaped particles of oval cross-sections. It was found that (i) the distribution of branches and contact normals are almost identical, (ii) the second rank fabric tensor does not adequately describe the microstructure of highly anisotropic samples, (iii) the density of contacts whose normals lie along the major and minor principal stress axes, varies sharply initially and then approaches a constant value in the course of deformation, and (iv) the density of contacts with planes parallel to the maximum shear stress plane remains practically constant throughout the deformation.

INTRODUCTION

A microscopic study of the evolution of the microstructure of an assembly of granular materials under a uniform confining pressure and, in addition, subjected to a pure shear is reported here. The granular material used in the study consisted of photoelastically sensitive rod-shaped particles of oval cross-sections.

The purpose of the experimental study, in general terms, was to establish guidelines for modeling the macroscopic behavior of granular materials. Specifically, the objective of this work was to investigate the correlation among certain microscopic fabric measures such as the distribution of contact normals, unit branches, etc., and to study the evolution of these quantities in the course of deformation.

Experiments on assemblies of photoelastic particles were pioneered by Dantu (ref. 1), Weber (ref. 2) and Wakabayashi (ref. 3), and later followed by several other investigators (refs. 4-15). Except for the works reported by Drescher (ref. 10),Konishi, et al. (ref. 13), Oda, et al. (ref. 11), Nemat-Nasser et al. (ref. 14), and Allersma (ref. 12), experiments reported in the literature have been performed on particles of circular cross-sections. The experiments of Drescher (ref. 10) and Allersma (ref. 12) were performed on crushed glass, and were mainly concerned with the overall behavior of the material and not the evolution of the fabric. The experiments reported by Konishi, et al. (ref. 13), Oda et al. (ref. 11) and Nemat-Nasser et al. (ref. 14), however, are on assemblies of rod-shaped particles with oval cross-sections subjected to biaxial compression. The results of these experiments show, among other things, that the fabric of a granular material depends at least on the distribution of contact normals, the shape of particles, and the shape of the associated voids. The microscopic deformation mechanism in simple shear tests is observed to be mostly in the form of frictional sliding, although some particle rolling may still take place. This is generally believed to be representative of the actual behavior of granular materials under typical loading conditions in the field.

MATERIAL AND APPARATUS

The granular material tested was composed of cylindrical rods of circular and elliptical cross-sections. Three different sizes of cross-sections with diameters ranging from 6mm to 15mm were used. The ratios of major to minor axis of the oval cross-sections of particles were 1.1 and 1.4. The rod-shaped particles were cast, using urethane prepolymer. These particles have been used by Konishi et al. (ref. 13) to study the behavior under biaxial loading conditions. The samples considered in the present experiments, were composed of mixed size particles and were relatively dense with an initial void ratio of 0.18.

A detailed description of the simple shear testing device used will appear elsewhere (ref. 16).

TESTING PROCEDURE AND ANALYSIS OF DATA

The loading frame is placed in the field of a circular polariscope consisting of a monochromatic light source, polarizer, two quarterwave plates, and an analyzer. After the assembly is completed, under a small constant isotropic confining load, the shear force is incrementally applied. At each stage of the loading, photographs of the isochromatic fringe pattern in the stressed assembly are taken for later analysis.

The photographs of the isochromatic fringe pattern in the stressed assembly were analyzed using a digital image analysis system consisting of a PC AT, a frame grabber (PC Vision Plus) from Imaging Technology Inc., a Vidicon camera (C1000) from Hamamatsu TV Co., and a RGB monitor. With the aid of programs Image Lab and Image Tool from Imaging Technology, a powerful program for obtaining histograms of several microscopic quantities was developed. After digitizing the photographs of the isochromatic fringe pattern obtained at each stage of loading, the program was employed to find the corresponding distributions of (i) contact normals, (ii) branches, and other related quantities.

REPRESENTATION OF DISTRIBUTION DENSITY FUNCTION

In works on crystallography, it is customary to represent the distribution density function of an orientation by spherical harmonic functions. Another alternative is the invariant formulation by Kanatani (ref. 17) which for two dimensions, is given by

$$E(n) = \frac{1}{2\pi}(1 + J_{ij}n_in_j + J_{ijkl}n_in_jn_kn_l + ...) , \tag{1}$$

where E(n) is the distribution density function of the unit vector n, and where

$$J_{ij} = 4\left[<n_in_j> - \frac{1}{2}\delta_{ij}\right], \tag{2}$$

$$J_{ijkl} = 16\ [<n_in_jn_kn_l> - \delta_{(ij} <n_kn_{l)}> + \frac{1}{8}\ \delta_{(ij}\delta_{kl)}]. \tag{3}$$

The angular brackets denote averages taken over all orientations. The symmetric, traceless second-rank tensor J_{ij} is closely related to various fabric or anisotropy tensors (ref. 18) previously introduced in the literature. The fourth-rank tensor J_{ijkl} is completely symmetric and traceless. The components of these two tensors can be represented, in two dimensions, in terms of four parameters, A, B, C and D, defined by

$$A = <\sin\theta\cos\theta>, \quad C = <\cos^2\theta>,$$
$$B = <\sin\theta\cos^3\theta>, \quad D = <\cos^4\theta>, \tag{4}$$

where θ is the orientation of n measured from the vertical. Note that the parameters A and C are related to the degree of concentration, J, and the preferred orientation, β, introduced by Konishi (ref. 6),

$$J^2 = \frac{1}{8}\ J_{ij}J_{ij} = 4A^2 + \ (1\text{-}2C)^2, \tag{5}$$

$$\tan 2\beta = \frac{2J_{12}}{J_{11}-J_{22}} = \frac{2A}{1-2C}. \qquad (6)$$

From the observed distribution of the six microscopic quantities mentioned in the previous section, the distribution density function, E(n), and the four parameters A, B, C and D were determined at selected points during loading and unloading. The plot of the observed distribution density function, E(n), was then compared with those of the zeroth-, second- and fourth-order approximations, $E_0(n)$, $E_2(n)$ and $E_4(n)$, respectively, where

$$E_0(n) = \frac{1}{2\pi},$$

$$E_2(n) = \frac{1}{2\pi}(1 + J_{ij}n_in_j) = \frac{1}{2\pi} + \frac{1}{\pi}[2A\sin 2\theta + (2C-1)\cos 2\theta],$$

$$E_4(n) = \frac{1}{2\pi}(1 + J_{ij}n_in_j + J_{ijkl}n_in_jn_kn_l) = \frac{1}{2\pi} + \frac{1}{\pi}[2A\sin 2\theta + (2C-1)\cos 2\theta +$$

$$+ (1-8C+8D)\cos 4\theta + 4(2B-A)\sin 4\theta]. \qquad (7)$$

The evolution of the parameters A, B, C and D during the shearing was also determined.

TEST RESULTS

The plots of stress ratio versus shear strain and volumetric strain versus shear strain for one of the tests, denoted by D1007, are shown in Figs. 1a and 1b, respectively. The numbered data points indicate the points where photographs of the isochromatic fringe pattern were taken. The plots shown in these figures are similar to the ones obtained for sand. Note that the sample has dilated after unloading to zero shear strain. This is thought to have been caused by loading the sample well beyond the peak stress.

The evolution of the degree of concentration, J, and the preferred orientation, β, in the course of shearing are shown in Figs. 1b and 1c, respectively. The results are similar to those reported earlier by Konishi (ref. 6). Initially, the sample was only slightly anisotropic, as is indicated by a small value of J equal to 0.079 (Fig. 1c). Roughly speaking, a value of 1 for J corresponds to the highest level of order and a value of zero corresponds to isotropy. J increases from its initial value by about 80% at the peak stress.

Most of the change in the preferred orientation, β, occurs at the very early stage of deformation, after which β changes gradually, approaching a value close to that of the inclination of the principal axis of stress at or immediately after the peak stress.

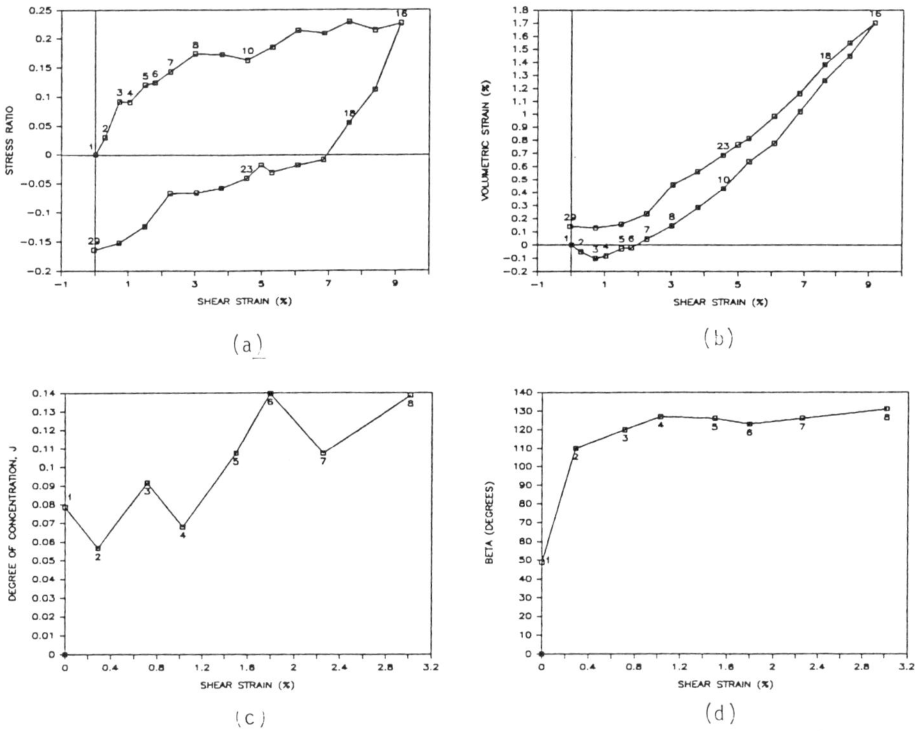

Fig. 1. Variation of (a) stress ratio, (b) volumetric strain, (c) degree of concentration, and (d) orientation of the major principal axis of the second-rank fabric tensor, J_{ij}, with shear strain.

The observed distribution of contact normals as well as its second and fourth order approximations (see Eqs. (7)) are shown in Fig. 2. Consistent with earlier observations, it is seen here also that some contacts with unit normals along the minor principal stress axis are lost while new contacts with unit normals along the major compressive principal stress axis are continually generated.

Clearly, the basic features of the actual distribution are well represented by the fourth-order approximation. In fact, it can be shown that the error in calculating E(n) by excluding the fourth- and higher-order terms can be more than ±30% along the major and minor principal stress axes. Thus, for a highly anisotropic microstructure, the second-rank tensor J_{ij} does not appear to be adequate, and, for a more reasonable representation of the microstructure, one must, at least, retain the fourth-order term involving the fabric tensor J_{ijkl}.

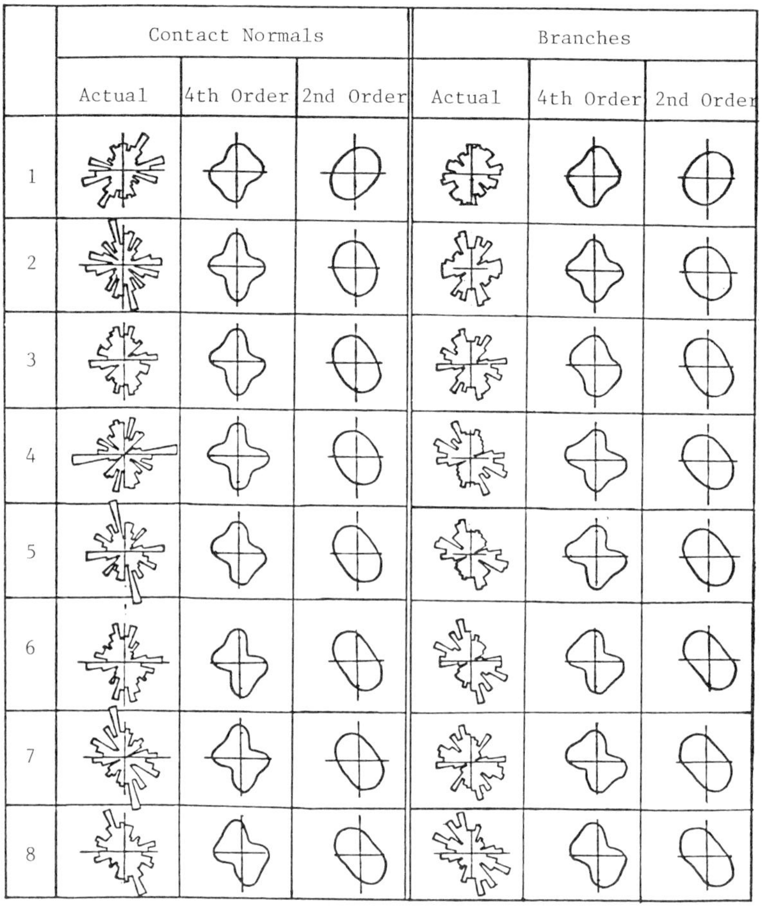

Fig. 2. Distribution density of contact normals and branches at the 8 stages of loading indicated in Fig. 1.

As seen from Fig. 2, the distribution of branches and its evolution is almost identical to that of contact normals. Hence, for granular materials composed of particles with an aspect ratio of about 1.4 (nearly the same as that of sand particles), the distribution density function E(n) of contact normals will also approximate the distribution density function of branches.

The evolution of the fabric parameters, A and B, is shown in Fig. 3a. It is observed that both of these quantities have the same order of magnitude and that they both decrease rapidly after the initial shearing, and gradually approach a constant value. The evolution of the parameters C and D is shown in Fig. 3b. Again the order of magnitude of these parameters is the same. However, these parameters seem to vary only slightly throughout the deformation. These results imply that, in the expression for E(n) (see Eqs. (7)), the sine terms (i.e., those involving the odd functions of θ) vary substantially in the course of shearing deformation, while the cosine terms (i.e., those that are even functions of θ) are practically constant throughout the deformation process.

The variation of A, B, C and D with the stress ratio is also found to be similar to the change of these parameters with the shear strain, shown in Fig. 3.

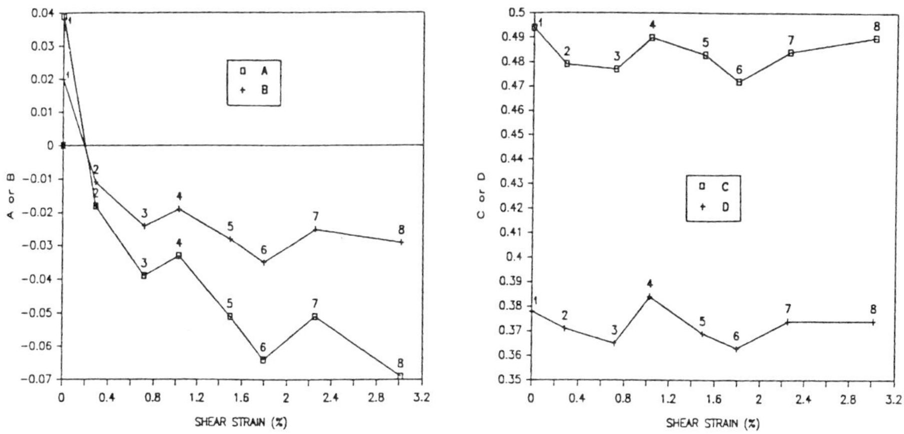

Fig. 3. Evolution of parameters A, B, C and D with shear strain.

As a corollary of the above observation, we can see that when the sine terms in the expression for E(n) are zero, i.e., when

$$\sin 2n\theta = 0 \quad (n=1,2,...)\ ,$$

or when $\theta=0^{\circ}, 90^{\circ},...,$ then

$$\cos 2n\theta = (-1)^{n}\ ,$$

and hence $E(0^{\circ})$ and $E(90^{\circ})$ are practically constant. Note that for $\theta=45^{\circ}, 135^{\circ},....,$ we have

$$\sin 2n\theta = (-1)^{n+1}\ , \quad \cos 2n\theta=0.$$

Therefore, $E(45^{\circ})$ and $E(135^{\circ})$, which are the distribution densities of contact normals along the major and minor principal stress axes, vary initially and then approach a constant value in the course of deformation; see Fig. 4.

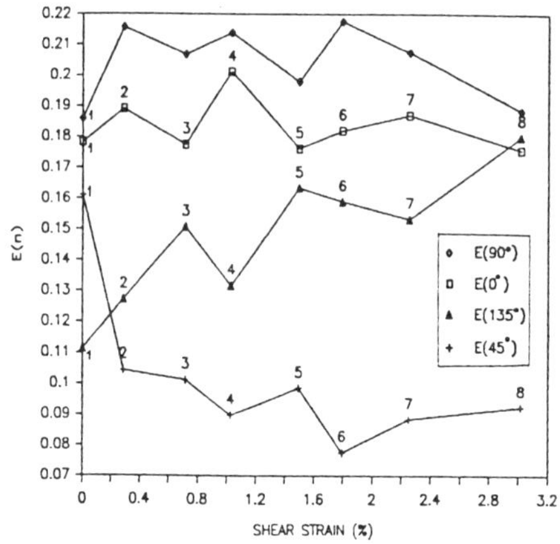

Fig. 4. Evolution of the distribution densities of contact normals along $\theta=0^{o}$, 45°, 90° and 135° orientations.

CONCLUSIONS

The results obtained in this work can be summarized as follows:

1. The fourth' and higher' order terms in the expansion of E(n) can be significant and thus the second-rank fabric tensor J does not adequately describe the microstructure. This is particularly true for highly anisotropic materials.

2. The distributions of branches and contact normals are almost identical.

3. The distribution densities of contacts whose normals lie along the major and minor principal stress axes, as shown in Fig. 4, vary initially and then approach a constant value in the course of shearing deformation. The density of contacts with normals along the normal to the maximum shear stress planes, however, remains practically constant throughout the deformation.

ACKNOWLEDGMENTS

This work has been supported by Air Force Office of Scientific Research under Grant No. AFOSR-87-0079 to the University of California, San Diego. The authors wish to express their appreciation to Mr. John Schmidt of Northwestern University, who designed and constructed the experimental apparatus, and to Dr. Y. Kishino, who helped Mr. Schmidt, and through whose efforts the system was assembled at Northwestern University and some preliminary experiments on the system were performed. Dr. Kishino was also instrumental in developing someearly versions of the computer programs which were modified and eventually used for data acquisition.

REFERENCES

1. Dantu, P., "Etude mechanique d'un milieu pulverulent forme de spheres egales de compacite maxima," Proc. 5th Int'l Conf. Soil Mech. Found. Eng., 1, 1961, 61-70.

2. Weber, J. D., "Recherches concernant les contraintes intergranulaires danse les milieux pulverulents," Cahiers du Groupe Francais de Rheologie, 3, 1966, 161-170.

3. Wakabayashi, T., "Photoelastic Method for Determination of Stress in Powdered Mass," Proc. 7th Japan National Cong. for Appl. Mech., 1957, 153-163.

4. Oda, M. and Konishi, J., "Microscopic Deformation Mechanism of Granular Material in Simple Shear, Soils and Foundations, Vol. 14, No. 4, 1974, 25-38.

5. Konishi, J., "Microscopic Consideration on Deformation and Strength Behavior of Granular Material Like Sand," Faculty of Engineering Report, Shinshu University, 1978 (in Japanese).

6. Konishi, J., "Microscopic Model Studies on the Mechanical Behavior of Granular Materials," Proceedings of the U.S.-Japan Seminar on Continuum Mechanical and Statistical Approaches in the Mechanics of Granular Materials, eds., Çowin, S. C. and Satake, M., Gakujutsu Bunken Fukyukai, Tokyo, Japan, 1978, 27-45.

7. Ishizuka, M., Satake, M., and Niizeki, S., "Analysis of Deformation Mechanism of Granular Material in Simple Shear Using Photoelasticity," Proc. of the 35th Annual Meeting of the JSCE, Vol. III, 1980, 33-34 (in Japanese).

8. Biarez J. and Wiendieck, K., "La comparison qualitative entre l'anisotropie mecanique et l'anisotropie de structure des milieux pulverulents," C. R. Acad. Sci., Vol. 256, 1217-1220.

9. Drescher, A. and de Josselin de Jong, G., "Photoelastic Verification of a Mechanical Model for the Flow of a Granular Material," J. Mech. Phys. Solids, Vol. 20, 1972, 337-351.

10. Drescher, A., "An Experimental Investigation of Flow Rules for Granular Materials Using Optically Sensitive Glass Particles," Geotechnique, Vol. 26, No. 4, 1976, 591-601.

11. Oda, M., Nemat-Nasser, S., and Konishi, J., "Stress-Induced Anisotropy in Granular Masses," Soils and Foundations, Vol. 25, No. 3, 1985, 85-97.

12. Allersma, H. G. B., "Photo-elastic Stress Analysis and Strains in Simple Shear," in Proc. of the IUTAM Conf. on Deformation and Failure of Granular Materials, eds., Vermeer, P. A., and Luger, H. J., Balkema, Rotterdam, 1982, 345-353.

13. Konishi, J., Oda, M., and Nemat-Nasser, S., "Induced Anisotropy in Assemblies of Oval Cross-Sectional Rods in Biaxial Compression," Mechanics of Granular Materials: New Models and Constitutive Relations, eds., Jenkins, J. T. and Satake, M., Elsevier, Amsterdam, 1983, 31-39.

14. Nemat-Nasser, S., Oda, M. and Konishi, J., Fabric and Micro-Deformation Mechanism of Granular Materials: Biaxial Compression Tests on Assemblies of Photoelastic Rods, Manuscript, 1986.

15. Supel, J. A., "Local Destruction of Granular Media Caused by Crushing a Single Grain," Arch. Mech., Vol. 37, 1985, 535-548.

16. Nemat-Nasser, S., Mehrabadi, M. M., Shodja, H., Subhash, G., "A Microscopic Study of the Granular Material Behavior in Pure Shear," in preparation.

17. Kanatani, K., "Distribution of Directional Data and Fabric Tensors," Int'l J. Eng. Sci., Vol. 22, No. 2, 1984, 149-164.

18. Satake, M., "Constitution of Mechanics of Granular Materials Through the Graph Theory," Proceedings of the U.S.-Japan Seminar on Continuum Mechanical and Statistical Approaches in the Mechanics of Granular Materials, eds., Cowin, S. C. and Satake, M., Gakujutsu Bunken Fukyukai, Tokyo, 1978, 47-62.

Micromechanics of Granular Materials, edited by M. Satake and J.T. Jenkins
Elsevier Science Publishers B.V., Amsterdam, 1988 — Printed in The Netherlands

CONTACT TENSOR IN CONSTITUTIVE MODEL FOR GRANULAR MATERIALS

Y. TOBITA[1] AND E. YANAGISAWA[2]

1 Department of Civil Engineering, Hachinohe Institute of Technology, Myo, Hachinohe, 031, Japan
2 Department of Civil Engineering, Tohoku University, Sendai, 980, Japan

SUMMARY

The roles of the contact tensor in the constitutive model for granular materials are considered. After deriving the micromechanically based contact tensor and the modified stress tensor, we give possible physical meanings of them. The constitutive model with these tensors and their rates can account for more features of deformation of granular materials. The anisotropic hardening feature and the non-coaxiality between the rate of deformation and the stress tensors are given special attention in this paper.

INTRODUCTION

The effects of inherent and induced anisotropies on the mechanical behavior of granular materials have received much attention in last two decades.(e.g., refs.1-4) In order to give mathematical expressions for the internal fabrics in a granular assembly, several kinds of "fabric tensors" were proposed and discussed.(refs. 5 to 7).

It seems natural to consider that the constitutive model with fabric tensors can account for more aspects of deformation features of granular materials than e.g., the mathematical plasticity theory, because of the fact that fabric tensors embody information about the density and anisotropic internal structures.

In this paper, we derive the contact tensor in a similar form to the relation between the traction vector and the stress tensor. Based on the contact tensor, we define the transformation tensor and the modified stress tensor. It is suggested that the (isotropic) yield condition in terms of the modified stresses leads to the anisotropic hardening nature of granular materials. It is also discussed by using the general constitutive model with the objectivity that the modified stress rate defined in this paper is closely related to the non-coaxial property between the stress and the "inelastic" deformation rate.

In the followings, we prefer to use the indicial notation in the Cartesian coordinate system x_i with unit base vectors $\mathbf{e}_i$. The usual summation convention with respect to repeated indices in a term is operated unless otherwise stated. In order to simplify the notation, we also use the symbolic notation. It suffices to know the following rule between the symbolic notation and indicial notation to understand the contents.

$$\mathrm{tr}(\mathbf{A}) = A_{kk},\ \mathbf{1} = \delta_{ij},\ \mathbf{A}^T = A_{ji}\ \mathbf{AB} = A_{ik}B_{kj},\ \mathrm{tr}(\mathbf{AB}) = A_{ik}B_{ki}$$

CONTACT TENSOR

In this section, we derive the contact tensor, defined in terms of spatial distribution of small contact areas between grain particles, so as to give a similar form to the relation between the traction vector and the stress tensor.(For more detailed discussion, see Tobita(ref.8))

We consider a microvolume V in which many particles are involved sufficient for the statistical treatment. Supposing that the contact area $dS^{(\alpha)}$ and its unit normal vector $\mathbf{n}^{(\alpha)}$ can be evaluated at each contact α, we define the spatial distribution function $C(\mathbf{n})$ being given as a summation of contact areas associated with unit normal $\mathbf{n}$. $C(\mathbf{n})$ is attached to an elementary solid angle $d\Omega$ with the unit normal $\mathbf{n}$.(see Fig.1) Next we consider the projection of $C(\mathbf{n})\mathbf{n}$ on to the imagined plane with unit normal $\underset{\sim}{\nu}$,

$$dS(\underset{\sim}{\nu}) = C(\mathbf{n})n_k\nu_k. \qquad (1)$$

The direction of the projected contact area $dS(\underset{\sim}{\nu})$ is taken to preserve its original direction $\mathbf{n}$. Thus we define a vector quantity

$$d\mathbf{S}(\underset{\sim}{\nu}) = C(\mathbf{n})n_k\nu_k n_i. \qquad (2)$$

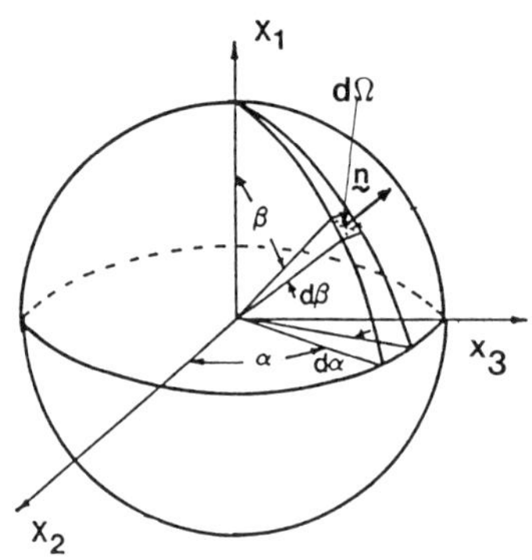

Fig.1 Elementary Solid Angle

Integrating eq.(2) over all solid angles satisfying the condition that $n_k\nu_k \geq 0$ in order to avoid the duplicated counting of the contact area and excluding the effect of microvolume as well as the effect of the integration, we arrive at the vector quantity whose magnitude denotes the contact area projected onto the $\underset{\sim}{\nu}$ plane in a unit volume and whose direction denotes the averaged direction over all contacts weighted by the projected area.

$$S_i^{(\underset{\sim}{\nu})} = \frac{1}{2\pi V}\int_{\Omega_{1/2}} C(\mathbf{n})n_k\nu_k n_i d\Omega, \qquad (3)$$

where $\Omega_{1/2}$ indicates the solid angles satisfying the condition that $n_k\nu_k \geq 0$ and $\int_{\Omega_{1/2}} d\Omega = 2\pi$ was used. We call this quantity of eq.(3) as <u>the contact vector</u>.

Employing the unit base vectors $\mathbf{e}_i$: i = 1,2,3, instead of the imagined plane, we arrive at

$$S_j^{(\mathbf{e}_i)} = \frac{1}{2\pi V}\int_{\Omega_{1/2}} C(\mathbf{n})n_i n_j d\Omega,$$

where $\mathbf{n}\mathbf{e}_i = n_i$ was used. Rewriting $S_j^{(\mathbf{e}_i)}$ by C_{ij}, C_{ij} denotes the j-th component on the $\mathbf{e}_i$ plane. Thus we have the definition of the contact tensor as

$$C_{ij} = \frac{1}{2\pi V}\int_{\Omega_{1/2}} C(\mathbf{n})n_i n_j d\Omega, \qquad (4)$$

Since $n_k n_k = 1$, C_{kk} is equivalent to the density of contact area defined by

$$\overline{S} = \frac{1}{2\pi V}\int \Omega_{1/2} C(\mathbf{n}) d\Omega. \tag{5}$$

Comparing eq.(3) with eq.(4), The contact vector $S^{(\nu)}$ can be expressed by

$$S_i^{(\nu)} = C_{ki}\nu_k. \tag{6}$$

This equation is, in a mathematical form , similar to the relationship between the traction vector f_i evaluated at the ν plane and the Chauchy stress tensor σ_{ij} given by

$$f_i = \sigma_{ji}\nu_j. \tag{7}$$

The similarity of eq.(6) to eq.(7) indicates that the contact tensor is really a second rank tensor quantity; namely, it satisfies the transformation rule of the second rank tensor. (see ,e.g.,refs.9,10)

The effectiveness of the fabric tensor in formulating constitutive models has received due attention in recent years. The elastic stiffness tensor E_{ijkl} in a linear elastic material:$\sigma_{ij} = E_{ijkl}\varepsilon_{kl}$, where$\varepsilon_{kl}$ denotes the strain tensor, was expressed as a function of the fabric tensor using the representation theorem of the isotropic function by Cowin(ref.11). In a similar approach to Cowin, Oda(ref.12) developed the mathematical expression of the anisotropic yield function for an elastic ideal-plastic material using the orientation tensor. The orientation tensor with respect to the spatial distribution of long axes of elongated particles can be assumed to be an unchanged quantity up to the failure state, which is strongly supported by experimental results.

It is a well known fact that some internal fabrics of granular materials changes to a great extent in the course of inelastic deformation, especially in the positive dilatant phase. The contact tensor is one of the "sensitive" fabric tensors to reflect the internal changes in a granular material. It is therefore expected that the constitutive formulation including the contact tensor can lead to an anisotropic hardening nature of granular materials. This will be discussed in a later section.

TRANSFORMATION TENSOR

Let us imagine an idealized isotropic granular material with the contact density of S_o, the transformation tensor F_{ij} is defined by

$$F_{ij} = C_{ij}/C_o \quad ; \quad C_o = (S_o/3). \tag{8}$$

We consider the transformation with respect to a small plane with unit normal ν_i and the area of it dA

$$dA_i^* = F_{ki}dA_k. \tag{9}$$

Selecting the coordinate axes to be coincident with those of the contact tensor, we give a schematic representation of the transformation rule: eq.(9), in Fig.2.

Since F_{ij} is proportional to C_{ij}:eq.(8), it is easily understood that the direction of the unit normal of the transformation plane ν_i^* :

$$\nu_i^* = dA_i^*/\|d\mathbf{A}^*\|, \quad \|d\mathbf{A}^*\| = \sqrt{dA_1^{*2} + dA_2^{*2} + dA_3^{*2}}, \tag{10}$$

is coincident with the contact vector: eq.(6).

MODIFIED STRESS TENSOR

Considering a small plane dA with unit normal $\underset{\sim}{\nu}$, we define the force vector **F** acting on the plane, which is given as a summation of interparticle forces transmitting through d**A**. Assuming the homogeneous state of stresses in a considered volume, the force vector F_i can be expressed in terms of the Chauchy stress σ_{ij} as

$$F_i = \sigma_{ji} dA_j. \tag{11}$$

Considering that the force vector is also acting on the transformed plane dA^*, F_i can also be expressed in a different way in terms of the new stress quantity T_{ij} as

$$F_i = T_{ji} dA_j^*. \tag{12}$$

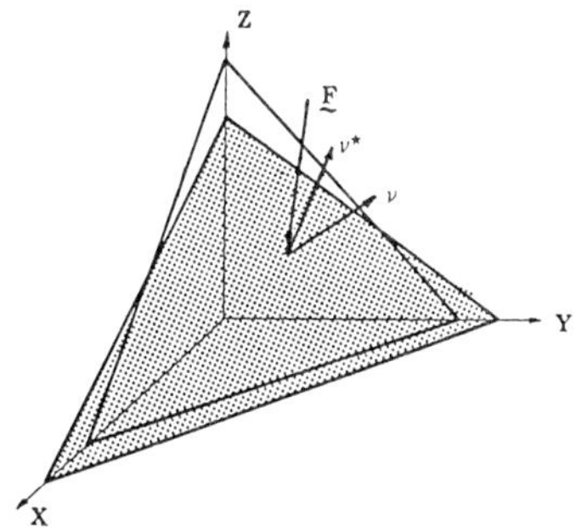

Fig.2 Transformed plane

Comparing eq.(11) with eq.(12), with the aid of eq.(7), we define the new stress quantity T_{ij} as

$$T_{ij} = \Phi_{ik}\sigma_{kj} \quad ; \quad \Phi_{ik} = \mathbf{F}_{ik}^{-1}. \tag{13}$$

It is easily verified that the new stress quantity behaves as a second rank tensor by noting that σ_{ij} and Φ_{ij} satisfies the transformation rule of the second rank tensor: $A_{\alpha\beta} = Q_{\alpha i}Q_{\beta j}A_{ij}$ and the property of the diagonal tensor defined as $Q_{\alpha i} = \mathbf{e}_\alpha^* \mathbf{e}_i^*$.[*1] We call this quantity as the <u>modified stress tensor</u>.

The rate of the modified stress tensor is given by

$$\dot{T}_{ij} = \Phi_{ik}\dot{\sigma}_{kj} + \dot{\Phi}_{ik}\sigma_{kj}. \tag{14}$$

The first term of the right hand side of eq.(14) denotes <u>the rate of the change of the modified stress tensor due to stress changes</u> and the second term denotes that due to associated fabric changes. In this decomposition of rate quantities, the first term due to stress changes is believed to be effective in characterizing the evolution rule of fabrics and the rate type constitutive formulation. We define <u>the modified stress rates</u> as the first term of eq.(14) and express it by

*1 **ab** indicates the inner product of vector quantities **a** and **b**, i.e., **ab** = $a_k b_k$.

$$\overset{\Delta}{T}_{ij} = \Phi_{ik}\dot{\sigma}_{kj}. \tag{15}$$

In next two sections, we briefly discuss the role of the modified stress tensor in the evaluation of anisotropic hardening nature of granular materials and the role of the modified stress rates in the evaluation of the non-coaxiality between the inelastic deformation rate and the stress tensor.

MODIFIED STRESSES FOR ANISOTROPIC HARDENING

In order to get a physical insight into the modified stress, we consider the case where the same traction **f** is acting on the imagined $\underset{\sim}{\nu}$ unit plane with different fabrics (a) and (b) as shown in Fig.3.

If we decompose the traction vector into two parts: one is taken along the unit normal of the transformed plane (being parallel to the contact vector) and the other is taken along the lying unit vector on the transformed plane. It is easily understood that slips in case (a) with negative dilatancy angle are of more occurrence than those in case (b) with positive dilatancy angle.

The modified stress tensor is effective to distinguish case(a) and case(b). Namely, the quantities T_s and T_n appeared in Fig.3 can be expressed in terms of the modified stress tensor as

$$T_s = T_{ji}\nu_j^*\mu_i^*, \quad T_n = T_{ji}\nu_j^*\nu_i^*. \tag{16}$$

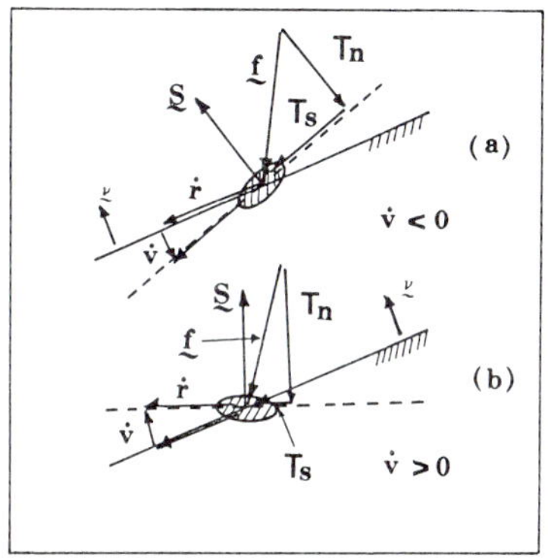

Fig.3

Eq.(16) can be easily verified by recalling the definition of the modified stress tensor: $f_i = T_{ji}\nu_j^*$ and by noting that $T_s = \mathbf{f}\underset{\sim}{\mu}^*$ and $T_n = \mathbf{f}\underset{\sim}{\nu}^*$.

From the simple example mentioned above, it may be expected that the yield condition given in terms of the modified stress tensor with respect to a suitably selected referential plane,

$$T_s/T_n = \mu(\gamma) \tag{17}$$

may lead to a nice feature in the evaluation of anisotropic hardening behavior of granular materials, which is believed to be stemming from anisotropic fabrics developed during inelastic deformation. The yield function of the type of eq.(17) was evaluated by Tobita(ref.13) for granular materials with anisotropic fabrics in the case that the coaxiality of the contact tensor and the stress tensor is held. The anisotropic hardening behavior evaluated by the shape of the yield surface given by eq.(17) in the usual stress space shows, at least from a qualitative point of view, good correspondences with experimental results as far as considered in ref.13.

It is therefore expected that the constitutive model in terms of the modified stress tensor can lead to the anisotropic hardening nature of granular materials with less mathematical complexity.

MODIFIED STRESS RATES AND NON-COAXIALITY

The modified stress rates defined by eq.(15) is closely related to the property of the non-coaxiality between the inelastic deformation rate **D** and the stress tensor $\underset{\sim}{\sigma}$. Without recourse to the "actual" constitutive model including the modified stress rate, we verify this fact in a rather general manner. Discussion is limited to the two dimensional case in the sequel.

For the discussion, we assume the constitutive equation in the form:

$$D_{ij} = H_{ijkl}(\underset{\sim}{\Phi})\,\dot{\sigma}_{kl}\,,\quad H_{ijkl} = H_{jikl} = H_{ijlk}. \tag{18}$$

In the followings, we closely follow Tokuoka(ref. 14). In eq.(18)[ ] was used to signify the linear dependence of H_{ijkl} on the stress rate $\dot{\sigma}_{kl}$. Infinitesimal strain is considered here. In eq.(18) the role of deformation rate and the stress rate was interchanged by comparing with the general formulation in which the stress rate is given by a function of deformation rate. This change does not lead to significant problems if the discussion is limited to the hardening regime, because one to one correspondence is assured for this regime. In order to satisfy the material objectivity, eq.(18) must satisfy

$$\mathbf{Q}\mathbf{H}(\underset{\sim}{\Phi})[\dot{\underset{\sim}{\sigma}}]\mathbf{Q}^T = \mathbf{H}(\mathbf{Q}\underset{\sim}{\Phi}\mathbf{Q}^T)[\mathbf{Q}\dot{\underset{\sim}{\sigma}}\mathbf{Q}^T], \tag{19}$$

where $\mathbf{Q} = Q_{\alpha i}$ was defined as $\mathbf{e}^*_\alpha\mathbf{e}_i$. In two dimensions, eq.(19) is equivalent to(ref.15)

$$\mathbf{D} = [h_1\mathrm{tr}(\dot{\underset{\sim}{\sigma}}) + h_2\mathrm{tr}(\underset{\sim}{\Phi}\dot{\underset{\sim}{\sigma}})]\mathbf{1} + [h_4\mathrm{tr}(\dot{\underset{\sim}{\sigma}}) + h_5\mathrm{tr}(\underset{\sim}{\Phi}\dot{\underset{\sim}{\sigma}})] + h_{10}\dot{\underset{\sim}{\sigma}} \tag{20}$$

where $h_\alpha : \alpha = 1,2,4,5$ and 10 are functions of invariants of Φ_{ij}.

Selecting the coordinate axes to be coincident with the principal axes of the contact tensor (the principal axes of Φ_{ij} is coincident with C_{ij}.), the fourth rank tensor H_{klpq} can be expressed by

$$\begin{aligned} H_{klpq} &= A_{kp}\delta_{kl}\delta_{pq} + \tfrac{1}{2}B_{kl}(\delta_{kl}\delta_{lq} + \delta_{kq}\delta_{lq})^{*1} \quad (\text{no sum on } k,l,p,q) \\ A_{kp} &= h_1 + h_4\Phi_k + h_2\Phi_p + h_5\Phi_k\Phi_p \\ B_{kl} &= h_{10} \end{aligned} \tag{21}$$

The matrix form of eq.(18) in two dimensions is given by

$$\begin{Bmatrix} D_1 \\ D_2 \\ D_3 \end{Bmatrix} = \begin{bmatrix} A_{11}+B_{11} & A_{12} & 0 \\ A_{21} & A_{22}+B_{22} & 0 \\ 0 & 0 & B_{12} \end{bmatrix} \begin{Bmatrix} \dot{\sigma}_1 \\ \dot{\sigma}_2 \\ \dot{\sigma}_3 \end{Bmatrix}, \tag{22}$$

where the indicial notation 11,22,12 were shifted to 1,2,3 respectively.

Two simple cases are considered in this paper.

(a) The principal axes of the contact tensor are coincident with those of the stress tensor after sufficient deformation is applied to the material by keeping the direction of principal stresses to be fixed. Then an abrupt change in the direction of principal stress rate is applied in a manner that $\dot{\sigma}_{11} = \dot{\sigma}_{22} = 0$ and $\dot{\sigma}_{12} \neq 0$.

---

*1 δ_{ij} indicates the Kronecker's delta, i.e., $\delta_{ij}=1$ ($i=j$) and $\delta_{ij} = 0$ ($i \neq j$)

In this case, $\dot{\sigma}_3 > 0$ leads to the non-coaxial component of $D_3 = D_{12}$,

$$D_{12} = B_{12}\dot{\sigma}_{12} = h_{10}\dot{\sigma}_{12}. \tag{23}$$

Expanding h_{10} in eq.(23) in a polynomial series with respect to the invariants of Φ_{ij}, we have

$$h_{10} = \hbar_{100} + \hbar_{101}I_{\Phi} + \hbar_{102}II_{\Phi} + ---, \tag{24}$$

where $\hbar_{100}$, $\hbar_{101}$,... are material constants and I_{Φ} and II_{Φ} are first and second invariants of Φ_{ij}. In small deformation theory, $\hbar_{100}$ can be interpreted as elastic constant, so that the inelastic deformation rate is given by

$$D^{i}_{12} = D_{12} - D^{e}_{12} = [\ \hbar_{101}\mathrm{tr}(\underset{\sim}{\Phi})\dot{\sigma}_{12} + ...]. \tag{25}$$

The quantity $\mathrm{tr}(\underset{\sim}{\Phi})\dot{\sigma}_{12} = (\Phi_{1k}\dot{\sigma}_{k2} + \Phi_{2k}\dot{\sigma}_{k2})$ can be interpreted as the modified stress rates. Thus, it is found that the modified stress rate is related to the non-coaxial deformation rate in this case.

(b) The principal directions of the contact tensor deviate from those of the stress tensor by the angle of θ as shown in Fig.4. The proportional loading in the sense that the principal stress rate is applied to the directions of the principal stress is further applied.

We first transform the stress tensor in $\mathbf{e}^*_\alpha$ coordinate system into the $\mathbf{e}_i$ system: $\dot{\sigma}_{ij} = Q_{i\alpha}Q_{j\beta}\dot{\sigma}_{\alpha\beta}$, we can evaluate the deformation rate D_{ij} by putting the results into eq.(22). The deformation rate in the coordinate system along the principal stress directions can be obtained by the transformation: $D^*_{\alpha\beta} = Q_{\alpha i}Q_{\beta j}D_{ij}$.

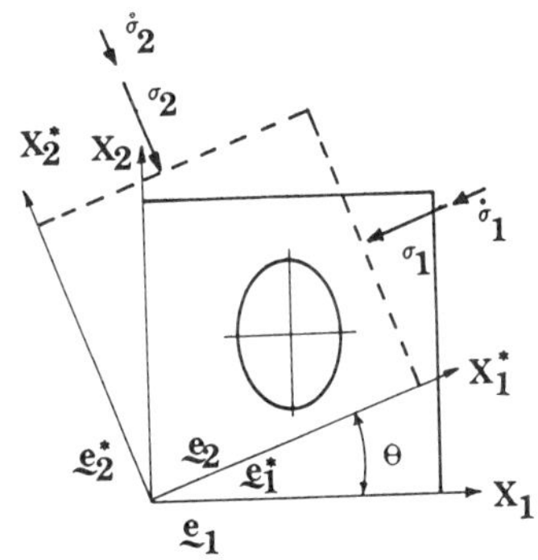

Fig.4

The non-coaxial component D^*_{12} in the principal stress coordinate system can be expressed in the form:

$$D^*_{12} = \frac{1}{2}\sin 2\theta\left[\left\{\frac{a+b}{2} + \frac{a-b}{2}\cos 2\theta\right\}\dot{\sigma}^*_1 + \left\{\frac{a+b}{2} - \frac{a-b}{2}\cos 2\theta\right\}\dot{\sigma}^*_2\right], \tag{26}$$

where $a = A_{21} - A_{11} - B_{11} + B_{22}$, $b = A_{22} + B_{22} - A_{12} - B_{12}$. Expanding h_α ;α = 1,.. 5 and 10 in a polynomial series with respect to the invariants of Φ_{ij}, we have following material constants for rank 1 materials (the order of $\underset{\sim}{\Phi}$ in constitutive equation is 1) to be specified.

$\hbar_{10}$, $\hbar_{11}I_\Phi$, $\hbar_{20}$, $\hbar_{40}$, $\hbar_{100}$, $\hbar_{101}I_\Phi$ (in two dimensions).

In this case

$$\frac{a-b}{2} = 0\ ,\ \frac{a+b}{2} = \hbar_{40}(\Phi_2 - \Phi_1)(\dot{\sigma}_1^* + \dot{\sigma}_2^*)$$

For rank 1 materials, we have $D_{12}^* = 0$ if $(\dot{\sigma}_1^* + \dot{\sigma}_2^*) = 0$; namely, if we apply the mean stress constant test to the specimen, the non-coaxial deformation rate will not generate.This may not be the case for granular materials. So we next consider rank 2 materials. For rank 2 materials, D_{12}^* can be expressed as

$$D_{12}^* = \frac{1}{2}\sin2\theta(\Phi_2-\Phi_1)[\{\hbar_{40} + \hbar_{41}I_\Phi + \hbar_{50}(\Phi_1 + \Phi_2)\}(\dot{\sigma}_1^* + \dot{\sigma}_2^*)$$
$$+ \hbar_{50}(\Phi_2 - \Phi_1)\cos2\theta(\dot{\sigma}_1^* - \dot{\sigma}_2^*)]$$

In this case we have $D_{12}^* \neq 0$ even when $(\dot{\sigma}_1^* + \dot{\sigma}_2^*) = 0$. From the calculation above, the modified stress rate is found to have close connection with the non-coaxial property of the deformation rate.

CONCLUSIONS

The roles of the contact tensor, the modified stress tensor and the modified stress rate were discussed in this paper. The contact tensor itself has been found to be very effective in formulating the non-linear constitutive model,e.g., the instantaneous elastic stiffness tensor can be, in a reasonable manner, represented as a function of the contact tensor with the help of the representation theorem of the isotropic function theory. As examples of the application of the fabric tensor, nice features of the modified stress tensor and the modified stress rate were discussed in some detail. It was found that:(a) the modified stress tensor was of primary importance in considering the anisotropic hardening nature of granular materials; (b) the modified stress rate was closely related to the property of non-coaxiality of the "inelastic" deformation rate and the stress tensors.

REFERENCES

1 Oda,M., Soils and Foundations, 12,1,(1972) 1-18
2 Arthur,J.R.F. and Menzies,B.K., Geotechnique, 22, 2,(1972), 115-128
3 Oda,M.,Koishikawa,I. and Higuchi,T., Soils and Foundations, 18,1 (1978), 25-38
4 Arthur,J.R.F. et al., Geotechnique, 27,1 (1977), 13-30
5. Satake,M., Deformation and Failure of Granular Materials Delft, A.A. Balkema,(1982) 63-68
6 Oda, M., Nemat-Nasser,S. and Mehrabadi,M.M., Int.J. Num. Ana. Methods in Geomechanics 6,(1982) 77-94
7 Kanatani,K., J.of JSSMFE, 23,4,(1983)171-177 (in Japanese)
8 Tobita,Y., The thesis submitted to Tohoku University for Dr.Engrg.,(1987)
9 Malvern,L.E. Introduction to the Mechanics of a continuous medium , Prentice-Hall,(1969)
10 Borisenko,A.I. and Tarapov,I.E.,Vector and Tensor Analysis with applications, Dover Publications Inc.(1968)
11 Cowin,S.C., Mechanics of Materials, 4, (1985) 137-147
12 Oda,M., Preprint of US-Japan Seminar on Micromechanics of Granular Materials (1987),70-88
13 Tobita,Y., Soils and Foundations (to appear)
14. Tokuoka , T., Introduction to rational mechanics(11) in Science of Machines (Kikai no Kenkyuu) 32, 11 (1980) 90-96(in Japanese)
15 Mehrabadi,M.M. Ph.D thesis to Tulane University, (1979)

Micromechanics of Granular Materials, edited by M. Satake and J.T. Jenkins
Elsevier Science Publishers B.V., Amsterdam, 1988 — Printed in The Netherlands

MICROMECHANICAL MODELLING OF CONSTITUTIVE RELATIONS FOR GRANULAR MATERIAL

C. S. CHANG

Department of Civil Engineering, University of Massachusetts,
Amherst, Massachusetts 01003

SUMMARY

A micromechanical approach to represent the stress-strain behavior of an assemblage of granular particles is developed and examples of its application for the cases of regular packings and random packings of elastic spheres are presented.

INTRODUCTION

The earliest attempt of using micromechanical approach to modelling of granular material was by Duffy and Mindlin in 1957 (ref. 1). They presented a stress-strain equation for the face centered cubic packing of equal sized elastic spheres. By taking a representative unit cube based on packing geometry they first defined the relationship between stress and contact forces according to equilibrium. Then the contact displacement is obtained from contact theory for two spheres. Finally the strain of the packing is determined from the deformed shape of the unit cube which is obtained based on these contact displacements.

This approach has been followed by other investigators (refs. 2,3,4) and stress-strain relationships have been developed for four additional regular packings namely: simple cubic, tetrahedral, cubical tetrahedral, and tetragonal sphenoidal. The effort has never been extended to other regular packings as for most packings the representative element is not a cubical shape, as indicated by Makhlouf and Stewart (ref. 4). For such cases, the definitions of stresses and strains are obscure in Duffy-Mindlin method.

This difficulty can be overcome by utilizing the definitions of stress as an volume average of contact forces developed recently in the field of micromechanics of granular soil (refs. 5,6,7). Then, with the use of energy conservation principle, similar to that used in ref.8, which requires that the work done described in terms of macro-field variables equal to the work done at all particle contacts. The obtained stress-strain relationship is applicable to all regular as well as irregular packings. Further extension of this method is outlined for the cases of Random packings.

STRESSES IN GRANULAR MEDIA

For a granular media, the average stress of a volume V can be expressed by the discrete contact forces on the surface of the volume (refs. 9,10,11). The stress can also be expressed by the forces on all contacts within the volume V in a general form

$$\sigma_{ij} = \frac{1}{V} \sum_{\alpha} z_i^{\alpha} f_j^{\alpha} \tag{1}$$

where z_i is fabric measure and f_j is the contact force between two particles.

Various investigators have used different fabric measures z_i to represent the stress tensor in terms of the volume average of the contact forces in a granular assembly. Kishino (ref. 5) and Chrisstoffersen, Mehrabadi and Nemat-Nasser (ref. 6) have derived a relationship between contact forces and overall stress based on the principle of virtual work. Rothenburg and Selvadurai (ref. 7) have another expression derived from the analogy of statistical average momentum for granular system. Table I summarizes the fabric measures used by these investigators.

TABLE I Summary of Fabric Representations in the definition of Stresses

z_i^{α} (fabric measure)	Reference
x_i^{α} (position vector)	Kishino (5)
l_i^{α} (branch vector)	Chrisstoffersen et.al.(6)
r_i^{α} (contact vector)	Rothenburg & Selvadurai (7)

The fabric measures given in Table I are illustrated in Fig. 1. The centroid of mass of two particles in contact are denoted as A and B and the contact point is denoted as C. Position vector is defined as the vector connecting C and the reference point of the coordinates, branch vector is the vector connecting the two centroids A and B and contact vector is that connecting the centroid to the contact point.

The branch vector between two particles in contact $\tilde{l}^{AB}$ and the contact vectors of the two particles, $\tilde{r}^{A}$ and $\tilde{r}^{B}$, are related to each other by

$$\tilde{l}^{AB} = \tilde{r}^{A} - \tilde{r}^{B} \tag{2}$$

Hence, the expression given by Chrisstoffersen et.al. is equivalent to the expression given by Rothenburg and Selvadurai.

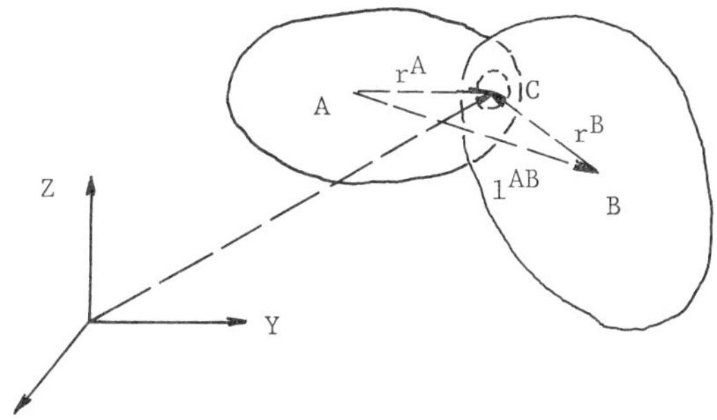

Fig. 1 Fabric representations in the definition of stress

LOCAL CONSTITUTIVE RELATIONSHIP

As shown in Fig. 2, the local coordinate system is constructed for each contact with three orthogonal unit vectors **n, s, t**. Where **n** is defined as the unit vector normal to the contact area, and in a chosen global axes XYZ, the angle β is measured on XY plane in a counter-clockwise direction from the positive X-axis and the angle α is measured from the positive Z axis to the normal vector. Both α and β range from 0 to 180 degrees. The force at contact α, f_j, can be transformed to t_i in the local coordinate system using the transformation matrix T_{ij}

$$t_i^{\alpha} = T_{ij}^{\alpha} f_j^{\alpha} \tag{3}$$

In the local coordinate system, the local rate constitutive relationship can be expressed in the following general form

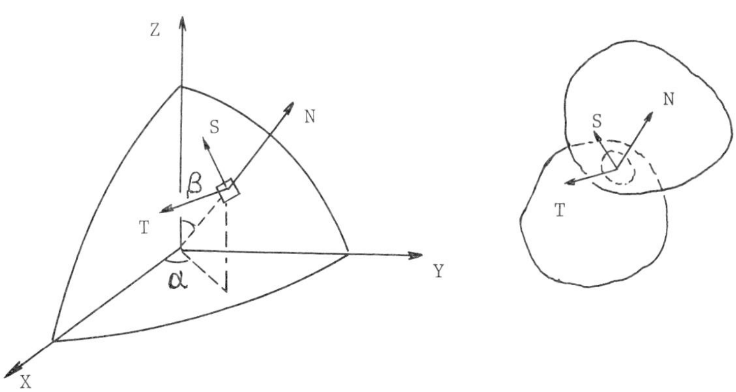

Fig. 2 Local coordinate system

$$t_i^\alpha = c_{ij}^\alpha \, \delta_j^\alpha \tag{4}$$

where δ_j^α is the displacement rate in the local coordinate system. It relates to the displacement rate d_j^α in the xyz coordinate system by

$$\delta_i^\alpha = T_{ij}^\alpha \, d_j^\alpha \tag{5}$$

STRAINS IN GRANULAR MEDIA

For granular material, the overall strain is interpreted in an average sense over a representative volume of packing. In the following derivation, the 'average' rate of strain, $\dot{\varepsilon}_{ij}$, is obtained from the consideration of the first law of thermodynamics, which is an assertion of 'conservation of energy'. Applying external stress on a packing while no heat is being transmitted to the system is an adiabatic process and under this condition the first law shows that the external work done on the system is equal to the change in the internal energy of the system. Thus

$$\sigma_{ij} \, \dot{\varepsilon}_{ij} = \frac{1}{V} \sum_{\alpha=1}^{N} f_i^\alpha \, d_i^\alpha \tag{6}$$

where N is the total number of contact points in volume V.
Substituting Eq. 1 into Eq. 6, it can be written that

$$\sum_{\alpha=1}^{N} z_i^\alpha \, f_j^\alpha \, \dot{\varepsilon}_{ij} = \sum_{\alpha=1}^{N} f_i^\alpha d_i^\alpha \tag{7}$$

From Eq.7, it can be reasoned that

$$z_i^\alpha \, \dot{\varepsilon}_{ij} = d_j^\alpha \tag{8}$$

From Eqs. 3, 4, 5 and 8, it follows

$$\sum_{\alpha=1}^{N} z_q^\alpha \, T_{pm}^\alpha \, c_{mk}^\alpha \, T_{kj}^\alpha \, z_i^\alpha \, \dot{\varepsilon}_{ij} = \sum_{\alpha=1}^{N} z_q^\alpha \, f_p^\alpha \tag{9}$$

Further, from Eq. 1, the stress-strain relationship of the granular media with a representative volume V can be established as follows

$$D_{ijqp} \, \dot{\varepsilon}_{ij} = \dot{\sigma}_{qp} \tag{10}$$

$$\text{where } D_{ijqp} = \frac{1}{V} \sum_{\alpha=1}^{N} z_q^\alpha \, T_{pm}^\alpha \, c_{mk}^\alpha \, T_{kj}^\alpha \, z_i^\alpha \tag{11}$$

EXAMPLE OF REGULAR PACKING

The stress-strain relationship for a regular packing of equal sized spheres is established following the method outlined above. The representative volume chosen here consists of a single sphere and its associated void space. The volume V is equal to the volume of one sphere V_s multiplied by a factor of (1 + e), where e is the void ratio of this packing. Such an unit cell is same as 'Voronoi Polyhedron' used for partitioning space occupied by a packing as discussed by Finney (ref. 12). It is noted that this representative volume is different from the unit cube chosen by other investigators who followed Duffy-Mindlin method (ref. 1).

For this representative volume, using the concept of contact vector (ref. 7), the stress tensor is related to the discrete contact forces by the following equation:

$$\sigma_{ij} = \frac{1}{V} \sum_{\alpha=1}^{N} r_i^\alpha f_j^\alpha \tag{12}$$

where N is the total number of contacts per sphere, σ_{ij} is the average stress of the packing represented by an unit cell, and r_i^α is the contact vector of αth contact point of the sphere.

For the contact of elastic spheres, the local constitutive matrix c_{ij} in Eq. 4 can be determined by Hertzian or non-Hertzian contact theory (refs. 13, 14). The normal and tangential compliances given by Mindlin and Deresiewicz (ref. 15) based on Hertzian theory is used. For detailed information on compliances under various loading condition the reader is referred to the summary in a review of the contact theory by Deresiewicz (refs. 16,17).

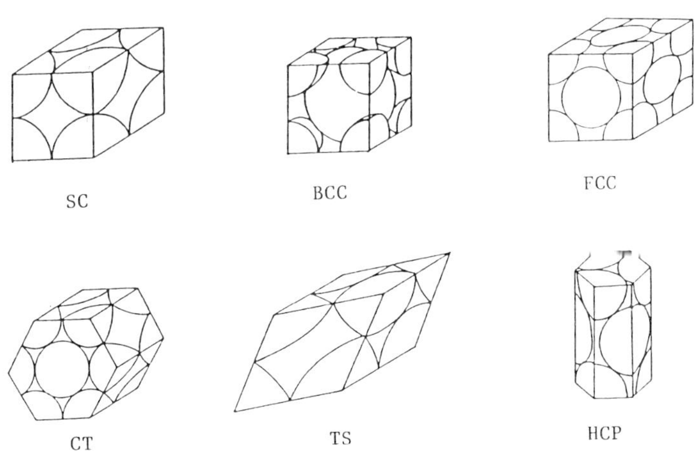

Fig. 3a Six regular packings of equal sized spheres

Using this present method it is possible to compute stress-strain relationship for any type of regular packing under arbitrary loading conditions. Examples of stress-strain curves are shown in Fig. 3 for FCC, SC, HCP, BCC, CT, TS packings, for $\phi_\mu = 30^\circ$, σ_3 = 10 psi, followed by a deviatoric stress loaded in the 45° direction for the six packings. The denser packing, in general, shows a stiffer behavior than the looser packings. Different packing structures show different stress-strain behavior even with same void ratio, for example, FCC and HCP.

Any of the 42 possible regular packings of equal spheres can be represented by a 'Voronoi Polyhedron'(ref. 18). Thus, following the above formulation, the stress-strain solutions for other packings can be easily obtained.

EXAMPLE OF RANDOM PACKING

The stress-strain relationship for a random packing of particles is also studied following the same method outlined above. To take into account of microstructure of a random granular assembly, the fabric measure z_i and contact forces f_j in Eq. 1 can be represented as function of the variables, $\Theta_1,..\Theta_k$, associated with microstructural characteristics of this random assembly, so that $z_i = z_i(\Theta_1,..\Theta_k)$ and $f_j = f_j(\Theta_1,..\Theta_k)$. Let $P(\Theta_1,..\Theta_k)$ be the probability density function of these variables of microstructural characteristics of this granular assembly. Therefore $N\,P(\Theta_1,..\Theta_k)\,\Delta\Theta_1..\Delta\Theta_k$ represents the number of contacts in the range of Θ_1 to $\Theta_1+\Delta\Theta_1,..,\Theta_k$ to $\Theta_k+\Delta\Theta_k$. Thus the average stress of this random assembly can be written as

$$\sigma_{ij} = \frac{N}{V} \sum_{\Theta_1} .. \sum_{\Theta_k} z_i(\Theta_1,..\Theta_k)\, f_j(\Theta_1,..\Theta_k)\, P(\Theta_1,..\Theta_k)\, \Delta\Theta_1..\Delta\Theta_k \tag{13}$$

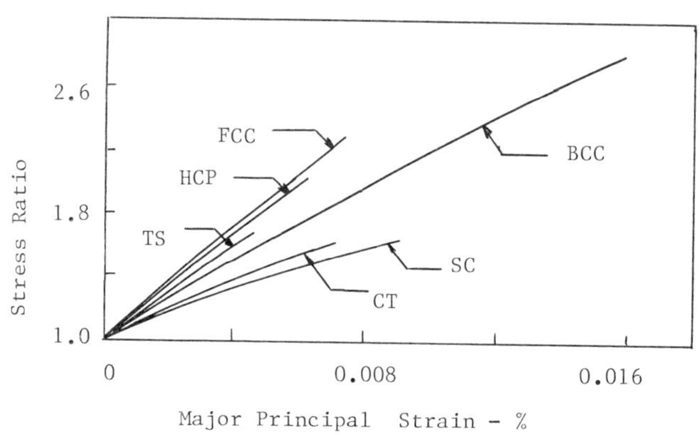

Fig. 3b Stress-strain curves for the six packings

If density function P is known for a random assembly, the stress-strain relationship can be readily established as follows

$$D_{ijqp}\ \dot{\varepsilon}_{ij} = \dot{\sigma}_{qp} \qquad (14)$$

where $D_{ijqp} = \frac{N}{V} \sum_{\Theta_1} .. \sum_{\Theta_k} z_q\ T_{pm}\ c_{mk}\ T_{kj}\ z_i\ P(\Theta_1, .. \Theta_k)\ \Delta\Theta_1 .. \Delta\Theta_k \qquad (15)$

It is noted that $z_q\ T_{pm}\ c_{mk}\ T_{kj}\ z_i$ are all functions of $\Theta_1, .. \Theta_k$.

For example, if the expression of branch vector orientations, γ, β, is used to represent the microstructure (ref. 6). Thus $z_i = z_i(\gamma, \beta)$ and $f_i = f_i(\gamma, \beta)$. Let $P(\gamma, \beta)$ be the probability density function. Then, $N\ P(\gamma, \beta)\ \Delta\gamma\ \Delta\beta$ is the number of contacts in the range of γ to $\gamma + \Delta\gamma$, β to $\beta + \Delta\beta$. The overall stress can be written by

$$\sigma_{ij} = \frac{N}{V} \sum_{\gamma} \sum_{\beta} z_i(\gamma, \beta)\ f_j(\gamma, \beta)\ P\ (\gamma, \beta)\ \Delta\gamma\ \Delta\beta \qquad (16)$$

Following the discussion above, the constitutive tensor may be given by

$$D_{ijqp} = \frac{N}{V} \sum_{\gamma} \sum_{\beta} z_q\ T_{pm}\ c_{mk}\ T_{kj}\ z_i\ P\ (\gamma, \beta)\ \Delta\gamma\ \Delta\beta \qquad (17)$$

where $z_q\ T_{pm}\ c_{mk}\ T_{kj}\ z_i$ are all functions of γ, β.

CONCLUDING REMARKS

A method to establish the stress-strain relationship of granular media has been described. For simple cubic packing this method yields the same result as that reported in literature. The method can be easily applied to establish the stress-strain relationships for other regular packings of equal sized spheres. In the case of random packings a general constitutive equation has been outlined. Future work on the areas of characterization of the microstructure of a granular assembly and the evolution of probability density function with strains should be important for further understanding the fundamentals of granular mechanics.

ACKNOWLEDGEMENT

This paper is based on the research supported by the Air Force office of Scietific Research under Grant No. AFOSR-86-0151 to the University of Massachusetts.

REFERENCES

1 Duffy, J. and Mindlin, R.D.: Stress-strain relation of a granular media. J. Appl. Mech., Trans ASME, 585-593 (1957).
2 Duffy, J.: A differential stress-strain relation for the hexagonal close packed array. J. Appl. Mech., Trans ASME, 88-94 (1959).
3 Deresiewicz, H.: Stress-strain relations for a simple model of a granular medium. J. Appl. Mech., Trans ASME, 402-406 (1958).
4 Makhlouf, H. and Stewart, J.J.: Elastic constants of cubical-tetrahedral and tetragonal sphenoidal arrays of uniform spheres. Proc. Intl. Sym. Wave Propagation and Dynamic Properties of Earth Materials, Albuquerque, NM, 825-837 (1967).
5 Kishino, Y.: Statistical Consideration on Deformation Characteristics of Granular Materials. Proc. U.S.-Japan Seminar on Continuum Mechanical and Statistical Approaches in the Mechanics of Granular Materials, Ed. S.C. Cowin and M. Satake, 114- 122, Tokyo: Gakujutsu Bunken Fukyukai, 1978.
6 Chrisstoffersen, J., Mehrabadi, M.M. and Nemat-Nasser, S.: A micromechanical description of granular material behavior. J. Appl. Mech., Trans ASME, 339-344 (1981).
7 Rothenburg, L. and Selvadurai, A.P.S. : Micromechanical Definition of the Cauchy·Stress Tensor for Particulate Media. Mechanics of Structured Media, Proc. Intl. Symp. on Mechanical behavior of Structured Media, Ed. A.P.S. Selvadurai, 469-486, Elsevier, 1981.
8 Kanatani, K.I. : Mechanical Properties of Ideal Granular Materials. Mechanics of Granular Materials, Proc. U.S.-Japan Seminar on New Models and Constitutive relations in the Mechanics of Granular Materials, Ed. J.T. Jenkins and M. Satake, 235-244, Elsevier, 1983.
9 Drescher, A. and de Josselin de Jong, G.: Photoelastic verification of a mechanical model for the flow of a granular material. J. Mech. Phys. Solids, Vol. 20, 337-351 (1972).
10 Satake, M : Fundamental Quantities in the Graph Approach to Granular Materials. Mechanics of Granular Materials, Proc. U.S.-Japan Seminar on New Models and Constitutive relations in the Mechanics of Granular Materials, Ed. J.T. Jenkins and M. Satake, 10-21, Elsevier, 1983.
11 Satake, M.: Constitution of mechanics of granular materials through the graph theory. Proc. U.S.-Japan Seminar on Continuum Mechanical and Statistical Approaches in the Mechanics of Granular Materials, Ed. S.C. Cowin and M. Satake, 47-62, Tokyo: Gakujutsu Bunken Fukyukai, 1978.
12 Finney, J.L.: Random packings and structure of simple liquids I. The geometry of random close packing. Proc. Roy. Soc. A, Vol. 319, 479-493, (1970).
13 Mindlin, R.D.: Compliance of elastic bodies in contact. J. Appl. Mech., Trans ASME, 259-268 (1949).
14 Johnson, K.L.: Contact Mechanics, 1st ed., Great Britain: Cambridge University Press 1985.
15 Mindlin, R.D. and Deresiewicz, H.: Elastic spheres in contact under varying oblique forces. J. Appl. Mech., Trans ASME, 327-344 (1953).
16 Deresiewicz, H.: Mechanics of granular matter. Advances in Applied Mechanics, Vol. 5, 233-306, New York: Academic Press (1958).
17 Deresiewicz, H.: Bodies in contact with applications to granular media. R.D. Mindlin and Applied Mechanics, Ed. G. Herrmann, New York: Pergamon Press 1973.
18 Shahinpoor, M.: Statistical mechanical considerations on storing bulk solids. Bulk Solids Handling, Vol. 1, 31-36 (1981).

Micromechanics of Granular Materials, edited by M. Satake and J.T. Jenkins
Elsevier Science Publishers B.V., Amsterdam, 1988 — Printed in The Netherlands

NEW OPTIMALITY THEORY IN NATURE AND COAXIALITY BETWEEN STRESS AND STRAIN INCREMENT TENSORS

S. NIISEKI

Deparment of Civil Engineering, Tohoku University, Sendai 980

SUMMARY

A new optimality theory in nature is formulated only through the first and second laws of thermodynamics and the fluctuation in nature. On the basis of this theory, the coaxiality between stress and strain increment tensors during the deformation of granular materials is considered.

INTRODUCTION

The idea that nature pursues economy in all her working is one of the oldest principle of theoretical science from the time of Greeks to the present day.

However, in nonequilibrium phenomena, we have not yet had such a useful general variational theory as Gibbs's one in equilibrium. This fact makes it difficult for us to understand the evolution of nonequilibrium phenomena in nature. In the light of the concept of stability, a new general variational theory for nonlinear nonequilibrium processes is formulated only through the first and second laws of thermodynamics and the fluctuation in nature.

Now, since St-Venant in 1870 proposed the hypothesis that there existed the coaxiality between stress and strain increment tensors, this hypothesis has widely been accepted in the metal plasticity through Reuss's relation, experimental data, etc. However, at present, there are various kinds of experimental data for this problem in the research of granular materials as follows.

(1) Arthur, et al.(refs. 1 & 2) and Kumasaka, et al.(ref. 3) asserted the existence of coaxiality between stress and strain increment tensors.

(2) Drescher, et al.(ref. 4) and Oda, et al.(ref. 5) asserted no existence of the coaxiality between them.

(3) Roscoe(ref.6) showed that the coaxility between them did not exist in the beginning stage of deformation but existed in the advanced stage of it

To clarify the above confused situation, this problem is theoretically considered by application of the new variational theory mentioned above.

BASIC CONCEPT FOR FORMULATION OF NEW THEORY

Here, based on the well-known concept of stability, the nonequilibrium phenomena are considered from the thermodynamical point of view below. The *first law of thermodynamics* states that the total amount of energy is always conserved in all the processes of energy transformation and/or flow. For example, the deformation of materials under a load is the transformation of the external mechanical work into the internal energy such as the strain energy, plastic energy, etc. Thus, the evolution of all natural phenomena can be regarded as the transformation and/or flow of energy. Therefore, the stability of nonequilibrium phenomena is completely equivalent to that of the processes of transformation and/or flow of energy, the relationship of which is expressed by the first law of thermodynamics.

On the other hand, the *second law of thermodynamics* expresses that all nonequilibrium phenomena always evolve towards the direction in which the available energy is dissipated and thus the entropy production always increases in nonequilibrium phenomena. Further, the *fluctuation* frequently arises from the thermal molecular motion or comes from the outer world of the system concerned and this has some amount of energy with infinitesmal motion. If the fluctuation in a nonequilibrium phenomenon spontaneously fades away by virtue of the property expressed by the second law of thermodynamics, this nonequilibrium phenomenon is stable by the *definition of stability*: if a nonequilibrium phenomenon to which any infinitesmal perturbation is applied never evolves into some other new nonequilibrium phenomenon qualitatively different from the one before the application of infinitesmal perturbation, this nonequilibrium phenomenon is stable.

It is noticed that the founders of thermodynamics in the 19th century had not yet recognized the existence of fluctuation in nature but since the Einstein theory of Brownian motion in 1905, the existence and importance of fluctuation have gradually been recognized in nonequilibrium phenomena. It is considered that in nonlinear nonequilibrium phenomena, this fluctuation in nature plays a role of the perturbation mentioned in the above definition of stability. Therefore, the nonequilibrium phenomena with no qualitative change for some finite length of time in spite of the existence of fluctuations, which we can often observe, should be stable by the definition of stability. Thus, the above consi-

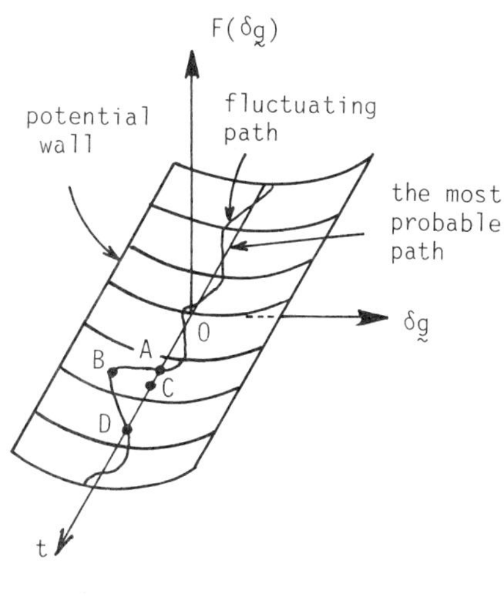

Fig. 1. Fluctuating path and potential wall in a nonequilibrium phenomenon

deration for stability implys that the potential wall governing the returnablity of a fluctuating path as illustrated in Fig. 1 exists in nature.

Then, it is considered that the maximum or minimum property of physical phenomena results from the potential wall governing the returnability in the definition of stability irrespective of the constitutive equations in each individual natural phenomenon and is the counterpart of stability like Gibbs's theory, which is both a variational theory and a stability one in equilibrium.

In addition, it is noted that if microscopic fluctuations in nature are taken into consideration, all the macroscopic laws become statistical like the Langevin equation but in the continuum mechanics, usually, they are described by use of the most probable values just from the historical point of view.

NEW OPTIMALITY THEORY IN NATURE

For simplicity, let us use the Eulerian description referred to the rectangular Cartesian co-ordinates X_i (i=1,2,3) and consider a thermomechanical system. The first law of thermodynamics in the local form is written as

$$\rho\dot{e} = \sigma_{ij}d_{ij} + q_{i,i} + \rho r \quad (1)$$

where $d_{ij} = (v_{i,j} + v_{j,i})/2$ (2)

ρ , e , σ_{ij} , q_i , r and v_i are the density, the internal energy per unit mass, the stress tensor, the heat flux vector, the heat radiation per unit mass and the velocity vector respectively and the superposed dot denotes the material derivative. The rate of total specific entropy $\dot{S}$ is composed of the rate of entropy production ${}_i\dot{S}$ which results from a nonequilibrium process in the system and the entropy flux ${}_e\dot{S}$ which flows into or out of the system:

$$\dot{S} = {}_i\dot{S} + {}_e\dot{S} \quad (3)$$

where, denoting the absolute temperature by θ , ${}_e\dot{S}$ is defined in this case by

$$\rho\,{}_e\dot{S} = (q_i/\theta)_{,i} + \rho(r/\theta) \quad (4)$$

Then, the second law of thermodynamics is expressed as

$${}_i\dot{S} \geqq 0 \quad (5)$$

Referring to Eq.(4), paying attention to the relationship between the rate of entropy production ${}_i\dot{S}$ and the physical quantities in Eq.(1) such as the dimension and combining Eqs.(1) and (5), we have

$$\rho\theta\,{}_i\dot{S} \geqq \rho\dot{e} - (\sigma_{ij}d_{ij} + q_{i,i} + \rho r) \quad (6)$$

or $$\rho\theta\,{}_i\dot{S} \geqq \sigma_{ij}d_{ij} + q_{i,i} + \rho r - \rho\dot{e} \quad (7)$$

It is noticed that these inequalities essentially consist only of the first and second laws of thermodynamics as is clear from the form of them.

Here, let us consider the consistency of the inequalities (6) and (7) with the above mentioned thermodynamics. From Eq.(4) and the inequality (7), we get

$$\rho({}_i\dot{S} - {}_e\dot{S} + \dot{e}/\theta) - \sigma_{ij}d_{ij}/\theta + q_i(\theta^{-1})_{,i} \geqq 0 \quad (8)$$

but this inequality is contradictory to the following Clausius-Duhem inequality which is derived from Eqs.(1),(3) and (4) and the inequality(5).

$$\rho({}_i\dot{S} + {}_e\dot{S} - \dot{e}/\theta) + \sigma_{ij}d_{ij}/\theta - q_i(\theta^{-1})_{,i} \geqq 0 \tag{9}$$

On the other hand, the result obtained from Eq.(4) and the inequality (6) is consistent with the inequality (9). Thus, the combined form of Eqs.(1) and (5) is limited to the inequality(6) only. It is noticed that the inequality(6) is also derived from the Clausius-Duhem inequality(9) and Eq.(4). However, the former procedure is considered to make it clear that the inequality(6) can be derived only from the first and second laws of thermodynamics.

Here, let us consider the effect of the microscopic fluctuation in nonequilibrium phenomena below. Thus, let us introduce the microscopic fluctuation into the first law of thermodynamics(1) in the *conventional macroscopic description* and rewrite Eq.(1) for a *fluctuating state* as

$$F(\delta\underset{\sim}{g}) = \rho\dot{e} - (\sigma_{ij}d_{ij} + q_{i,i} + \rho r) \tag{10}$$

where $\delta\underset{\sim}{g}$ denotes some perturbation from the path ACD without fluctuations in Fig. 1 which is included in various physical quantities in the right hand of Eq.(10) and $F(\delta\underset{\sim}{g})$ defined by Eq.(10) is named the *fluctuation function*. It is noticed that Eq.(10) is the description for the first law of thermodynamics in a fluctuating state and along the most probable or average path ACD with no fluctuation, by virtue of the first law of thermodynamics(1) in the conventional macroscopic description, we should have

$$F(\underset{\sim}{0}) = \rho\dot{e} - (\sigma_{ij}d_{ij} + q_{i,i} + \rho r) = 0 \tag{11}$$

As explained in the previous section, nonequilibrium phenomena in nature are frequently stable in spite of the existence of microscopic fluctuation and from the definition of stability, this fact means that the fluctuating energy is always dissipated by virtue of the property expressed by the second law of thermodynamics. Further, though there always existed microscopic fluctuations, the nonequilibrium phenomenon with the negative rate of entropy production has never been observed since the establishment of the second law of thermodynamics. Therefore, the second law of thermodynamics(5) is still valid for such a fluctuating state. It is noticed that the relationship between the right and left hand sides of inequality(6) essentially depends only on the second law of thermodynamics(5) and so the inequality(6) is considered to be applicable to a fluctuating state like the above. Requiring that even if any internal or external perturbation such as a microscopic fluctuation acts upon a nonequilibrium phenomenon, the rate of entropy production should always be positive, the following thermodynamical inequality is obtained from the inequality(6).

$$\rho\dot{e} - (\sigma_{ij}d_{ij} + q_{i,i} + \rho r) > 0 \tag{12}$$

Taking account of Eq.(10), this inequality is rewritten as

$$\rho\dot{e} - (\sigma_{ij}d_{ij} + q_{i,i} + \rho r) = F(\delta\underset{\sim}{g}) > 0 \tag{13}$$

It is noticed that we have selected the case of positive rate of entropy production from the two possibility in Eq.(5), i.e. positive rate or zero and as long as the above inequality holds, the possibility that the rate of entropy production becomes negative or zero is completely denied through the inequality (6), i.e. the rate of entropy production is always positive. This fact means that the available energy and fluctuating one in a nonequilibrium phenomenon are always dissipated and therefore this phenomenon is stable. Furthermore, considering the first law of thermodynamics(10) in the *microscopic description* along a fluctuating path ABD in Fig. 1 under the condition (13), we have

$$\Delta F_{AB} + \Delta F_{BD} = 0 \tag{14}$$

This fact means that Eqs.(10) and (13) are consistent with the first law of thermodynamics(1) in the conventional macroscopic description. Combining Eqs.(11) and (13), we get

$$\rho\dot{e} \geq \sigma_{ij}d_{ij} + q_{i,i} + \rho r \tag{15}$$

Next, let us consider the purely dissipative process. Firstly, the stress tensor σ_{ij} and the deformation rate one d_{ij} are divided into the reversible parts ${}_R\sigma_{ij}$ and ${}_Rd_{ij}$ and the irreversible parts ${}_D\sigma_{ij}$ and ${}_Dd_{ij}$ respectively as

$$\sigma_{ij} = {}_R\sigma_{ij} + {}_D\sigma_{ij} \tag{16}$$

$$d_{ij} = {}_Rd_{ij} + {}_Dd_{ij} \tag{17}$$

Then, the rate of irreversible mechanical work D is given by

$$D = \sigma_{ij}d_{ij} - {}_R\sigma_{ij}\,{}_Rd_{ij} \tag{18}$$

and the equation of equilibrium of total entropy is expressed as

$$\rho\theta\dot{S} = D + q_{i,i} + \rho r \tag{19}$$

From Eqs.(3), (4) and (19), we have

$$\rho\theta\,{}_i\dot{S} = D + (\ln\theta)_{,i}q_i \tag{20}$$

By use of Eqs.(1) and (19) and the relationship among the Helmholtz free energy ϕ, e and S : $e = \phi + \theta S$ (21)

we have

$$\rho\dot{\phi} = -\rho\dot{\theta}S + \sigma_{ij}d_{ij} - D \tag{22}$$

Then, from Eqs.(3), (4), (6), (21) and (22), we get

$$\rho\theta\,{}_i\dot{S} \geq \rho\theta\,{}_i\dot{S} - \{D + (\ln\theta)_{,i}q_i\} \tag{23}$$

Considering Eq.(23) in a fluctuating state with the positive rate of entropy production, in the similar manner to the case deriving Eq.(15) from Eq.(6), we obtain

$$\rho\theta\,{}_i\dot{S} \geq D + (\ln\theta)_{,i}q_i \tag{24}$$

Now, if the left and right hand sides in this inequality take the most probable value and a fluctuating one respectively, we have

$$[\rho\theta\,{}_i\dot{S}]^* \geq D + (\ln\theta)_{,i}q_i \tag{25}$$

where $[\rho\theta\,{}_i\dot{S}]^*$ is the most probable value of $[\rho\theta\,{}_i\dot{S}]$. The inequality(25) is rewritten into the following equivalent form:

$$[\rho\theta\,{}_i\dot{S}]^* = [D + (\ln\theta)_{,i}q_i]_{max.} \tag{26}$$

For the case that the right and left hand sides in the inequality(24) have the most probable value and a fluctuating one repectively, giving the similar consideration to the above, we get

$$[\rho\theta_i\dot{S}]_{min.} = [D + (\ln\theta),_i q_i]^* \tag{27}$$

As, by Eq.(20), the most probable values of the left and right hand sides in the inequality(24) should be equal to each other, from Eqs.(26) and (27), we have

$$[\rho\theta_i\dot{S}]_{min.} = [D + (\ln\theta),_i q_i]_{max.} \tag{28}$$

For the purely dissipative mechanical process, Eq.(28) reduces to

$$[\rho\theta_i\dot{S}]_{min.} = [D]_{max.} \tag{29}$$

This equation means that *in the most probable or average state, the external and internal mechanical work should be maximum and minimum respectively.*

CONSIDERATION FOR COAXIALITY BETWEEN STRESS AND STRAIN INCREMENT TENSORS

Here, let us consider the coaxility between stress and strain increment tensors in the granular materials in the plane strain state. Suppose that the directions of axes of stress tensor are identical with those of x, y and z axes in the rectangular Cartesian co-ordinates and the directions of principal axes of strain increment tensor deviate from those of stress tensor only in the x-z plane in the above co-ordinates by an angle θ as shown in Fig. 2. Then, the increment dW of external mechanical work calculated in the above co-ordinates is expressed by

$$dW = \sigma_{ij} d\varepsilon_{ij} \tag{30}$$

$$\text{where } \sigma_{ij} = \begin{pmatrix} \sigma_1 & & 0 \\ & \sigma_2 & \\ 0 & & \sigma_3 \end{pmatrix} \tag{31}$$

$$d\varepsilon_{ij} = \begin{pmatrix} d\varepsilon_x & 0 & d\varepsilon_{xz} \\ 0 & 0 & 0 \\ d\varepsilon_{xz} & 0 & d\varepsilon_z \end{pmatrix} \tag{32}$$

$$\left.\begin{aligned} d\varepsilon_x &= d\varepsilon_1\cos^2\theta + d\varepsilon_3\sin^2\theta \\ d\varepsilon_z &= d\varepsilon_1\sin^2\theta + d\varepsilon_3\cos^2\theta \\ d\varepsilon_{xz} &= -(d\varepsilon_1 - d\varepsilon_3)\sin\theta\cos\theta \end{aligned}\right\} \tag{33}$$

$$0 \leqq \theta < \pi/2 \tag{34}$$

σ_1, σ_2 and σ_3 are the principal values of stress tensor respectively, $d\varepsilon_1$ and $d\varepsilon_3$ are the principal values of strain increment tensor respectively and they satisfy the following conditions:

$$\sigma_1 > \sigma_2 > \sigma_3 \tag{35}$$

$$\text{and } d\varepsilon_1 > d\varepsilon_3 \tag{36}$$

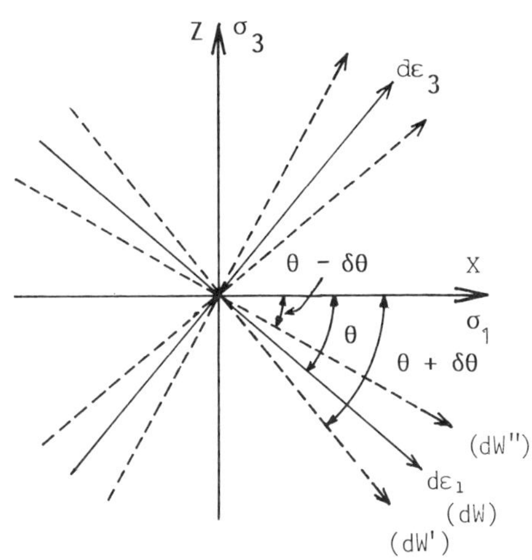

Fig. 2. Relationship between principal axes of stress and strain increment tensors

It is noticed that the conditions (35) and (36) never decrease the generality for the consideration below. Substituting Eqs.(31) through (33) into Eq.(30), Eq.(30) is rewritten by

$$dW = \sigma_1 d\varepsilon_x + \sigma_3 d\varepsilon_z$$
$$= (\sigma_1 d\varepsilon_1 + \sigma_3 d\varepsilon_3) - (\sigma_1 - \sigma_3)(d\varepsilon_1 - d\varepsilon_3)\sin^2\theta \qquad (37)$$

Usually, there are two types of deformation, i.e. stress controlled deformation and strain controlled one. Here, let us consider the stress controlled deformation in which the directions of principal axes and principal values of stress tensor are specified, the angle θ between principal axes of stress and strain increment tensors increases or decreases by an infinitesmal angle $\delta\theta$ and the values $d\varepsilon_1$ and $d\varepsilon_3$ change by infinitesmal values $\delta d\varepsilon_1$ and $\delta d\varepsilon_3$ respectively. It may be supposed that $\delta\theta$, $\delta d\varepsilon_1$ and $\delta d\varepsilon_3$ correspond to the microscopic fluctuations of θ, $d\varepsilon_1$ and $d\varepsilon_3$ in nature respectively.

If the directions of principal axes of strain increment tensor rotate in the clockwise direction by an infinitesmal angle $\delta\theta$ and the principal values of strain increment tensor change into $d\varepsilon_1 + \delta d\varepsilon_1$ and $d\varepsilon_3 + \delta d\varepsilon_3$ respectively as shown in Fig. 2. In the similar manner to the case of Eq.(37), the increment dW' of external mechanical work in this case is expressed by

$$dW' = (\sigma_1 d\varepsilon_1 + \sigma_3 d\varepsilon_3) + (\sigma_1 \delta d\varepsilon_1 + \sigma_3 \delta d\varepsilon_3)$$
$$-[(\sigma_1 - \sigma_3)(d\varepsilon_1 - d\varepsilon_3) + (\sigma_1 - \sigma_3)(\delta d\varepsilon_1 - \delta d\varepsilon_3)]\sin^2(\theta + \delta\theta) \qquad (38)$$

$\delta d\varepsilon_1$ and $\delta d\varepsilon_3$ are the fluctuating quantities and very small compared with $d\varepsilon_1$ and $d\varepsilon_3$. Therefore, neglecting these terms, Eq.(43) reduces to

$$dW' = \sigma_1 d\varepsilon_1 + \sigma_3 d\varepsilon_3 - (\sigma_1 - \sigma_3)(d\varepsilon_1 - d\varepsilon_3)\sin^2(\theta + \delta\theta) \qquad (39)$$

From Eqs.(37) and (39), we have

$$dW - dW' = (\sigma_1 - \sigma_3)(d\varepsilon_1 - d\varepsilon_3)[\sin^2(\theta + \delta\theta) - \sin^2\theta] > 0 \qquad (40)$$

It is noticed that because of the conditions(34), (35) and (36), the right hand side of the above equation is positive.

In the following, if we consider the case that the directions of principal axes of strain increment tensor rotate in the counterclockwise direction by an infinitesmal angle $\delta\theta$ in the similar manner to the case of Eq.(39), the increment dw'' of external mechanical work is expressed by

$$dW'' = (\sigma_1 d\varepsilon_1 + \sigma_3 d\varepsilon_3) - (\sigma_1 - \sigma_3)(d\varepsilon_1 - d\varepsilon_3)\sin^2(\theta - \delta\theta) \qquad (41)$$

From Eqs.(37) and (41), we have

$$dW'' - dW = (\sigma_1 - \sigma_3)(d\varepsilon_1 - d\varepsilon_3)[\sin^2\theta - \sin^2(\theta - \delta\theta)] > 0 \qquad (42)$$

It is noticed that because of the conditions(34), (35) and (36), the right hand side of the above equation is positive. From the inqualities(40) and (42), the following inequality is obtained.

$$dW'' > dW > dW' \qquad (43)$$

In addition, in the similar manner to the above, considering the case that $\theta = 0$, we have

$$\left.\begin{aligned} dW &> dW' \\ dW &> dW'' \end{aligned}\right\} \qquad (44)$$

The inequality(43) expresses that the smaller is the angle between the principal axes of stress and strain increment tensors, the bigger is the external mechanical work and the inequality(44) explains that when the directions of principal axes of stress tensor are identical with those of strain increment tensor, the external mechanical work is maximum.

Then, if we consider the strain controlled deformation in the similar manner to the case of stress controlled deformation, we can obtain the completely same results as those in the case of stress controlled deformation.

According to Eq.(29) in our new optimality theory, the external mechanical work should be maximum in the most probable or average state. Thus, from this fact, it is concluded that *even if the directions of principal axes of stress and strain increment tensors are not identical at an initial stage of deformation, the probability decreasing the angle between them is higher than the one increasing it and the most probable directions of them should become identical with deformation as soon as possible and once the most probable directions of them become identical, they statistically never deviate from each other.*

However, it is noted that the above conclution is applicable only to the *materials with fully irreversible deformation surrounding which the boundaries never prevent the essential deformation* of them.

CONCLUDING REMARKS

The coaxiality between stress and strain increment tensors was considered by use of a new general variational theory and this theoretical result supported the exprimental result of Arthur, et al. and Kumasaka, et al. The new general theory presented here has already been applied to solve various long-pending problems(ref. 7).

REFERENCES

1 J.R.F. Arthur and B.K. Menzies, Inherent Anisotropy in a Sand, Geotechnique, 22(1) (1972) 115-128

2 J.R.F. Arthur, K.S. Chua and T. Dunstan, Induced Anisotropy in a Sand, Geotechnique, 27(1) (1977) 13-30

3 H. Kumasaka, M. Satake and S. Niiseki, On Coaxiality Between Stress and Strain Increment Tensors(in Japanese), Proc. of the 36th National Conference of JSCE, III (1981) 3-4

4 A. Drescher and G. de Josselin de Jong, Photoelastic Verification of a Mechanical Model for the Flow of a Granular Material, J. of Mech. Phys. Solids, 20 (1972) 337-351

5 M. Oda and J. Konishi, Rotation of Principal Stresses in Granular Material During Simple Shear, Soils and Foundations, 14(4) (1974) 39-53

6 K.H. Roscoe, The Influence of Strains in Soil Mechanics, in: Milestones in Soil Mechanics, Tomas Telford Ltd, England, 1975, pp.281-329

7 S. Niiseki, New Optimality Theory, in: S. Ishizaka(Eds.), Science on Form, KTK Science Publishers, Tokyo, 1986, pp.593-600

Micromechanics of Granular Materials, edited by M. Satake and J.T. Jenkins
Elsevier Science Publishers B.V., Amsterdam, 1988 — Printed in The Netherlands

A CONSTITUTIVE MODEL FOR GRANULAR MATERIALS EVALUATING PRINCIPAL STRESS ROTATION

H.Matsuoka[1], K.Sakakibara[2] and Y.Suzuki[1]

[1]Dept. of Civil Eng., Nagoya Institute of Technology, Nagoya 466 (Japan)

[2]Japan Highway Public Corporation, Kasumigaseki, Tokyo 102 (Japan)

SUMMARY

A constitutive model for granular materials by which the general strain increments ($d\varepsilon_x$, $d\varepsilon_y$ and $d\gamma_{xy}$) are directly related to the general stress increments ($d\sigma_x$, $d\sigma_y$ and $d\tau_{xy}$) is proposed. The stress-strain matrix between the strain increments and the stress increments is expressed in general coordinates. In order to evaluate the influence of rotation of the principal stress axes on strains, "principal stress rotation tests", in which the stress path is the circumference of a Mohr's circle, are carried out by a "two-dimensional general stress apparatus" using a stack of aluminium rods, and analyzed by the proposed model. The effect of discontinuous change in the principal stress direction is also considered from the fabric change of granular materials. The model is extended to a three-dimensional one.

INTRODUCTION

It is natural to consider that the change in stresses on an arbitrary plane causes strains in every kind of material. In the theory of plasticity, however, such an idea is not satisfied because the constitutive relation is usually formulated by the stress invariants such as the principal stresses. For example, the strains caused by the "principal stress rotation test" along a Mohr's stress circle under constant principal stresses cannot be explained by the usual theory of plasticity. Here, a constitutive model for granular materials by which the general strain increments are directly related to the general stress increments is presented. The model predicts the strains not only due to "consolidation" and "shear", but also due to "principal stress rotation".

A CONSTITUTIVE MODEL FOR GRANULAR MATERIALS EXPRESSED IN GENERAL COORDINATES

The following equation has been derived from assuming the hyperbolic relationship between the shear-normal stress ratio (τ_{xy}/σ_x or τ_{xy}/σ_y) and the shear strain (γ_{xy}) expressed in general coordinates (refs. 1-4).

$$\gamma_{xy} = \frac{1}{G_o} \cdot \frac{\sin\phi \cdot \sin\phi_{mo} \cdot \sin 2\alpha}{\sin\phi - \sin\phi_{mo}} \tag{1}$$

where G_o is the gradient of the initial tangent of the hyperbolic relationship, ϕ is the internal friction angle, ϕ_{mo} is the mobilized internal friction angle

($\sin\phi_{mo}=(\sigma_1-\sigma_3)/(\sigma_1+\sigma_3)$) and α is the angle of the principal stress direction. As $1/G_o$ ($\equiv k_s$) is approximately proportional to the logarithm of the mean principal stress σ_m, γ_{xy} in Eq.(1) is considerd to be the function of ϕ_{mo}, α and σ_m. By totally differentiating γ_{xy} with respect to ϕ_{mo}, α and σ_m, [1]$d\gamma_{xy}{}^s$: shear strain increment due to shear ($d\phi_{mo}$), [2]$d\gamma_{xy}{}^r$: shear strain increment due to principal stress rotation ($d\alpha$) and [3]$d\gamma_{xy}{}^{ac}$: shear strain increment due to anisotropic consolidation ($d\sigma_m$) have been derived as follows(refs. 1-4):

$$d\gamma_{xy}^{s} = k_s\cdot\frac{\sin^2\phi\cdot\cos\phi_{mo}\cdot\sin 2\alpha}{(\sin\phi-\sin\phi_{mo})^2}\cdot d\phi_{mo} \qquad \text{"shear"} \tag{2}$$

$$d\gamma_{xy}^{r} = 2k_s\cdot\frac{\sin\phi\cdot\sin\phi_{mo}\cdot\sin 2(\alpha+\delta)}{\sin\phi-\sin\phi_{mo}}\cdot d\alpha \qquad \text{"principal stress rotation"} \tag{3}$$

$$d\gamma_{xy}^{ac} = k_c\cdot\frac{\sin\phi\cdot\sin\phi_{mo}\cdot\sin 2\alpha}{\sin\phi-\sin\phi_{mo}}\cdot\frac{d\sigma_m}{\sigma_m} \qquad \text{"anisotropic consolidation"} \tag{4}$$

It is considered in $d\gamma_{xy}{}^r$ that the principal stress direction deviates by δ from the principal strain increment direction. In order to obtain the normal strain increments $d\varepsilon_x$ and $d\varepsilon_y$ from those shear strain increments, the stress ratio vs. strain increment ratio relation, which corresponds to the flow rule in the theory of plasticity, is neccessary. So, the following stress ratio-strain increment ratio relationship proposed by Matsuoka (ref. 5) has been introduced.

$$\frac{\tau}{\sigma_N} = \lambda\cdot\left(-\frac{d\varepsilon_N}{d\gamma}\right) + \mu \tag{5}$$

where τ/σ_N is the shear-normal stress ratio on the mobilized plane ($\tau/\sigma_N=\tan\phi_{mo}$), $d\varepsilon_N/d\gamma$ is the normal-shear strain increment ratio on the mobilized plane, and (λ and μ) are the material constants. If Eq.(5) is expressed in general coordinates, we can get the following expression (refs. 1-3).

$$\left.\begin{matrix} d\varepsilon_x/d\gamma_{xy} \\ d\varepsilon_y/d\gamma_{xy} \end{matrix}\right\} = \frac{2\cdot\frac{\mu-\tan\phi_{mo}}{\lambda}\cdot\cos\phi_{mo}+\sin\phi_{mo}\pm\cos 2\alpha}{2\cdot\sin 2\alpha} \tag{6}$$

Combining Eqs.(2), (3) and (4) with Eq.(6), the normal strain increments $d\varepsilon_x$ and $d\varepsilon_y$ for these three kinds of shear strain increments can be obtained. It should

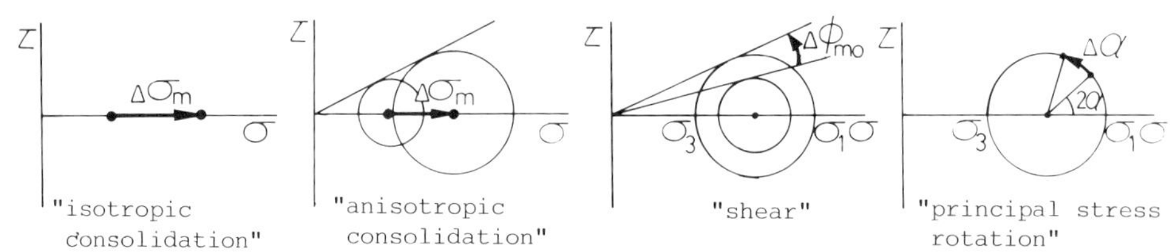

Fig. 1. Stress changes in Mohr's circle corresponding to isotropic consolidation, anisotropic consolidation, shear and principal stress rotation.

be noted that 2α in Eq.(6) is changed for $2(\alpha+\delta)$ when the strains due to "principal stress rotation" are calculated by using Eq.(3). In addition to the above-mentioned strain increments, [4]$d\varepsilon_x{}^{ic}=d\varepsilon_y{}^{ic}$: normal strain increments due to isotropic consolidation should be taken into account. They are written in a two-dimensional case as follows:

$$\left.\begin{matrix} d\varepsilon_x^{ic} \\ d\varepsilon_y^{ic} \end{matrix}\right\} = \frac{1}{2}\cdot\frac{0.434C_c}{1+e_o}\cdot\frac{d\sigma_m}{\sigma_m} \quad \text{or} \quad \frac{1}{2}\cdot\frac{0.434C_s}{1+e_o}\frac{d\sigma_m}{\sigma_m} \tag{7}$$

As seen from Eqs.(2)-(7), this constitutive model can predict the strains not only due to "isotropic consolidation", "anisotropic consolidation" and "shear", but also the strains due to "principal stress rotation" (see Fig.1). As $d\phi_{mo}$, $d\alpha$ and $d\sigma_m$ can be expressed by the general stress increments ($d\sigma_x$, $d\sigma_y$ and $d\tau_{xy}$), the general strain increments can be directly related to the general stress increments as follows (refs. 1-4):

$$\{d\varepsilon_x,\ d\varepsilon_y,\ d\gamma_{xy}\}^T = [D]^{-1}\cdot\{d\sigma_x,\ d\sigma_y,\ d\tau_{xy}\}^T \tag{8}$$

where [D] is the stress-strain matrix.

COMPONENTS OF THE STRESS-STRAIN MATRIX

The components of the matrix $[D]^{-1}$ can be written in the four domains (see Fig.2) as follows (refs. 1-4):

$$[D]^{-1} = \begin{Bmatrix} D_{11} & D_{12} & D_{13} \\ D_{21} & D_{22} & D_{23} \\ D_{31} & D_{32} & D_{33} \end{Bmatrix} \tag{9}$$

I.($d\phi_{mo} \geqq 0,\ d\sigma_m \geqq 0$)

D_{11} = FF × SS × A1 + FR × RR × B1 + FF × CA × C1 + CCI × C1
D_{12} = FF × SS × A2 + FR × RR × B2 + FF × CA × C1 + CCI × C1
D_{13} = FF × SS × A3 + FR × RR × B3
D_{21} = GG × SS × A1 + GR × RR × B1 + GG × CA × C1 + CCI × C1
D_{22} = GG × SS × A2 + GR × RR × B2 + GG × CA × C1 + CCI × C1
D_{23} = GG × SS × A3 + GR × RR × B3
D_{31} = HH × SS × A1 + HR × RR × B1 + HH × CA × C1
D_{32} = HH × SS × A2 + HR × RR × B2 + HH × CA × C1
D_{33} = HH × SS × A3 + HR × RR × B3

II.($d\phi_{mo} \geqq 0,\ d\sigma_m < 0$)

D_{11} = FF × SS × A1 + FR × RR × B1 + CSI × C1
D_{12} = FF × SS × A2 + FR × RR × B2 + CSI × C1
D_{13} = FF × SS × A3 + FR × RR × B3
D_{21} = GG × SS × A1 + GR × RR × B1 + CSI × C1
D_{22} = GG × SS × A2 + GR × RR × B2 + CSI × C1
D_{23} = GG × SS × A3 + GR × RR × B3
D_{31} = HH × SS × A1 + HR × RR × B1
D_{32} = HH × SS × A2 + HR × RR × B2
D_{33} = HH × SS × A3 + HR × RR × B3

III.($d\phi_{mo} < 0,\ d\sigma_m < 0$)

D_{11} = FFU × SSU × A1 + FR × RR × B1 + CSI × C1
D_{12} = FFU × SSU × A2 + FR × RR × B2 + CSI × C1
D_{13} = FFU × SSU × A3 + FR × RR × B3
D_{21} = GGU × SSU × A1 + GR × RR × B1 + CSI × C1
D_{22} = GGU × SSU × A2 + GR × RR × B2 + CSI × C1
D_{23} = GGU × SSU × A3 + GR × RR × B3
D_{31} = HH × SSU × A1 + HR × RR × B1
D_{32} = HH × SSU × A2 + HR × RR × B2
D_{33} = HH × SSU × A3 + HR × RR × B3

IV.($d\phi_{mo} < 0,\ d\sigma_m \geqq 0$)

D_{11} = FFU × SSU × A1 + FR × RR × B1 + CCI × C1
D_{12} = FFU × SSU × A2 + FR × RR × B2 + CCI × C1
D_{13} = FFU × SSU × A3 + FR × RR × B3
D_{21} = GGU × SSU × A1 + GR × RR × B1 + CCI × C1
D_{22} = GGU × SSU × A2 + GR × RR × B2 + CCI × C1
D_{23} = GGU × SSU × A3 + GR × RR × B3
D_{31} = HH × SSU × A1 + HR × RR × B1
D_{32} = HH × SSU × A2 + HR × RR × B2
D_{33} = HH × SSU × A3 + HR × RR × B3

where

$$FF = \frac{1}{2}\{2\frac{\mu - \tan\phi_{mo}}{\lambda}\cdot\cos\phi_{mo} + \sin\phi_{mo} + \cos 2\alpha\}$$

$$GG = \frac{1}{2}\{2\frac{\mu - \tan\phi_{mo}}{\lambda}\cdot\cos\phi_{mo} + \sin\phi_{mo} - \cos 2\alpha\}$$

$$HH = \sin 2\alpha$$

$$FFU = \frac{1}{2}\{2\frac{-\mu - \tan\phi_{mo}}{\lambda}\cdot\cos\phi_{mo} + \sin\phi_{mo} + \cos 2\alpha\}$$

$$GGU = \frac{1}{2}\{2\frac{-\mu - \tan\phi_{mo}}{\lambda}\cdot\cos\phi_{mo} + \sin\phi_{mo} - \cos 2\alpha\}$$

$$FR = \frac{1}{2}\{2\frac{\mu - \tan\phi_{mo}}{\lambda}\cdot\cos\phi_{mo} + \sin\phi_{mo} + \cos 2(\alpha + \delta)\}$$

$$GR = \frac{1}{2}\{2\frac{\mu - \tan\phi_{mo}}{\lambda}\cdot\cos\phi_{mo} + \sin\phi_{mo} - \cos 2(\alpha + \delta)\}$$

$$HR = \sin 2(\alpha + \delta)$$

$$SS = k_s\cdot\frac{\sin^2\phi\cdot\cos\phi_{mo}}{(\sin\phi - \sin\phi_{mo})^2}$$

$$SSU = k_s\cdot\frac{\sin^2\phi\cdot\cos\phi_{mo}}{(\sin\phi + \sin\phi_{mo})^2}$$

$$RR = 2k_s\cdot\frac{\sin\phi\cdot\sin\phi_{mo}}{\sin\phi - \sin\phi_{mo}}$$

$$CA = k_c\cdot\frac{\sin\phi\cdot\sin\phi_{mo}}{\sin\phi - \sin\phi_{mo}}$$

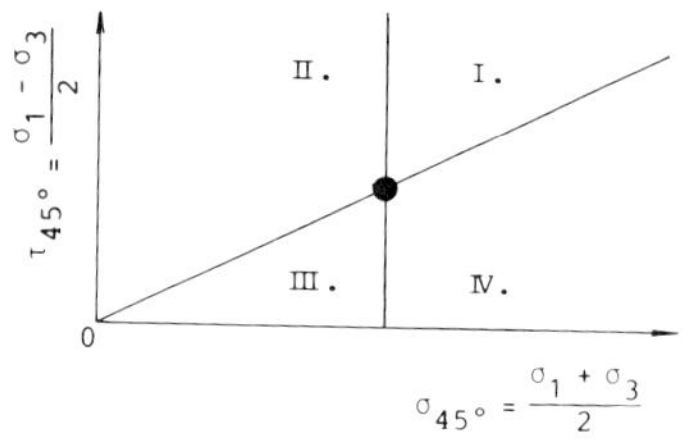

Fig. 2. Four domains of stress changes.

$$CCI = \frac{0.434}{2}\cdot\frac{C_c}{1 + e_o} \qquad CSI = \frac{0.434}{2}\cdot\frac{C_s}{1 + e_o}$$

$$A_1 = \frac{I_1/I_2}{8\tan\phi_{mo}}\cdot\cos^2\phi_{mo}\cdot\{2 - (I_1/I_2)\cdot\sigma_y\}$$

$$A_2 = \frac{I_1/I_2}{8\tan\phi_{mo}}\cdot\cos^2\phi_{mo}\cdot\{2 - (I_1/I_2)\cdot\sigma_x\}$$

$$A_3 = \frac{I_1/I_2}{8\tan\phi_{mo}}\cdot\cos^2\phi_{mo}\cdot\{2(I_1/I_2)\cdot\tau_{xy}\}$$

$$B_1 = \cos^2 2\alpha\cdot\frac{(-\tau_{xy})}{(\sigma_x - \sigma_y)^2} \qquad B_2 = \cos^2 2\alpha\cdot\frac{\tau_{xy}}{(\sigma_x - \sigma_y)^2}$$

$$B_3 = \cos^2 2\alpha\cdot\frac{1}{(\sigma_x - \sigma_y)} \qquad C_1 = \frac{1}{(\sigma_x + \sigma_y)}$$

$$I_1 = \sigma_x + \sigma_y \qquad I_2 = \sigma_x\sigma_y - \tau_{xy}^2$$

In the above equations the letter U means the unloading process. The stress-strain parameters in this constitutive model are k_s, k_c, λ, μ, ϕ, $C_c/(1+e_o)$ and $C_s/(1+e_o)$. ϕ denotes the internal friction angle in terms of effective stresses, C_c the compression index, C_s the swell index and e_o the initial void ratio. k_s is a parameter to be found in Eq.(1) ($1/G_o \equiv k_s$) or in SS, SSU and RR, which decides the magnitude of the shear strains due to shear and principal stress rotation. It is seen from Eq.(1) that k_s is the gradient of the initial tangent of the relation between $(\varepsilon_1-\varepsilon_3)$ and $(\sigma_1-\sigma_3)/(\sigma_1+\sigma_3)=\sin\phi_{mo}$ (in the case of $\alpha=45°$) under a constant mean principal stress σ_m. k_c is a parameter in CA, which decides the magnitude of the shear strain due to anisotropic consolidation. It can be estimated by the following equation which is derived from a microscopic study of the compression mechanism of granular materials.

$$k_c = 0.44\frac{C_c}{1 + e_o} \tag{10}$$

λ and μ are parameters in FF, GG, FFU, GGU, FR and GR, which correspond to the gradient and the ordinate intersection of the linear relation between the shear-normal stress ratio and the normal-shear strain increment ratio on the mobilized plane respectively (see Fig.8).

Even if the coordinate axes are rotated, the strain increments calculated by

Eqs.(2) and (3) have the same values at the same stress state and stress change. So, the calculated strain increments satisfy the characteristics as a tensor.

EXTENSION TO THREE DIMENSIONS

The constitutive relation mentioned here is a two-dimensional one. In order to extend it to three dimensions, the superposition of the "two-dimensional" principal strain increments is assumed as follows (refs. 5-6):

$$\left.\begin{aligned} d\varepsilon_1 &= d\varepsilon_{1(12)} + d\varepsilon_{1(13)} \\ d\varepsilon_2 &= d\varepsilon_{2(12)} + d\varepsilon_{2(23)} \\ d\varepsilon_3 &= d\varepsilon_{3(23)} + d\varepsilon_{3(13)} \end{aligned}\right\} \tag{11}$$

where $d\varepsilon_{1(12)}$, for example, denotes the "two-dimensional" major principal strain increment caused under the two principal stresses σ_1 and σ_2 (see Fig.3). The terms in the right sides of Eq.(11) mean such "two-dimensional" principal strain increments as obtained by Eq.(8).

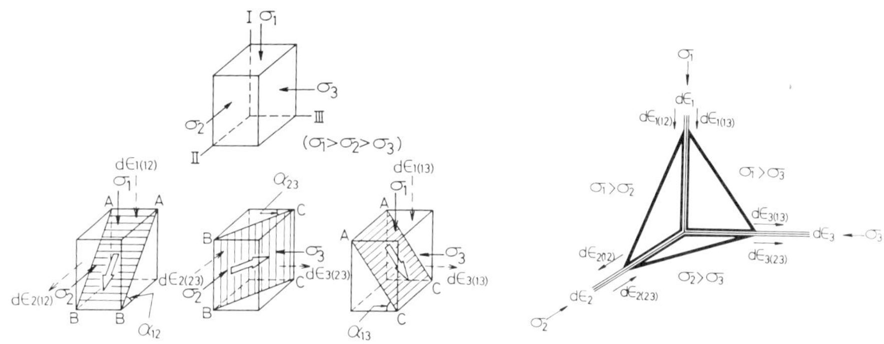

Fig. 3. Superposition of "two-dimensional" principal strain increments under respective pairs of principal stresses.

TWO-DIMENSIONAL GENERAL STRESS APPARATUS

A general stress apparatus which can apply the general stresses (σ_x, σ_y and τ_{xy}) independently has been made for a stack of aluminium rods (mixture of Φ1.6mm and Φ3.0mm, 50mm in length) as a two-dimensional model of granular materials. The normal stress σ_x in the vertical direction and the normal stress σ_y in the horizontal direction are applied by the wires through pulleys, and the shear stress τ_{xy} is applied by the loading plate on the sample. These three stresses σ_x, σ_y and τ_{xy} can be measured by the load cells independently (see Fig.4). The

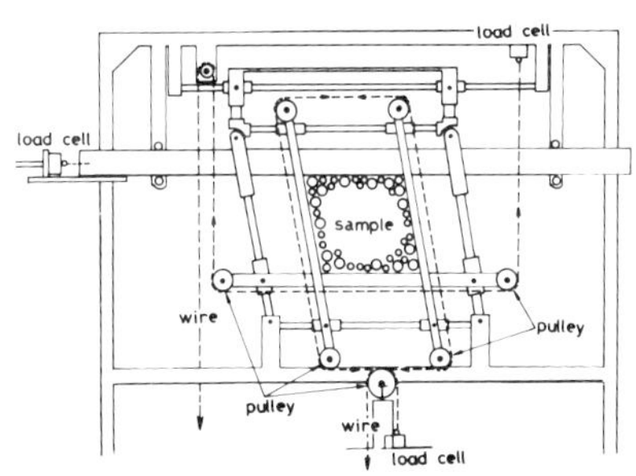

Fig. 4. Main part of two-dimensional general stress apparatus for rod mass.

general strains ε_x, ε_y and γ_{xy} can be also measured by dial gauges independently.

PRINCIPAL STRESS ROTATION TESTS WITH SEVERAL ROTATION TIMES

Fig.5(a) show three photoelastic pictures taken by a cyclic simple shear test on photoelastic rod mass. We can see the change in the principal stress direction under cyclic shearing. Fig.5(b) show the histograms of interparticle contact normals for the three pictures, expressed in a radial direction. They indicate the concentration of contact normals to the principal stress direction. These histograms are considered to be one of the expressions of fabric.

Fig.6(a) shows the qualitative pattern of a histogram of interparticle contact normals expressed in a radial direction, which changes from a circular distribution to an elliptical distribution during shear. Such an elliptical distribution rotates continuously during rotation of the principal stress axes, as shown in Fig.6(b). So, it is easy to understand the origin of the strains caused by principal stress rotation. From $\alpha=0°$ to $\alpha=180°$ the elliptical distribution passes in the "virgin" zone in which such fabric change has never been experienced, and after $\alpha=180°$ the experienced fabric change will be repeated. Paying attention to the fact that the shear strain increment due to principal stress rotation $d\gamma_{xy}{}^r$ has the same parameter k_s as the shear strain increment due to shear $d\gamma_{xy}{}^s$ (see Eqs.(2) and (3)), the value of k_s is reduced in the experienced fabric by measuring the ratio of the tangential gradient Ⓐ to Ⓑ of the cyclic stress-strain relationship at the stress ratio $\sin\phi_a=0.25$ in Fig. 7, which is the same stress ratio in the principal stress rotation. This is based on the similarity of the change in structures during cyclic shearing and principal stress rotation (see Fig.6).

Fig.8 shows the shear-normal stress ratio vs. normal-shear strain increment ratio relationship on the mobilized plane ($\alpha=45°+\phi_{mo}/2$) when the stress state circles four times along a Mohr's circle. The broken line in the figure denotes the line of no volumetric strain increment ($d\varepsilon_v=0$). It is seen from Fig.8 that the measured values converge to the line of no volumetric strain increment with the number of rotation times. Therefore, in the analysis the value of the strain increment ratio is assumed to be the intermediate value between the present value and the value at $d\varepsilon_v=0$ of the strain increment ratios.

Fig.9(a)-(d) show the results of the principal stress rotation test on the stack of aluminium rods in which the stress state circles four times along a Mohr's circle, and the analytical values by the proposed model. It is seen from this figure that the measured strains in the 2nd to 4th cycles are smaller than in the 1st cycle, and the measured volumetric strain ε_v becomes smaller with the number of rotation times. The analytical values based on the above-mentioned ideas explain well such tendency of the measured values.

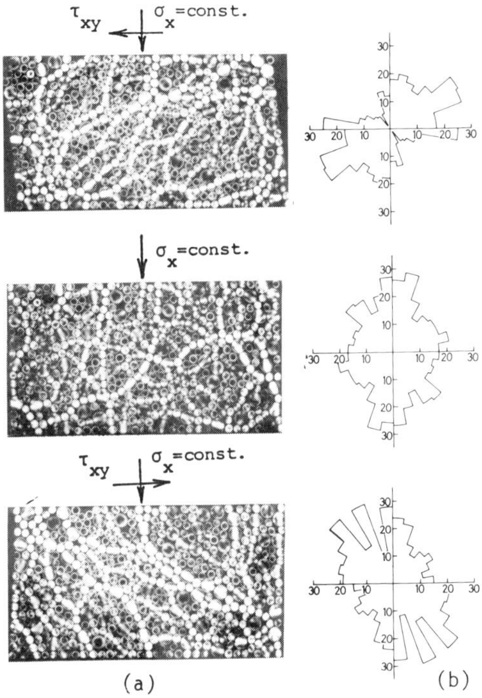

Fig. 5. Photoelastic pictures taken by a cyclic simple shear test on photoelastic rod mass and the histograms of interparticle contact normals.

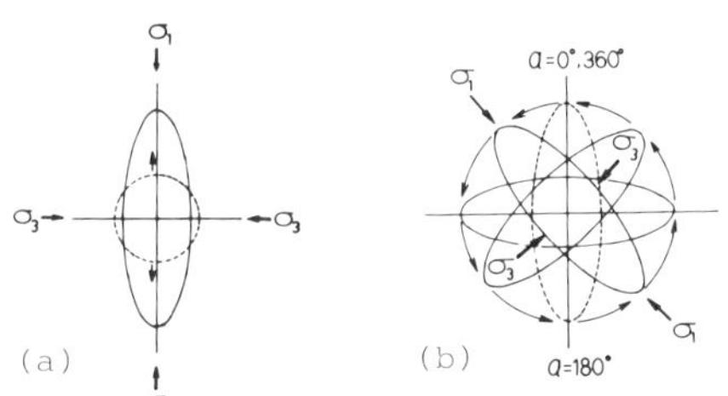

Fig. 6. Pattern of histogram of inter-particle contact normals expressed in radial direction (a) under shear and (b) under principal stress rotation.

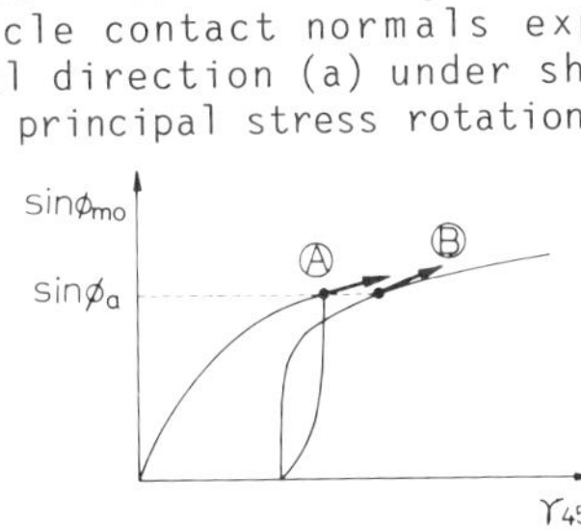

Fig. 7. Tangential gradients of shear-normal stress ratio vs. shear strain relation on 45° plane under virgin and cyclic loadings.

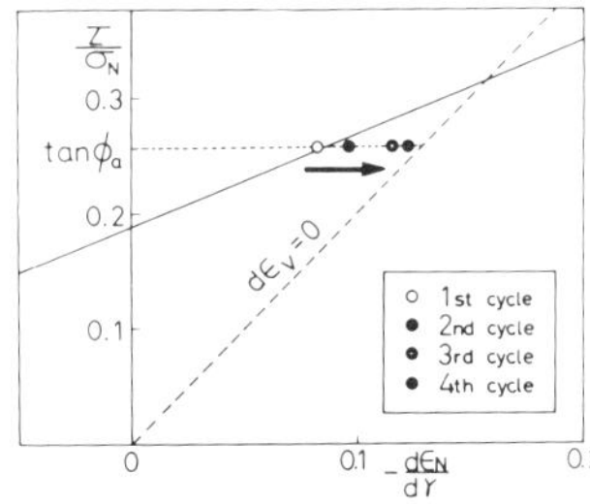

Fig. 8. Shear-normal stress ratio vs. normal-shear strain increment ratio relation on mobilized plane under four times of principal stress rotation.

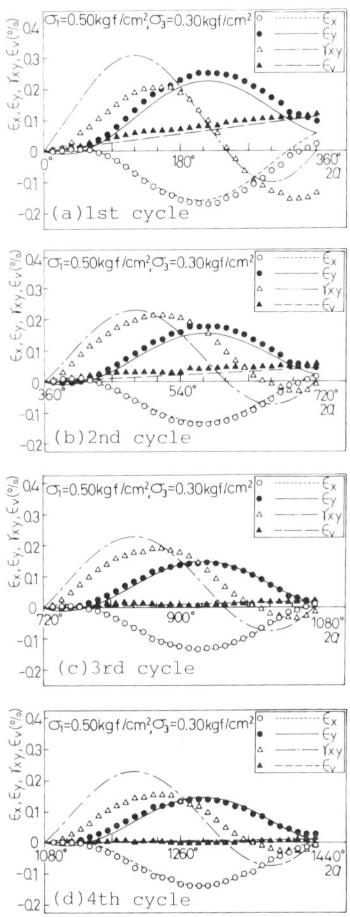

Fig. 9. Comparison between measured strains of principal stress rotation test on rod mass and calculated strains.

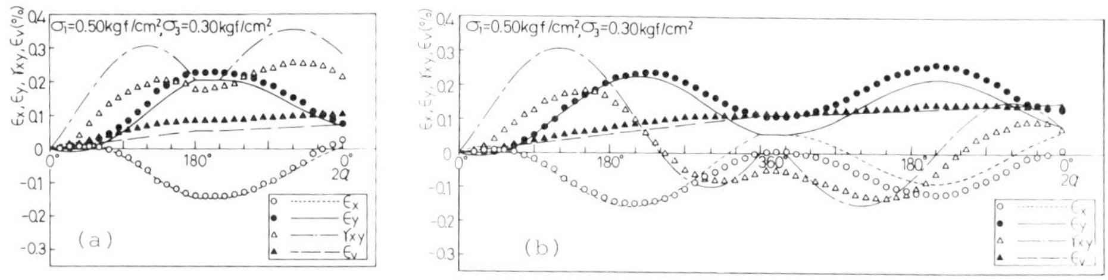

Fig. 10. Comparison between measured strains of principal stress rotation test on rod mass and calculated strains, when rotated reversely at (a)$2\alpha = 180°$ and (b)$2\alpha = 360°$.

Fig.10(a) and (b) show the results of the principal stress rotation tests on the same sample in which the stress states circle from $2\alpha = 0°$ to $2\alpha = 180°$ and 360° in the same direction and then the stress states circle reversely to $2\alpha = 0°$, and the analytical values by the proposed model. It is assumed in this analysis that the strains are not produced during $\Delta(2\alpha) = 30°$ after the reverse rotation of the principal stress direction, because the measured strains do not change sensitively just after the reverse rotation.

SIMPLE SHEAR TESTS

Fig.11(a), (b) and (c) show the comparison between the measured values of a simple shear test on the stack of aluminium rods and the calculated values. The calculated values are obtained from the following conditions of no vertical stress increment and no lateral strain increment.

$$d\sigma_x = 0 \quad (\sigma_x = \text{const.}) \tag{12}$$

$$d\varepsilon_y = d\varepsilon_y^s + d\varepsilon_y^r + d\varepsilon_y^{ac} + d\varepsilon_y^{ic} = 0 \tag{13}$$

where $d\varepsilon_y^s$, $d\varepsilon_y^r$, $d\varepsilon_y^{ac}$ and $d\varepsilon_y^{ic}$ are the lateral strain increments due to shear, principal stress rotation, anisotropic consolidation and isotropic consolidation respectively. The broken lines in Fig.11 are the calculated values neglecting the strain increment due to principal stress rotation. Fig.11(b) shows the measured and calculated values of the changes in

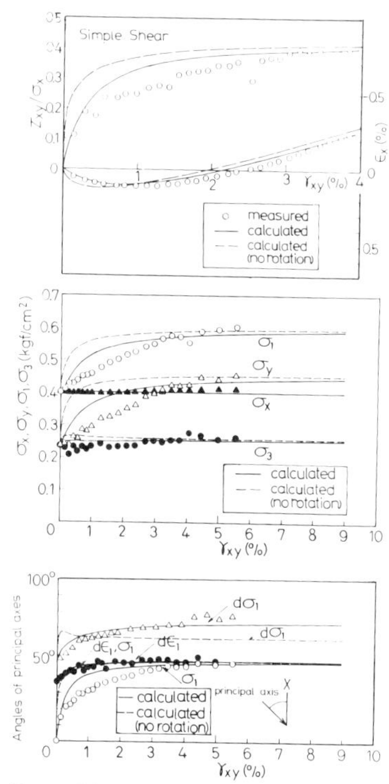

Fig. 11. Comparison between measured values of simple shear test on rod mass and calculated values.

stresses (σ_x, σ_y, σ_1 and σ_3) during simple shear. Fig.11(c) also represents the measured and calculated values of the changes in angles of the principal axes (σ_1, $d\sigma_1$ and $d\varepsilon_1$) during simple shear.

The stress-strain parameters used for all these analyses are as follows. For the stack of aluminium rods: k_s=0.31%, λ=0.8, μ=0.19, ϕ=24°, $C_c/(1+e_o)$=1.2% and δ=30°.

EFFECT OF DISCONTINUOUS CHANGE IN PRINCIPAL STRESS DIRECTION

Next, we consider the case that the direction of the principal stresses is changed in reloading process. Fig.12(a)-(c) show the elliptical patterns of the qualitative histograms of contact normals expressed in a radial direction from the origin. Fig.12(a) shows the case that the principal stress direction in reloading (③) is different by 90° from that in the first "virgin" loading (②). Fig.12(b) represents the case that the principal stress direction in reloading (③) is different by 45° from that in the first loading (②). Fig.12(c) shows the case that the principal stress direction in reloading (③) is the same as in the first loading (②). Fig.13(a)-(c) show the measured stress-strain curves corresponding to the above three cases, obtained by the general stress apparatus of aluminium rod mass. They are expressed by the relationship among the stress ratio $(\sigma_1-\sigma_3)/(\sigma_1+\sigma_3)$, the shear strain $(\varepsilon_1-\varepsilon_3)$ and the volumetric strain $(\varepsilon_1+\varepsilon_3)$.

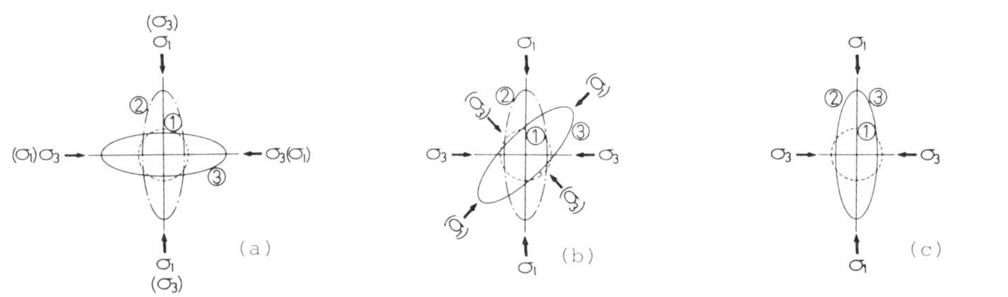

Fig. 12. Qualitative patterns of the histograms of contact normals when the principal stress direction is changed in reloading process.

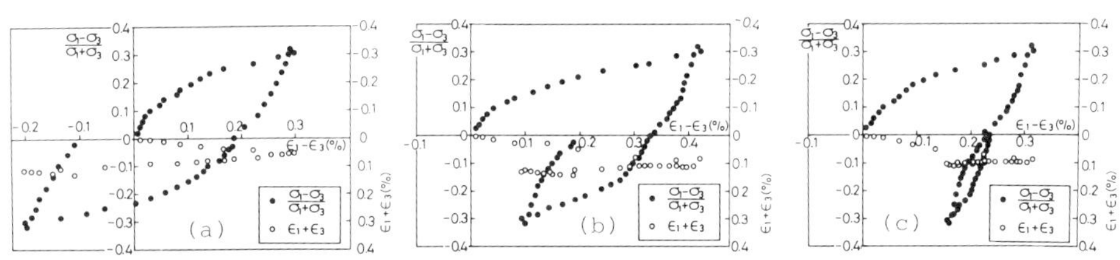

Fig. 13. Measured stress-strain relations corresponding to three cases in Fig. 12.

As shown in the elliptical patterns of Fig.12, the change in the distribution of contact normals is biggest in case (a), smallest in case (c) and intermediate in case (b). It is seen from Fig.13(a)-(c) that the big change in fabric causes big strains and the small change causes small strains. We usually consider case (a) and case (c), for example, in compression and extension conditions, but we should consider the more general case such as case (b).

CONCLUSIONS

The main results are summarized as follows:

1)A constitutive model for granular materials by which the general strain increments ($d\varepsilon_x$, $d\varepsilon_y$ and $d\gamma_{xy}$) are directly related to the general stress increments ($d\sigma_x$, $d\sigma_y$ and $d\tau_{xy}$) is proposed. The stress-strain matrix is expressed in general coordinates. The model predicts the strains not only due to "consolidation" and "shear", but also due to "principal stress rotation".

2)The model explains well the results of "principal stress rotation tests" whose stress states are rotated several times along a Mohr's stress circle under constant principal stresses. The influence of rotation of the principal stress axes on strains is evaluated by the proposed model.

3)As an example, the stress-strain behaviour under simple shear is analyzed by the proposed model.

4)The effect of discontinuous change in principal stress direction in reloading process is also considered from the fabric change of granular materials.

ACKNOWLEDGEMENTS

The authors wish to thank Prof. T. Nakai at Nagoya Institute of Technology for helpful discussions. They are also indebted to Messrs. Y. Iwata and J. Nozoe for their assistance.

REFERENCES

1 H. Matsuoka, Y. Iwata, and K. Sakakibara, Proc. of 2nd Int. Symp. on Numerical Models in Geomechanics, Ghent, 1986, pp.67-78.
2 H. Matsuoka, Y. Iwata, K. Sakakibara, and Y. Hara, Proc. of 2nd Int. Conf. on Constitutive Laws for Engineering Materials, Theory and Applications, 1, 1987, pp.413-420.
3 H. Matsuoka and K. Sakakibara, Soils and Foundations, Vol.27, No.4, 1987, pp.73-88.
4 H. Matsuoka, Y. Suzuki and K. Sakakibara, Proc. of 6th Int. Conf. on Numerical Methods in Geomechanics, Innsbruck, 1988 (to appear).
5 H. Matsuoka, Soils and Foundations, Vol.14, No.2, 1974, pp.47-61.
6 H. Matsuoka, Proc. of 4th Int. Conf. on Numerical Methods in Geomechanics, Edmonton, Vol.1, 1982, pp.223-233.

Micromechanics of Granular Materials, edited by M. Satake and J.T. Jenkins
Elsevier Science Publishers B.V., Amsterdam, 1988 — Printed in The Netherlands

A UNIFIED MECHANICAL QUANTITY FOR GRANULAR MATERIALS IN THREE-DIMENSIONAL STRESSES

T. NAKAI

Dept. of Civil Eng., Nagoya Institute of Technology, Gokiso-cho, Showa-ku, Nagoya 466 (Japan)

ABSTRACT

A mechanical quantity is presented to describe uniquely the behavior of granular materials in three-dimensional stresses. The validity of this quantity is discussed on the basis of the results of triaxial compression and extension tests and true triaxial tests under constant mean principal stresses. A way to apply this quantity t_{ij} to ordinary elastoplastic theory is then mentioned.

INTRODUCTION

Many elastoplastic models have been developed to describe stress-strain behavior of soils. Most of them employ the stress and strain increment parameters of the Mises type in formulating the models: the well-known Cambridge model (refs.1 and 2) is a typical example. However, these models using the Mises type parameters cannot describe the behavior of granular materials such as soil in three-dimensional stress in a uniform manner. Although some attempts have been done to overcome this disadvantage (e.g., modifying the yield and/or plastic potential functions), these are not essential.

From a point of view that the shear behavior of granular materials is governed by the frictional law in three-dimensional stresses, Matsuoka and Nakai (refs.3 and 4) proposed the concept of "Spatially Mobilized Plane" (SMP). By extending this concept, a new concept (named SMP*) was developed by which the three-dimensional behavior of granular materials are explained more precisely (refs.5 and 6). Further, from the reconsideration of the concept of SMP*, a mechanical quantity (second order symmetric tensor) has been presented (refs.7 and 8). In the present paper, the validity of this mechanical quantity is examined using various test results in three-dimensional stresses, and its significance and a way for applying to elastoplastic theories are discussed.

SHEAR BEHAVIOR OF GRANULAR MATERIALS IN THREE-DIMENSIONAL STRESSES

Fig.1 shows results of the constant mean principal stress tests on Toyoura sand (initial void ratio $e_0 \simeq 0.68$) under triaxial compression and extension conditions in terms of the relation between q/p and $d\varepsilon_v/d\varepsilon_d$, according to the stress and strain increment parameters of the Mises type which are used in most of constitutive models such as Cambridge model (refs.1 and 2). Here, these stress and strain increment parameters are defined as

$$p = \frac{1}{3}(\sigma_1 + \sigma_2 + \sigma_3) = \frac{1}{3}J_1 \tag{1}$$

$$q = \frac{1}{\sqrt{2}}\sqrt{(\sigma_1 - \sigma_2)^2 + (\sigma_2 - \sigma_3)^2 + (\sigma_3 - \sigma_1)^2} = \sqrt{J_1^2 - 3J_2} \tag{2}$$

$$d\varepsilon_v = d\varepsilon_1 + d\varepsilon_2 + d\varepsilon_3 \tag{3}$$

$$d\varepsilon_d = \frac{\sqrt{2}}{3}\sqrt{(d\varepsilon_1 - d\varepsilon_2)^2 + (d\varepsilon_2 - d\varepsilon_3)^2 + (d\varepsilon_3 - d\varepsilon_1)^2} \tag{4}$$

in which J_1, J_2 and J_3 are the first, second and third effective stress invariants

$$J_1 = \sigma_1 + \sigma_2 + \sigma_3 \quad , \quad J_2 = \sigma_1\sigma_2 + \sigma_2\sigma_3 + \sigma_3\sigma_1 \quad , \quad J_3 = \sigma_1\sigma_2\sigma_3 \tag{5}$$

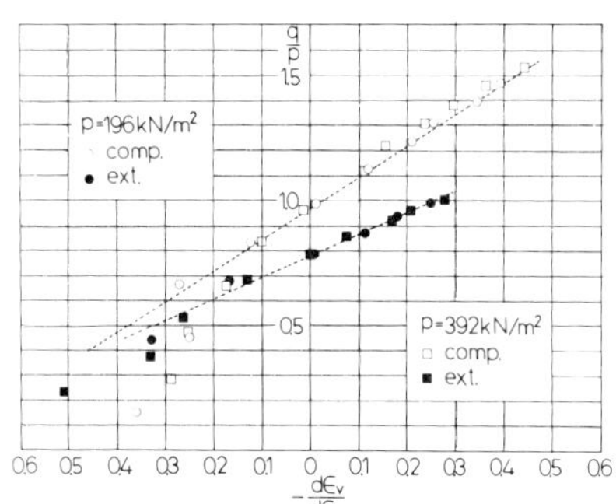

Fig. 1. Relation between q/p and $d\varepsilon_v/d\varepsilon_d$ in triaxial compression and extension tests

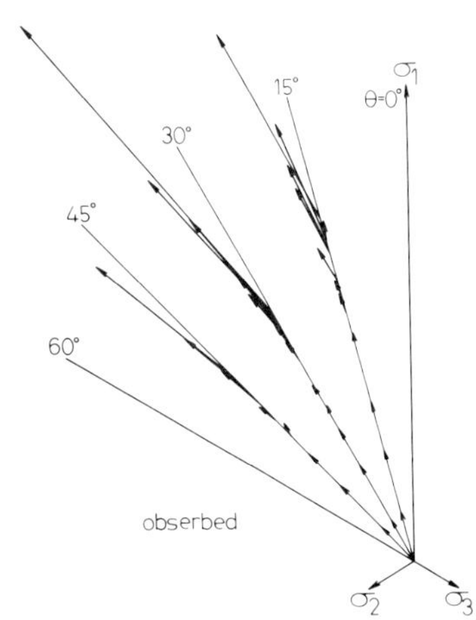

Fig. 3. Observed strain increment vectors on octahedral plane in true triaxial tests

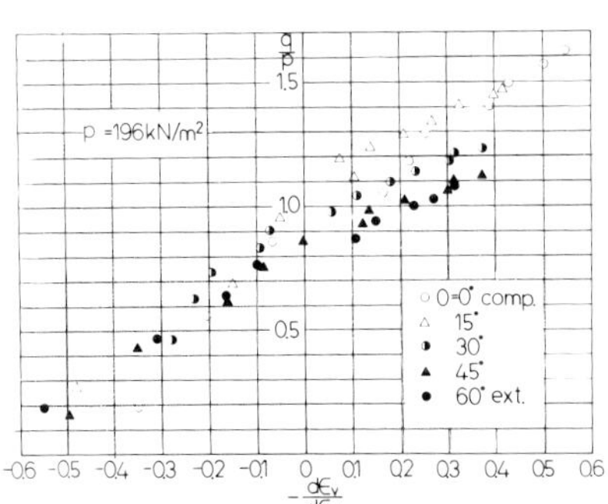

Fig. 2. Relation between q/p and $d\varepsilon_v/d\varepsilon_d$ in true triaxial tests

Fig.2 also shows results of the true triaxial tests on Toyoura sand, arranged in terms of the same relation as in Fig.1. In this figure, θ denotes the angle between σ_1-axis and the corresponding radial stress path on the octahedral plane: the angle θ=0°and θ=60°denote the stress paths under triaxial compression and extension conditions, and θ=15°, θ=30°and θ=45°those under the three different principal stresses. Further, Fig.3 shows the directions of observed strain increment vectors on the octahedral plane in true triaxial tests under three different principal stresses (θ=15°, 30° and 45°).

It is obvious from Figs.1 and 2 that the stress ratio-strain increment ratio relation in three-dimensional stresses cannot uniquely be described by the stress and strain increment parameters of the Mises type in Eqs.(1) to (4). Fig.3 also indicates that the direction of shear strain increment does not coincide with that of shear stress on the octahedral plane so that the shape of cross-section of the plastic potential function is not a circle on the octahedral plane. Therefore, we can understand that the plastic potential function g in three-dimensional stresses is not given by the function of (p and q).

MECHANICAL QUANTITY FOR GRANULAR MATERIALS IN THREE-DIMENSIONAL STRESSES

Definition of mechanical quantity t_{ij}

In order to explain uniquely the stress-strain behavior of granular materials in three-dimensional stresses described in the preceding section, Nakai and Matsuoka (refs.5 and 6) proposed the extended concept of "Spatially Mobilized Plane" (briefly SMP*). This concept of SMP* is based on the idea that there exist unique relations between the shear-normal stress ratio on the SMP (τ_{SMP}/σ_{SMP}) and the amounts of strain increments ($d\varepsilon_{SMP}{}^*$ and $d\gamma_{SMP}{}^*$). Here, SMP is a plane where granular materials are most mobilized on the average in three-dimensional space, and indicated by the plane ABC in Fig.4, where I, II and III axes represent the principal directions of stress (refs.3 and 4). Since the values at the points where the SMP intersects these three axes are proportional to the roots of the respective principal stresses, the direction cosines (a_1, a_2 and a_3) of the normal to the SMP are given by

$$a_i = \sqrt{\frac{J_3}{\sigma_i J_2}} \qquad (i = 1,\ 2,\ 3) \tag{6}$$

The normal and shear stresses on the SMP (τ_{SMP} and σ_{SMP}) are defined as

$$\sigma_{SMP} = \sigma_1 a_1^2 + \sigma_2 a_2^2 + \sigma_3 a_3^2 \tag{7}$$

$$\tau_{SMP} = \sqrt{(\sigma_1 - \sigma_2)^2 a_1^2 a_2^2 + (\sigma_2 - \sigma_3)^2 a_2^2 a_3^2 + (\sigma_3 - \sigma_1)^2 a_3^2 a_1^2} \tag{8}$$

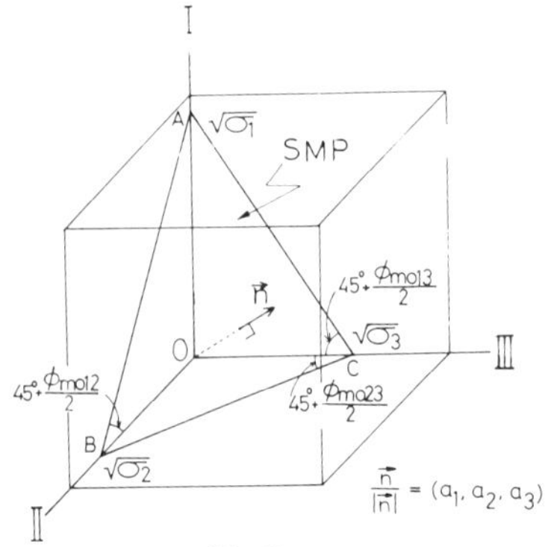

$$\phi_{moij} = \sin^{-1}\frac{\sigma_i - \sigma_j}{\sigma_i + \sigma_j} \quad (i, j = 1, 2, 3,\ i < j)$$

$$\tan(45° + \frac{\phi_{moij}}{2}) = \sqrt{\frac{1 + \sin\phi_{moij}}{1 - \sin\phi_{moij}}} = \sqrt{\frac{\sigma_i}{\sigma_j}}$$

Fig. 4. A cubical element and Spatially Mobilized Plane (plane ABC) in three-dimensional space

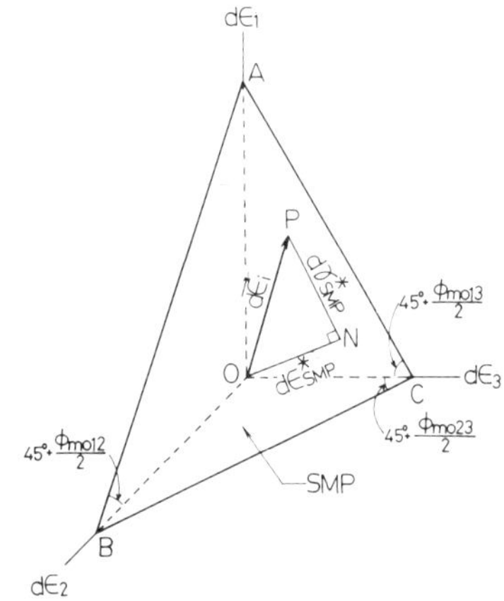

Fig. 5. Strain increment parameters ($d\varepsilon_{SMP}{}^*$ and $d\gamma_{SMP}{}^*$) represented in principal strain increment space

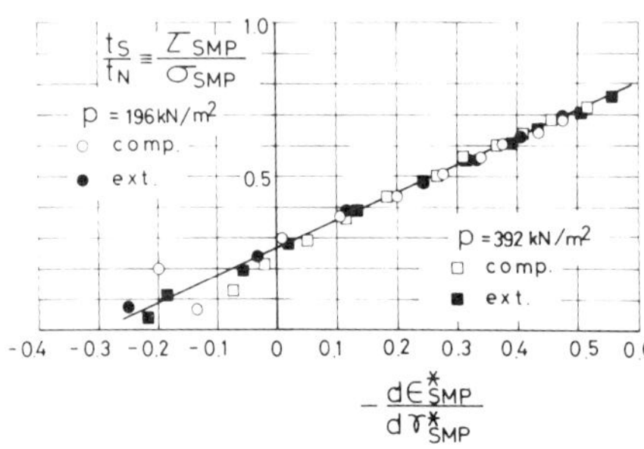

Fig. 6. Relation between t_S/t_N $\equiv \tau_{SMP}/\sigma_{SMP}$ and $d\varepsilon_{SMP}{}^*/d\gamma_{SMP}{}^*$ in triaxial compression and extension tests

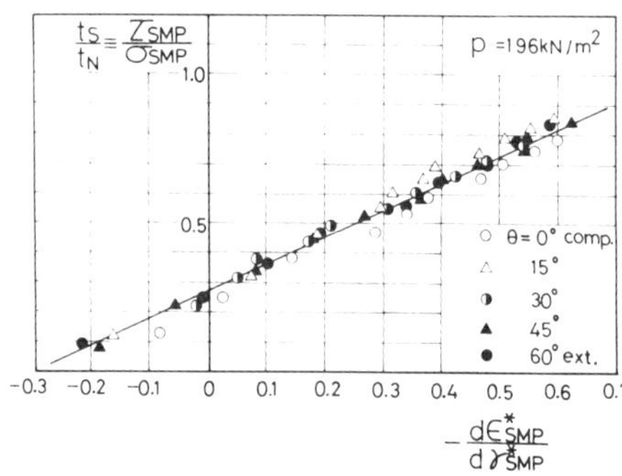

Fig. 7. Relation between t_S/t_N $\equiv \tau_{SMP}/\sigma_{SMP}$ and $d\varepsilon_{SMP}{}^*/d\gamma_{SMP}{}^*$ in true triaxial tests

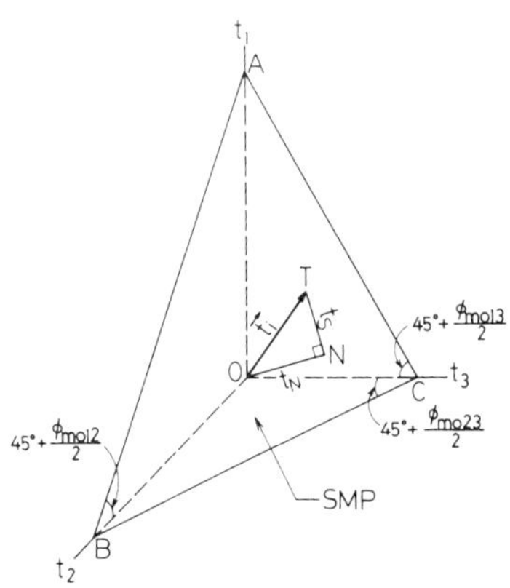

Fig. 8. Stress parameters (t_N and t_S) represented in principal value space of t_{ij}

Also, the amounts of strain increments ($d\varepsilon_{SMP}^*$ and $d\gamma_{SMP}^*$) are given by the normal component and parallel component of the principal strain increment vector to the SMP (see Fig.5)

$$d\varepsilon^{\star}_{SMP} = d\varepsilon_1 a_1 + d\varepsilon_2 a_2 + d\varepsilon_3 a_3 \tag{9}$$

$$d\gamma^{\star}_{SMP} = \sqrt{d\varepsilon_1^2 + d\varepsilon_2^2 + d\varepsilon_3^2 - (d\varepsilon_1 a_1 + d\varepsilon_2 a_2 + d\varepsilon_3 a_3)^2} \tag{10}$$

Figs.6 and 7 show the same test results as those of Figs.1 and 2 as rearranged in terms of the relation between τ_{SMP}/σ_{SMP} and $d\varepsilon_{SMP}^*/d\gamma_{SMP}^*$, according to the concept of SMP*. It is noticed from these figures that the concept of SMP* is interpreted as the theory in which unique relations hold between the stress parameters of Eqs.(7) and (8) and the strain increment parameters of Eqs.(9) and (10). In the present study, we will generalize the concept of SMP* so as to apply this concept to theory of plasticity. For this purpose, the following quantities are introduced (ref.7).

$$t_1 = \sigma_1 a_1 \quad , \quad t_2 = \sigma_2 a_2 \quad , \quad t_3 = \sigma_3 a_3 \tag{11}$$

Then, if we define amounts of t_i (i=1, 2, 3) which are given by the same forms of transformation as ($d\varepsilon_{SMP}^*$ and $d\gamma_{SMP}^*$) in Eqs.(9) and (10) (see Fig.8), these amounts coincide with σ_{SMP} and τ_{SMP}.

$$t_N = t_1 a_1 + t_2 a_2 + t_3 a_3 = \sigma_{SMP} \tag{12}$$

$$t_S = \sqrt{t_1^2 + t_2^2 + t_3^2 - (t_1 a_1 + t_2 a_2 + t_3 a_3)^2} = \tau_{SMP} \tag{13}$$

Therefore, the concept of SMP* implies that there are unique relations between the normal and parallel components of $\vec{t}_i=(t_1, t_2, t_3)$ to the SMP and those of $\vec{d\varepsilon}_i=(d\varepsilon_1, d\varepsilon_2, d\varepsilon_3)$. Extending this interpretation, we introduce a tensor t_{ij} whose principal value are (t_1, t_2 and t_3) and principal directions coincide with those of σ_{ij}. The tensor t_{ij} is, therefore, a symmetric tensor expressed as

$$t_{ij} = \sigma_{ik} a_{kj} \tag{14}$$

where σ_{ik} is the stress tensor, and a_{ik} is the symmetric tensor whose principal values are given by (a_1, a_2 and a_3) in Eq.(6) and its principal directions coincide with those of σ_{ij}.

Application of t_{ij} to theory of plasticity

In ordinary isotropic hardening elastoplastic models of soils such as Cambridge model, the plastic potential function g (it is identical to the yield

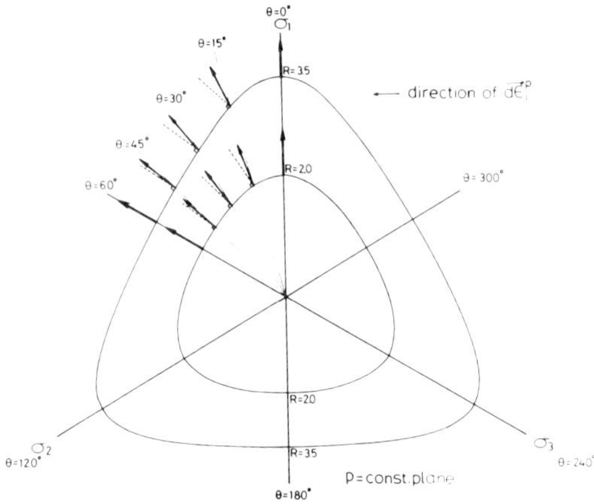

Fig. 9. Shape of plastic potential function represented on octahedral plane in principal stress space and directions of plastic strain increments

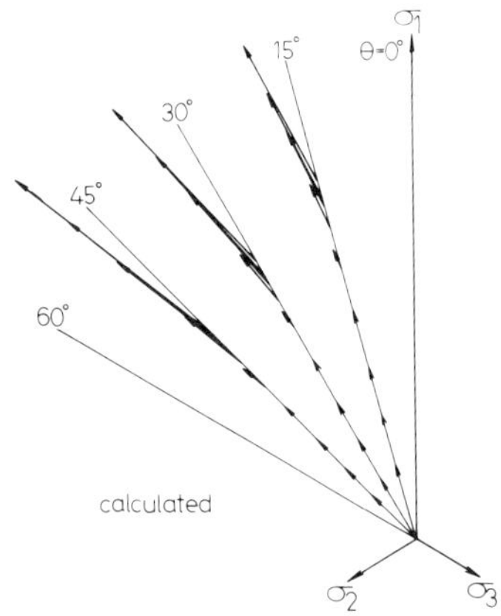

Fig. 10. Calculated strain increment vectors on octahedral plane in true triaxial tests

function in the associated model) is given by a function of (p and q)

$$g = g(p, q) \tag{15}$$

and the plastic strain increment tensor $d\varepsilon_{ij}^p$ is expressed using the following flow rule in σ_{ij}-space.

$$d\varepsilon_{ij}^{p} = \Lambda\frac{\partial g}{\partial \sigma_{ij}} \qquad (\Lambda\text{: a scalar}) \tag{16}$$

On the other hand, according to the t_{ij}-concept presented here, we formulate the plastic potential function (or the yield function) in the form of

$$g = g(t_N, t_S) \tag{17}$$

and the flow rule is assumed in t_{ij}-space as follows:

$$d\varepsilon_{ij}^{p} = \Lambda\frac{\partial g}{\partial t_{ij}} \tag{18}$$

The results calculated in such a way not only satisfies naturally the unique relation between $t_S/t_N \equiv \tau_{SMP}/\sigma_{SMP}$ and $d\varepsilon_{SMP}^*/d\gamma_{SMP}^*$ in Figs.6 and 7 but also explain the observed directions of shear strain increments in Fig.3. Fig.9 shows the function given in the form of Eq.(17) on the octahedral plane in the principal stress space for two cases of $R \equiv (\sigma_1/\sigma_3)_{(comp.)} \equiv (\sigma_1/\sigma_3)_{(ext.)} = 2$ and 3.5. The shape of the function on the octahedral plane is in agreement with that of the Matsuoka-Nakai criterion (SMP criterion) (refs.3 and 4). In this figure, the vectors indicate the direction of the plastic strain increment on the octahedral plane under the stress condition of $\theta=0°$, 15°, 30°, 45° and 60°. Under the three different principal stresses (θ=15°, 30° and 45°) the direction of each vector deviates from that of the corresponding deviatoric stress vector (radial direction) with a definite trend as the stress ratio increases. It is also noticed that its direction is not normal to the curve and is between the radial direction and the direction normal to the curve (direction of broken lines), because the flow rule is assumed not in σ_{ij}-space but in t_{ij}-space. Fig.10 shows the calculated directions of strain increments corresponding to the observed directions in Fig.3.

Correlation to "second stress tensor" by Satake

Satake (ref.9) proposed the following "second stress tensor" $\overset{*}{\sigma}_{ij}$ so as to describe the behavior of granular materials.

$$\overset{*}{\sigma}_{ij} = \frac{1}{3}\sigma_{ik}\phi_{kj}^{-1} \tag{19}$$

Here, ϕ_{kj} denotes what he called the fabric tensor and, e.g., is characterized by the frequency distribution of the interparticle contact angles originated in microscopic studies on granular materials (refs.10 and 11). This is discussed in reference to the fabric change due to a change in stress ratio, so-called induced anisotropy. If we now assume that ϕ_{ij} coincides with a_{ij}^{-1}, it is obvious from Eqs.(14) and (19) that t_{ij} corresponds to $\overset{*}{\sigma}_{ij}$. In this case, it is seen from Eq.(6) that there is the following relation between the principal values of ϕ_{ij} and the principal stresses:

$$\phi_1:\phi_2:\phi_3 = \sqrt{\sigma_1}:\sqrt{\sigma_2}:\sqrt{\sigma_3} \tag{20}$$

Eq.(20) shows that the concept of t_{ij} is qualitatively compatible with the experimental observation by Oda (ref.10) that the normals to the interparticle contacts gradually concentrate on the average in the direction of the major principal stress as the stress ratio increases. Also, Satake (ref.12) shows that the fabric tensor ϕ_{ij} for the induced anisotropy is expressed as

$$\phi_{ij} = \frac{\sigma_{ij}^{\alpha}}{\sigma_{kk}^{\alpha}} \tag{21}$$

and confirms experimentally by biaxial compression tests and simple shear tests on stack of photoelastic rods that the value of α in Eq.(21) is between 0.4 and 0.6. This result almost correspond to Eq.(20) (when α =0.5 Eqs.(20) and (21) are identical).

From the above discussion, the tensor t_{ij} is considered to be the quantity reflecting the induced anisotropy of such granular materials as soils.

CONCLUSIONS

From reconsideration of the extended concept of "Spatially Mobilized Plane" (SMP*), a mechanical quantity t_{ij} to describe uniquely the behavior of granular materials in three-dimensional stresses is obtained. This quantity t_{ij} is a symmetric tensor which are given by the inner product of stress tensor σ_{ik} and the symmetric tensor a_{kj} whose principal values are the direction cosines of the SMP. The validity of t_{ij} is checked by the results of triaxial tests and true triaxial tests. Further, a way to apply t_{ij} to elastoplastic theory is mentioned, and a correlation to Satake's "second stress tensor" is discussed.

ACKNOWLEDGMENT

The author wishes to thank Prof. H. Matsuoka at Nagoya Institute of Technology for his helpful supports and discussions. The author is also indebted to Prof. M. Satake at Tohoku University for his kind advice.

REFERENCES

1 K.H. Roscoe, A.N. Schofield and A. Thurairajah, Yielding of Clays in Stress Wetter than Critical, Geotechnique, Vol.13, No.3, 1963, pp. 211-240.

2 A.N. Schofield and C.P. Wroth, Critical State Soil Mechanics, McGraw-Hill, London, 1968.

3 H. Matsuoka and T. Nakai, Stress-deformation and Strength Characteristics of Soil under Three Different Principal Stress, Proceedings of Japan Society of Civil Engineers, No.232, 1974, pp. 59-70.

4 H. Matsuoka and T. Nakai, Stress-strain Relationship of Soil Based on the SMP, Proceedings of Specialty Session 9, 9th ICSMFE, 1977, pp. 153-162.

5 T. Nakai and H. Matsuoka, A Unified law for Soil Shear Behavior under Three-dimensional Stress Conditions, Proceedings of Japan Society of Civil Engineers, No.303, 1980, pp. 65-77 (in Japanese).

6 T. Nakai and H. Matsuoka, Shear Behaviors of Sand and Clay under Three-dimensional Stress Condition, Soils and Foundations, Vol.23, No.4, 1983, pp. 87-105.

7 T. Nakai and H. Mihara, A New Mechanical Quantity for Soils and Its Application to Elastoplastic Constitutive Models, Soils and Foundations, Vol.24, No.2, 1984, pp. 82-94.

8 T. Nakai and H. Matsuoka, A Generalized Elastoplastic Constitutive Model for Clay in Three-Dimensional Stresses, Soils and Foundations, Vol.26, No.3, 1986, pp. 81-98.

9 M. Satake, Fabric Tensor in Granular Materials, Proceedings of IUTAM-Conference on Deformation and Failure of Granular Materials, Delft, 1982, pp. 63-68.

10 M. Oda, The Mechanism of Fabric Changes during Compressional Deformation of Sand, Soils and Foundations, Vol.12, No.2, 1972, pp. 1-18.

11 H. Matsuoka, A Microscopic Study on Shear Mechanism of Granular Materials, Soils and Foundations, Vol.14, No.1, 1974, pp. 29-43.

12 M. Satake, Consideration on Induced Anisotropy by Use of the Fabric Tensor, Proceeding of 19th Japan National Conference on Soil Mechanics and Foundation Engineering, Vol.2, 1984, pp. 301-302 (in Japanese).

Micromechanics of Granular Materials, edited by M. Satake and J.T. Jenkins
Elsevier Science Publishers B.V., Amsterdam, 1988 — Printed in The Netherlands

EXTENDED SUBLOADING SURFACE MODEL FOR ELASTOPLASTIC DEFORMATION

K. HASHIGUCHI
Dept. Agr. Eng., Kyushu Univ., Fukuoka, Japan

ABSTRACT

The subloading surface model does not assume a yield surface enclosing a purely elastic domain, and instead it assumes a subloading surface which always passes through a current stress point even in the unloading state. Thus, it describes a continuous stress rate-strain rate relation and its loading criterion does not require the judgement whether a current stress lies on a yield surface or not. In this paper, a mathematically exact formulation of this model is presented so as to be capable of describing a continuous stress rate-strain rate relation, an anisotropic hardening/softening and a hysteresis effect consistently, formulating a reasonable translation rule of the center of similarity of the normal-yield and the subloading surfaces and a plastic strain rate equation based on the extended consistency condition and deriving the associated flow rule for materials without a purely elastic domain.

INTRODUCTION

The classical theory of plasticity is concerned only with a description of the remarkable plastic deformation in the yield state, ignoring a plastic deformation due to a stress change within the yield surface by assuming its interior to be a purely elastic domain. In order to extend the classical theory so as to describe also the plastic deformation due to a stress change within the yield surface, the author proposed a "subloading surface model", and refined or extended it so as to describe also an induced anisotropy (refs.1-6). In this extension the classical yield state in which a stress lies on the yield surface and the state within that surface are called a "normal-yield state" and a "subyield state", respectively, and the classical yield surface is called a "normal-yield surface". The salient feature of this model is the assumption of the "subloading surface" which expands/contracts passing always through a current stress point even in the unloading state and retaining a similarity to the normal-yield surface, and the plastic modulus depends on the ratio of the size of the subloading surface to that of the normal-yield surface. Since a purely elastic domain is not assumed in this model, a continuous stress rate-strain rate relation is described bringing about a smooth elastic-plastic transition and the loading criterion does not require the judgement whether a stress lies on the yield surface or not.

In this paper, a complete formulation of the subloading surface model is presented refining the previous equations (ref.6) and deriving the associated flow rule for materials without a purely elastic domain with the generalized loading criterion applicable to not only hardening but also softening states.

BASIC CONSTITUTIVE EQUATIONS FOR THE NORMAL YIELD STATE

Constitutive equations for the normal-yield state in which a current stress lies on the normal-yield surface are formulated below, which will be extended to the subloading surface model formulation in the next section.

First, assume that the normal-yield surface is described by the following equation:

$$f(\hat{\boldsymbol{\sigma}}) - F(H) = 0 \tag{1}$$

where we set

$$\hat{\boldsymbol{\sigma}} \equiv \boldsymbol{\sigma} - \hat{\boldsymbol{\alpha}} \tag{2}$$

The second-order tensor $\boldsymbol{\sigma}$ is a stress, and the scalar $\hat{\boldsymbol{\alpha}}$ and the second-order tensor H are parameters to describe, respectively, the expansion/contraction and the translation of the surface. For simplicity, we assume that the surface described by (1) expands/contracts retaining a similarity in a stress space. Therefore, the function f is to be a homogeneous function of degree $\hat{\boldsymbol{\sigma}}$.

Let $\dot{H}$ where a superposed dot designates a material-time derivative be a function of plastic strain rate $\dot{\boldsymbol{\varepsilon}}^{p}$ (homogeneous degree one by dimensional invariance of time) and some plastic internal state variables to describe a history of plastic deformation.

Further, let $\dot{\hat{\boldsymbol{\alpha}}}$ be given as

$$\dot{\hat{\boldsymbol{\alpha}}} = \dot{A}\,\boldsymbol{I} + \dot{B}\,\frac{\hat{\boldsymbol{\sigma}}}{\|\hat{\boldsymbol{\sigma}}\|} \tag{3}$$

where $\dot{A}$ and $\dot{B}$ are functions of $\dot{\boldsymbol{\varepsilon}}^{p}$ (homogeneous degree one) and some plastic internal state variables depending on a plastic deformation.

By differentiating (1) and substituting the relation

$$\frac{\partial f(\hat{\boldsymbol{\sigma}})}{\partial \hat{\boldsymbol{\sigma}}} = \frac{nF}{\mathrm{tr}(\hat{\boldsymbol{n}}\hat{\boldsymbol{\sigma}})}\hat{\boldsymbol{n}} \qquad \text{when } f(\hat{\boldsymbol{\sigma}}) = F \tag{4}$$

$$\hat{\boldsymbol{n}} \equiv \frac{\partial f(\hat{\boldsymbol{\sigma}})}{\partial \hat{\boldsymbol{\sigma}}} \Big/ \left\| \frac{\partial f(\hat{\boldsymbol{\sigma}})}{\partial \hat{\boldsymbol{\sigma}}} \right\| \tag{5}$$

which results from (1) noting the homogeneity of the function ?, we have the consistency condition

$$\mathrm{tr}\{\hat{\boldsymbol{n}}(\dot{\hat{\boldsymbol{\sigma}}} - \frac{\dot{F}}{nF}\hat{\boldsymbol{\sigma}})\} = 0 \tag{6}$$

Here, assume that the associated flow rule holds for the normal-yield state:

$$\dot{\boldsymbol{\varepsilon}}^{p} = \dot{\hat{\lambda}}\,\hat{\boldsymbol{n}} \qquad (\dot{\hat{\lambda}} > 0) \tag{7}$$

where $\dot{\hat{\lambda}}$ is a proportionality factor.

By substituting (7) into (6) $\dot{\hat{\lambda}}$ is given as follows:

$$\dot{\hat{\lambda}} = \frac{\mathrm{tr}(\hat{\boldsymbol{n}}\,\dot{\boldsymbol{\sigma}})}{\hat{D}} \tag{8}$$

where

$$\hat{D} \equiv \mathrm{tr}\{\,\hat{\boldsymbol{n}}\,(\frac{F'}{n\,F}\,\hat{h}\,\hat{\boldsymbol{\sigma}} + \hat{\boldsymbol{a}}\,)\} \tag{9}$$

$$F' \equiv \mathrm{d}F \,/\, \mathrm{d}H \tag{10}$$

$\dot{H}$ and $\dot{\hat{\boldsymbol{\alpha}}}$ involve $\dot{\boldsymbol{\varepsilon}}^{p}$ in homogeneous degree one so that we can write

$$\dot{H} = \dot{\hat{\lambda}}\,\hat{h} \tag{11}$$

$$\dot{\hat{\boldsymbol{\alpha}}} = \dot{\hat{\lambda}}\,\hat{\boldsymbol{a}} \tag{12}$$

$\hat{h}$ and $\hat{\boldsymbol{a}}$ are scalar and second-order tensor functions of $\hat{\boldsymbol{n}}$ (homogeneous degree one) and some plastic internal state variables. $\hat{D}$ is called a plastic (or hardening) modulus in conformity with the similarity to the elastic modulus in a uniaxial loading state.

The constitutive equation (7) with (8) falls within the framework of the classical theory of plasticity in which the interior of the normal-yield surface is assumed to be a purely elastic domain. Therefore,

1) The discontinuous stress rate/strain rate relation is predicted, which changes abruptly at the moment when the stress reaches the normal-yield surface.
2) The loading criterion requires the judgement whether the current stress lies on the normal-yield surface or not.
3) Of course, the hysteresis loop for the partial unloading-reloading and the mechanical ratchetting phenomenon cannot be predicted.

SUBLOADING SURFACE MODEL

In order to overcome the above-mentioned shortcomings of the classical theory, we extend it so as to describe the plastic deformation in the subyield state. We incorporate a surface, called a "subloading surface", which expands /contracts, always passing through a current stress point regardless of whether a material lies in a loading or an unloading state and retaining a similarity and lying in a position of similarity to the normal-yield surface. And postulate that the plastic modulus depends on the ratio of the size of the subloading surface to that of the normal-yield surface. Further, we assume that the similarity-center of normal-yield and subloading surfaces moves with a plastic deformation. According to these basic concept, we formulate the explicit constitutive equation below.

Let the subloading surface be described by the equation (see Fig.1).

$$f(\bar{\boldsymbol{\sigma}}) = \mathscr{F} \quad (\leq F) \tag{13}$$

in setting

$$\bar{\boldsymbol{\sigma}} \equiv \boldsymbol{\sigma} - \bar{\boldsymbol{\alpha}} \tag{14}$$

where $\bar{\boldsymbol{\alpha}}$ is a parameter to describe a transition of the surface, and f is a value of $f(\bar{\boldsymbol{\sigma}})$ calculated by substituting current values of $\boldsymbol{\sigma}$ and $\bar{\boldsymbol{\alpha}}$, while the function $f(\bar{\boldsymbol{\sigma}})$ has the same form as the homogeneous function $f(\hat{\boldsymbol{\sigma}})$ in (1) in accordance with the assumption of the similarity of the subloading surface to the normal-yield surface.

Also, we introduce the parameter

$$R \equiv \left\{ \frac{f(\bar{\boldsymbol{\sigma}})}{F} \right\}^{1/n} \quad (0 \leqq R \leqq 1) \tag{15}$$

which represents the ratio of the size of the subloading surface to that of the normal-yield surface. Here, let the conjugate point on the normal-yield surface having the same outer-normal direction as that of the subloading surface at the current stress $\boldsymbol{\sigma}$ be denoted as $\boldsymbol{\sigma}_y$. Then, it holds that

$$\hat{\boldsymbol{\sigma}}_y = \bar{\boldsymbol{\sigma}} / R \tag{16}$$

$$\hat{\boldsymbol{n}}_y = \bar{\boldsymbol{n}} \tag{17}$$

where

$$\hat{\boldsymbol{\sigma}}_y \equiv \boldsymbol{\sigma}_y - \hat{\boldsymbol{\alpha}} \tag{18}$$

$$\hat{\boldsymbol{n}}_y \equiv \frac{\partial f(\hat{\boldsymbol{\sigma}}_y)}{\partial \hat{\boldsymbol{\sigma}}_y} \Big/ \left\| \frac{\partial f(\hat{\boldsymbol{\sigma}}_y)}{\partial \hat{\boldsymbol{\sigma}}_y} \right\| \tag{19}$$

$$\bar{\boldsymbol{n}} \equiv \frac{\partial f(\bar{\boldsymbol{\sigma}})}{\partial \bar{\boldsymbol{\sigma}}} \Big/ \left\| \frac{\partial f(\bar{\boldsymbol{\sigma}})}{\partial \bar{\boldsymbol{\sigma}}} \right\| \tag{20}$$

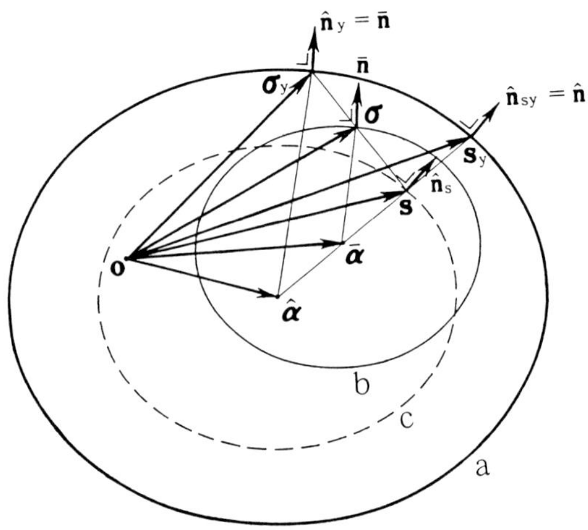

Fig.1 Normal yield surface: a and subloading surface: b.

Further, assume that the translation rule (3) of the normal-yield surface holds even in the subyield state, regarding a stress $\boldsymbol{\sigma}$ in (3) as a conjugate stress $\boldsymbol{\sigma}_y$. Hence, noting the relation (16), we have

$$\dot{\hat{\boldsymbol{\alpha}}} = \dot{A}\, \boldsymbol{I} + \dot{B} \frac{\bar{\boldsymbol{\sigma}}}{\| \bar{\boldsymbol{\sigma}} \|} \tag{21}$$

Now, we formulate the translation rule of the similarity-center $\boldsymbol{s}$ of the normal-yield and subloading surfaces.

The similarity-center $\boldsymbol{s}$ must not go out from the normal-yield surface. Also, if $\boldsymbol{s}$ lies on the normal-yield surface, a singularity of plastic modulus occurs in this point. Then, assume that

$$R_s \leqq \chi \tag{22}$$

where

$$R_s \equiv \left\{ \frac{f(\hat{\boldsymbol{s}})}{F} \right\}^{1/n} \quad (0 \leqq R_s < 1) \tag{23}$$

$$\hat{\boldsymbol{s}} \equiv \boldsymbol{s} - \hat{\boldsymbol{\alpha}} \tag{24}$$

χ $(0 \leqq \chi < 1)$ is a material constant.

(22) is rewritten as

$$\mathrm{tr}\{\hat{\boldsymbol{n}}_s(\dot{\hat{\boldsymbol{s}}} - \frac{\dot{F}}{nF}\hat{\boldsymbol{s}})\} \leqq 0 \quad \text{when } R_s = \chi \tag{25}$$

in a differential form, using the relation

$$\frac{\partial f(\hat{\boldsymbol{s}})}{\partial \hat{\boldsymbol{s}}} = \frac{n\chi^n F}{\mathrm{tr}(\hat{\boldsymbol{n}}_s\hat{\boldsymbol{s}})}\hat{\boldsymbol{n}}_s \quad \text{when } R_s = \chi \tag{26}$$

where

$$\hat{\boldsymbol{n}}_s \equiv \frac{\partial f(\hat{\boldsymbol{s}})}{\partial \hat{\boldsymbol{s}}} \Big/ \left\| \frac{\partial f(\hat{\boldsymbol{s}})}{\partial \hat{\boldsymbol{s}}} \right\| \tag{27}$$

Let (22) or (25) be called an "enclosing condition of similarity-center". In order to satisfy (25), referring to Fig.1, we assume that

$$\dot{\hat{\boldsymbol{s}}} - \frac{\dot{F}}{nF}\hat{\boldsymbol{s}} = C\dot{\varepsilon}^p(\boldsymbol{\sigma}_y - \boldsymbol{s}_y) \quad \text{when } R_s = \chi \tag{28}$$

where $\boldsymbol{s}_y$ designates an intersecting point of the normal-yield surface and the straight (radial) line starting from the point $\hat{\boldsymbol{\alpha}}$ and passing through the point $\boldsymbol{s}$ in the stress space, i.e.,

$$\boldsymbol{s}_y = \hat{\boldsymbol{\alpha}} + \frac{\hat{\boldsymbol{s}}}{R_s} \tag{29}$$

and we put

$$\dot{\varepsilon}^p \equiv \| \dot{\boldsymbol{\varepsilon}}^p \| \tag{30}$$

C ($\geqq 0$) is a material constant which controls the rate of translation of the similarity-center. Further, assume that

$$\dot{\hat{\boldsymbol{s}}} - \frac{\dot{F}}{nF}\hat{\boldsymbol{s}} = C\dot{\varepsilon}^p(\boldsymbol{\sigma}_y - \hat{\boldsymbol{\alpha}}) \quad \text{when } R_s = 0 \tag{31}$$

In Fig.1, $\hat{\boldsymbol{n}}_{sy}$ is the outer-normal of the normal-yield surface in the point $\boldsymbol{s}_y$, i.e.,

$$\hat{\boldsymbol{n}}_{sy} \equiv \frac{\partial f(\hat{\boldsymbol{s}}_y)}{\partial \hat{\boldsymbol{s}}_y} \Big/ \left\| \frac{\partial f(\hat{\boldsymbol{s}}_y)}{\partial \hat{\boldsymbol{s}}_y} \right\| \tag{32}$$

where we set

$$\hat{\boldsymbol{s}}_y \equiv \boldsymbol{s}_y - \hat{\boldsymbol{\alpha}} \tag{33}$$

For (28) and (31) to be satisfied, we assume the following linear equation of R_s.

$$\dot{\hat{\boldsymbol{s}}} - \frac{\dot{F}}{nF}\hat{\boldsymbol{s}} = C\dot{\varepsilon}^p\{\boldsymbol{\sigma}_y - \hat{\boldsymbol{\alpha}} + \frac{R_s}{\chi}(\hat{\boldsymbol{\alpha}} - \boldsymbol{s}_y)\} \tag{34}$$

which is rewritten as

$$\dot{\hat{\boldsymbol{s}}} - \frac{\dot{F}}{nF}\hat{\boldsymbol{s}} = C\dot{\varepsilon}^p(\frac{\bar{\boldsymbol{\sigma}}}{R} - \frac{\hat{\boldsymbol{s}}}{\chi}) \tag{35}$$

noting (16) and (29).

From (35) we have the translation rule of $\boldsymbol{s}$ as follows:

$$\dot{\boldsymbol{s}} = \dot{\hat{\boldsymbol{\alpha}}} + \frac{\dot{F}}{nF}\hat{\boldsymbol{s}} + C\dot{\varepsilon}^p(\frac{\bar{\boldsymbol{\sigma}}}{R} - \frac{\hat{\boldsymbol{s}}}{\chi}) \tag{36}$$

Next, we formulate the evolution equation of R, i.e., $\dot{R}$.

Substituting the relation

$$\hat{\boldsymbol{\sigma}}_y = \frac{\tilde{\boldsymbol{\sigma}}}{R} + \hat{\boldsymbol{s}} \tag{37}$$

into the relation $f(\hat{\boldsymbol{\sigma}}_y) = F$, we have

$$f\left(\frac{\tilde{\boldsymbol{\sigma}}}{R} + \hat{\boldsymbol{s}}\right) = F \tag{38}$$

where

$$\tilde{\boldsymbol{\sigma}} \equiv \boldsymbol{\sigma} - \boldsymbol{s} \tag{39}$$

We can calculate the value of R from (38), since $\boldsymbol{\sigma}$, $\boldsymbol{s}$, $\hat{\boldsymbol{\alpha}}$ and F are already known in general. Further, given the value of R, we can calculate $\bar{\boldsymbol{\alpha}}$ by the equation

$$\bar{\boldsymbol{\alpha}} = \boldsymbol{s} - R\,\hat{\boldsymbol{s}} \tag{40}$$

Differentiating (38) and noting (17) and the relation

$$\frac{\partial f(\hat{\boldsymbol{\sigma}}_y)}{\partial \hat{\boldsymbol{\sigma}}_y} = \frac{nF}{\mathrm{tr}(\hat{\boldsymbol{n}}_y\hat{\boldsymbol{\sigma}}_y)}\hat{\boldsymbol{n}}_y \tag{41}$$

the rate of R is derived as follows:

$$\dot{R} = \frac{\mathrm{tr}\{\bar{\boldsymbol{n}}(\dot{\tilde{\boldsymbol{\sigma}}} + R\,\dot{\hat{\boldsymbol{s}}} - \frac{\dot{F}}{nF}\bar{\boldsymbol{\sigma}})\}}{\mathrm{tr}(\bar{\boldsymbol{n}}\,\tilde{\boldsymbol{\sigma}})}R \tag{42}$$

(42) describes the relation of $\dot{R}$ to $\dot{\boldsymbol{\sigma}}$, $\dot{\hat{\boldsymbol{\alpha}}}$, $\dot{\boldsymbol{s}}$ and $\dot{F}$. On the other hand, in order to formulate the plastic modulus, let a scalar relation between a stress rate and a plastic strain rate be assumed. Then, postulate that in the loading state the current stress always goes away from the similarity-center and approaches asymptotically to the normal-yield surface, and thus R increases with a plastic deformation, i.e.,

$$\dot{R}/\dot{\varepsilon}^p \geqq 0 \quad \text{for } \dot{\boldsymbol{\varepsilon}}^p \neq \boldsymbol{0} \tag{43}$$

satisfying the conditions

$$\left.\begin{aligned} R = 0: \quad \dot{R}/\dot{\varepsilon}^p &= +\infty \\ R = 1: \quad \dot{R}/\dot{\varepsilon}^p &= 0 \end{aligned}\right\} \quad \text{for } \dot{\boldsymbol{\varepsilon}}^p \neq \boldsymbol{0} \tag{44}$$

$(44)_1$ is required in order to describe the smooth transition from the unloading to the loading state, since $\dot{R}/\dot{\varepsilon}^p = -\infty$ in the unloading state ($\dot{\boldsymbol{\varepsilon}}^p = \boldsymbol{0}$ and $\dot{R} < 0$). $(44)_2$ is obvious since it must hold that $R \leqq 1$.

Thus, we postulate the following relation between $\dot{R}$ and $\dot{\varepsilon}^p$.

$$\dot{R} = U\,\dot{\varepsilon}^p \quad \text{for } \dot{\boldsymbol{\varepsilon}}^p \neq \boldsymbol{0} \tag{45}$$

where U is a monotonically decreasing function of R and

$$R_{sc} \equiv \left\{\frac{f(\tilde{\boldsymbol{\sigma}})}{F}\right\}^{1/n} \tag{46}$$

with conditions

$$\begin{aligned} R = 0: \quad U &= +\infty \\ R = 1: \quad U &= 0 \end{aligned} \tag{47}$$

Examples of the function U are

$$U = u_1 (1 - R^m) / R_{sc} \qquad (48)$$

$$U = -u_2 \ln R / R_{sc}$$

where u_1, u_2 and m are material constants. R_{sc} plays a role to express Masing effect (ref.7).

Equating (42) and (45), we have

$$\mathrm{tr}\,[\bar{\boldsymbol{n}}\{\dot{\boldsymbol{\sigma}} - \frac{\dot{F}}{nF}\bar{\boldsymbol{\sigma}} - R\dot{\hat{\boldsymbol{\alpha}}} - (1-R)\dot{\boldsymbol{s}} - \frac{U}{R}\dot{\varepsilon}^{p}\tilde{\boldsymbol{\sigma}}\}] = 0 \qquad (49)$$

Further, substituting (36) into (49), we have the consistency condition for this extended model:

$$\mathrm{tr}\,[\bar{\boldsymbol{n}}\,(\dot{\hat{\boldsymbol{\sigma}}} - \frac{\dot{F}}{nF}\hat{\boldsymbol{\sigma}} - \{C(1-R)(\frac{\bar{\boldsymbol{\sigma}}}{R} - \frac{\hat{\boldsymbol{s}}}{\chi}) + \frac{U}{R}\tilde{\boldsymbol{\sigma}}\}\,\dot{\varepsilon}^{p})] = 0 \qquad (50)$$

Here, we assume that the associated flow rule holds also for the subloading surface:

$$\dot{\boldsymbol{\varepsilon}}^{p} = \dot{\bar{\lambda}}\,\bar{\boldsymbol{n}} \qquad (51)$$

By substituting (51) into (50), $\dot{\bar{\lambda}}$ is given as

$$\dot{\bar{\lambda}} = \frac{\mathrm{tr}(\bar{\boldsymbol{n}}\,\dot{\boldsymbol{\sigma}})}{\bar{D}} \qquad (52)$$

where

$$\bar{D} \equiv \mathrm{tr}\,[\bar{\boldsymbol{n}}\{\frac{F'}{nF}\bar{h}\,\hat{\boldsymbol{\sigma}} + \bar{\boldsymbol{a}} + C(1-R)(\frac{\bar{\boldsymbol{\sigma}}}{R} - \frac{\hat{\boldsymbol{s}}}{\chi}) + \frac{U}{R}\tilde{\boldsymbol{\sigma}}\}] \qquad (53)$$

$\dot{H}$ and $\dot{\hat{\boldsymbol{\alpha}}}$ involve $\dot{\boldsymbol{\varepsilon}}^{p}$ in homogeneous degree one so that we can write

$$\dot{H} = \dot{\bar{\lambda}}\,\bar{h} \qquad (54)$$

$$\dot{\hat{\boldsymbol{\alpha}}} = \dot{\bar{\lambda}}\,\bar{\boldsymbol{a}} \qquad (55)$$

$\bar{h}$ and $\bar{\boldsymbol{a}}$ are scalar and second-order tensor functions of $\bar{\boldsymbol{n}}$ (degree one) and some plastic internal state variables.

ON THE ASSOCIATED FLOW RULE AND A LOADING CRITERION

A physical meaning of the associated flow rule was given first by Drucker (ref.8). His interpretation of this rule premises the existence of the yield surface enclosing a purely elastic domain so that the maximum plastic work holds, however. Therefore, a more general interpretation of this rule is required when it is adopted to the extended constitutive equations which does not assume a purely elastic domain as in the subloading surface model. We consider the applicablity of this rule and the loading criterion below.

Ilyushin (ref.9) postulated that the work done during a strain cycle is non-negative, i.e.,

$$\oint_{\boldsymbol{\varepsilon}} \mathrm{tr}(\boldsymbol{\sigma}\,d\boldsymbol{\varepsilon}) \geq 0 \qquad (56)$$

For an infinitesimally small strain cycle, (56) is written as

$$\mathrm{tr}(\, d\boldsymbol{\sigma}^{r}\, d\boldsymbol{\varepsilon}\,) \geqq 0 \quad \text{or} \quad \mathrm{tr}(\, \dot{\boldsymbol{\sigma}}^{r}\, \dot{\boldsymbol{\varepsilon}}\,) \geqq 0 \tag{57}$$

where $\dot{\boldsymbol{\sigma}}^{r}$ is a relaxation stress rate (ref.10), i.e.,

$$\dot{\boldsymbol{\sigma}}^{r} = \mathbf{E}\,\dot{\boldsymbol{\varepsilon}}^{p} \tag{58}$$

referring to Fig.2 illustrated for a uniaxial loading case (ε_a: axial strain, σ_a^r: axial stress, σ_a: axial relaxation stress) in which the elastic lines ab and cd are regarded to be parallel since we consider an infinitesimally small strain cycle. The fourth-order tensor $\mathbf{E}$ which is a function of stress in general is an elastic modulus, i.e.,

$$\dot{\boldsymbol{\varepsilon}}^{e} = \mathbf{E}^{-1}\,\dot{\boldsymbol{\sigma}} \quad \text{or} \quad \dot{\boldsymbol{\sigma}} = \mathbf{E}\,\dot{\boldsymbol{\varepsilon}}^{e} \tag{59}$$

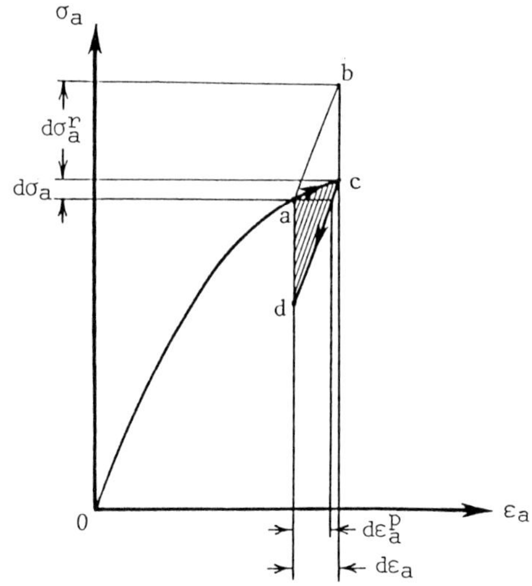

Fig.2 A positive work done by a strain cycle.

Let the integral form of (59) be described as

$$\boldsymbol{\sigma} = \mathbf{D}(\boldsymbol{\varepsilon}^{e}) \tag{60}$$

where $\mathbf{D}$ is a second-order tensor function of $\boldsymbol{\varepsilon}^{e}$.

On the other hand, we assume that a plastic deformation occurs only when a loading criterion expands:

$$\begin{aligned} \dot{\boldsymbol{\varepsilon}}^{p} &\neq \mathbf{0} \quad \text{for} \quad \mathrm{tr}\left(\frac{\partial \phi}{\partial \boldsymbol{\varepsilon}}\,\dot{\boldsymbol{\varepsilon}}\right) > 0 \\ \dot{\boldsymbol{\varepsilon}}^{p} &= \mathbf{0} \quad \text{for} \quad \mathrm{tr}\left(\frac{\partial \phi}{\partial \boldsymbol{\varepsilon}}\,\dot{\boldsymbol{\varepsilon}}\right) \leqq 0 \end{aligned} \tag{61}$$

where ϕ is a loading function of strain and some plastic internal state variables.

Further, assume that the direction of the relaxation stress rate $\dot{\boldsymbol{\sigma}}^{r}$ depends on a strain and some plastic internal state variables but not on their increments. Then, in order to satisfy (57) and (61), a direction of relaxation stress rate must be outer-normal to the loading surface $\phi = 0$:

$$\dot{\boldsymbol{\sigma}}^{r} = \dot{\Lambda}\,\frac{\partial \phi}{\partial \boldsymbol{\varepsilon}} \qquad (\dot{\Lambda} > 0) \tag{62}$$

where $\dot{\Lambda}$ is a proportionality factor.

From (58) and (62), a plastic strain rate is given by

$$\dot{\boldsymbol{\varepsilon}}^{p} = \dot{\lambda}\,\mathbf{E}^{-1}\,\frac{\partial \phi}{\partial \boldsymbol{\varepsilon}} \tag{63}$$

Here, noting

$$\partial \boldsymbol{\sigma} / \partial \boldsymbol{\varepsilon}^{e} = \mathbf{E} \tag{64}$$

and

$$\boldsymbol{\varepsilon}^{e} = \boldsymbol{\varepsilon} - \boldsymbol{\varepsilon}^{p} \tag{65}$$

we have

$$\partial \boldsymbol{\sigma} / \partial \boldsymbol{\varepsilon} = \mathbf{E} \tag{66}$$

Let the function which is obtained by replacing the argument $\boldsymbol{\varepsilon}$ to $\mathbf{D}^{-1}(\boldsymbol{\sigma}) + \boldsymbol{\varepsilon}^{p}$ in the function ϕ be denoted by f. The relation of the gradients of these functions is given by (66) as

$$\frac{\partial \phi}{\partial \boldsymbol{\varepsilon}} = \frac{\partial f}{\partial \boldsymbol{\sigma}} \frac{\partial \boldsymbol{\sigma}}{\partial \boldsymbol{\varepsilon}} = \frac{\partial f}{\partial \boldsymbol{\sigma}} \mathbf{E} \tag{67}$$

From (63) and (67), we have the associated flow rule in a stress space:

$$\dot{\boldsymbol{\varepsilon}}^{p} = \dot{\lambda} \frac{\partial f}{\partial \boldsymbol{\sigma}} \tag{68}$$

Also, the loading criterion (61) in a strain space is rewritten in a stress space as follows:

$$\begin{aligned} \dot{\boldsymbol{\varepsilon}}^{p} &\neq \mathbf{0} \quad \text{for} \quad \mathrm{tr}(\boldsymbol{n}\, \mathbf{E}\, \dot{\boldsymbol{\varepsilon}}) > 0 \\ \dot{\boldsymbol{\varepsilon}}^{p} &= \mathbf{0} \quad \text{for} \quad \mathrm{tr}(\boldsymbol{n}\, \mathbf{E}\, \dot{\boldsymbol{\varepsilon}}) \leqq 0 \end{aligned} \tag{69}$$

where

$$\boldsymbol{n} \equiv \frac{\partial f}{\partial \boldsymbol{\sigma}} \Big/ \left\| \frac{\partial f}{\partial \boldsymbol{\sigma}} \right\| \tag{70}$$

The plastic strain rate obeying the associated flow rule (68) is generally written by the following equation as shown (7) with (8) and (51) with (52).

$$\dot{\boldsymbol{\varepsilon}}^{p} = \dot{\lambda}\, \boldsymbol{n} \tag{71}$$

$$\dot{\lambda} = \frac{\mathrm{tr}(\boldsymbol{n}\, \dot{\boldsymbol{\sigma}})}{D} \tag{72}$$

where D is a plastic modulus which is a function of stress and some plastic internal state variables.

From (59) and (71) with (72) into the equation

$$\dot{\boldsymbol{\varepsilon}} = \dot{\boldsymbol{\varepsilon}}^{e} + \dot{\boldsymbol{\varepsilon}}^{p} \tag{73}$$

we have

$$\dot{\boldsymbol{\sigma}} = \mathbf{E} \left\{ \dot{\boldsymbol{\varepsilon}} - \frac{\mathrm{tr}(\boldsymbol{n}\, \mathbf{E}\, \dot{\boldsymbol{\varepsilon}})}{D + \mathrm{tr}(\boldsymbol{n}\, \mathbf{E}\, \boldsymbol{n})} \boldsymbol{n} \right\} \tag{74}$$

$$\dot{\lambda} = \frac{\mathrm{tr}(\boldsymbol{n}\, \mathbf{E}\, \dot{\boldsymbol{\varepsilon}})}{D + \mathrm{tr}(\boldsymbol{n}\, \mathbf{E}\, \boldsymbol{n})} \tag{75}$$

If there exists the state $D + \mathrm{tr}(\boldsymbol{n}\, \mathbf{E}\, \boldsymbol{n}) = 0$, there occurs the contradiction $\| \dot{\boldsymbol{\sigma}} \| = \infty$ from (74). Then, there does not exist the state $D + \mathrm{tr}(\boldsymbol{n}\, \mathbf{E}\, \boldsymbol{n}) = 0$. Also, since $D + \mathrm{tr}(\boldsymbol{n}\, \mathbf{E}\, \boldsymbol{n})$ is to be a continuous function of its arguments, it can never be negative. Thus, it holds that

$$D + \mathrm{tr}(\boldsymbol{n}\, \mathbf{E}\, \boldsymbol{n}) > 0 \tag{76}$$

(76) was suggested by A.C. Chan, Inst. Numer. Meth. Eng., Univ. College Swansea as a private correspondence.

Then, if the loading condition $\mathrm{tr}(\boldsymbol{n}\,\mathbf{E}\,\dot{\boldsymbol{\varepsilon}}) > 0$ is fulfilled, the collateral condition $\dot{\lambda} > 0$ is automatically fulfilled by (76) into (75).

In the unloading state it holds that

$$\mathrm{tr}(\boldsymbol{n}\dot{\boldsymbol{\sigma}}) \leqq 0 \tag{77}$$

which is obtained by substituting $\dot{\boldsymbol{\sigma}} = \mathbf{E}\,\dot{\boldsymbol{\varepsilon}}$ $(\dot{\boldsymbol{\varepsilon}}^{e} = \dot{\boldsymbol{\varepsilon}})$ into $(69)_2$. Here, consider the state $D < 0$ which occurs in a softening state (a falling portion of stress-strain curve). The quantity $\mathrm{tr}(\boldsymbol{n}\dot{\boldsymbol{\sigma}})/D$ is positive in the unloading from this state, since $\mathrm{tr}(\boldsymbol{n}\dot{\boldsymbol{\sigma}}) < 0$ and $D < 0$. In other words, $\mathrm{tr}(\boldsymbol{n}\dot{\boldsymbol{\sigma}})/D > 0$ is a necessary condition but not a sufficient condition to be a loading state.

In the above, the applicability of the associated flow rule was interpreted for the subloading surface too for which the loading criterion is given from (69) as

$$\dot{\boldsymbol{\varepsilon}}^{p} \neq \boldsymbol{0} \quad \text{for} \quad \mathrm{tr}(\bar{\boldsymbol{n}}\,\mathbf{E}\,\dot{\boldsymbol{\varepsilon}}) > 0$$

$$\dot{\boldsymbol{\varepsilon}}^{p} = \boldsymbol{0} \quad \text{for} \quad \mathrm{tr}(\bar{\boldsymbol{n}}\,\mathbf{E}\,\dot{\boldsymbol{\varepsilon}}) \leqq 0 \tag{78}$$

REFERENCES

1 K. Hashiguchi and M. Ueno, Elastoplastic constitutive equations of soils, in: S. Murayama and A.N. Schofield (Eds.), Constitutive Equations of Soils: Proc. 9th Int. Conf. on Soil Mech. and Found. Eng., Spec. Session 9, Tokyo, 1977, pp.73-82.

2 K. Hashiguchi, Plastic constitutive equations of granular materials, in: C.S. Cowin and M. Satake (eds.), Proc. US-Japan Seminar on Contin. Mech. and Statist. Appr. in Mech. of Granular Materials, Sendai, Gakujutsu-Bunken Fukyu-Kai, Tokyo, 1978, pp.321-329.

3 K. Hashiguchi, Constitutive equations of elastoplastic materials with elastic-plastic transition, ASME J. of Appl. Mech., 47 (1980), 266-272.

4 K. Hashiguchi, Anisotropic hardening model for granular media, in: G.N. Pande (eds.), Proc. Int. Symp. on Soils under Cyclic and Transient Loading, Swansea, A.A. Balkema Publ., Rotterdam, 1980, pp.469-474.

5 K. Hashiguchi, A mathematical description of elastoplastic deformation in normal-yield and subyield states. in: G.N. Pande and W.F. Van Impe (eds.), Proc. 2nd Int. Conf. on Numerical Methods in Geomechanics, Ghent (Belgium), M. Jackson & Son Publ. Ltd., Rendruth, 1986, pp.17-24.

6 K. Hashiguchi, Elastoplastic constitutive modekl with a subloading surface, in: G. Yagawa ans S.N. Atluri (eds.), Proc. Int. Conf. on Computational Mechanics, Tokyo, Springer-Verlag, Tokyo, 1986, pp.IV65-70.

7 Masing, G., 1926, Eigenspannungen und Verfestigung heim Messing, Proc. 2nd Int. Congr. Appl. Mech., Zurich, pp.332-335.

8 D.C. Drucker, A more fundamental approach to plastic stress-strain relations, in: Proc. 1st U.S. Natl. Congr. on Applied Mechanics, ASME, 1951, pp.487-491.

9 A.A. Ilyushin, On the postulate of plasticity, J. of Appl. Math. Mech. (translation of Prikladnaya Mathematika i Mekhanika), 25 (1961), pp.746-752.

10 P.J. Yorder and W.D. Iwan, On the formulation of strain space plasticity with multiple loading surfaces, ASME J. of Appl. Mech., 48 (1981), 773-778.

Micromechanics of Granular Materials, edited by M. Satake and J.T. Jenkins
Elsevier Science Publishers B.V., Amsterdam, 1988 — Printed in The Netherlands

A COMBINED HARDENING MODEL OF ANISOTROPICALLY CONSOLIDATED CLAY

H. Hirai
Department of Civil Engineering, Kumamoto University, Kumamoto (Japan)

ABSTRACT

A model introduced in the present paper is capable of describing the mechanical behaviour of anisotropically consolidated clay reasonably well. This model provides the non-associative flow rule based on combined isotropic-kinematic hardening. The good agreement is observed between simulation and experimental data.

INTRODUCTION

Various anisotropic behaviours of clay can be observed in many fields. Experiments of clay under anisotropically consolidated stress have showed that the stress-strain relationships are influenced by the initial degree of anisotropic consolidation. Specimens trimmed at different angle for the clay deposit consolidated one-dimensionally exhibit different undrained failure strength.

Various constitutive models of clay have been developed to incorporate the influence of anisotropic initial condition on deformation and strength. Pietruszczak et al. (ref.1) simulated the anisotropic behaviour on K_0-consolidated clays by using multisurface model based on combined isotropic-kinematic hardening. Using an anisotropic tensor to represent the rotation of yield surface, Kavvadas (ref.2) proposed a non-associative stress-strain relationship of clays. The constitutive models to account for the initial anisotropy and subsequent evolution of induced anisotropy of clay are presented by Banerjee et al. (ref.3) and Anandarajah et al. (ref.4) on the basis of the boundary surface plasticity. An anisotropic hardening model is proposed by Hirai (refs. 5-9) in order to describe the induced anisotropy of sand. This model can be easily applied to the simulation of clay behaviour.

The objective of this paper is to present a constitutive model that is capable of describing the mechanical behaviour of clay subjected to shear deformation subsequent to anisotropically consolidated stress. The initial yield surface rotates from isotropic stress line when subjected to anisotropic consolidation and the subsequent yield surface expands and translates as the shear stresses are applied to a sample. The clay behaviour is assumed to be simulated by combination of isotropic and kinematic hardening for which the expansion of yield surface is prescribed by conventional isotropic hardening and the translation of yield surface is specified by the evolution of kinematic hardening. The model simulations are compared with some experimental data obtained under undrained triaxial tests of clay.

YIELD SURFACE AND PLASTIC POTENTIAL

An important feature of cyclic behaviour of sand is induced anisotropy due to plastic deformation. The deformation-induced plastic anisotropy is described by the combination of isotropic and kinematic hardening. The yield surface in general takes the following form:

$$f(\sigma_{ij}, \alpha_{ij}, \beta_{ij}, k) = 0 \tag{1}$$

where σ_{ij} is stress, α_{ij} and k denote kinematic hardening and isotropic hardening respectively and β_{ij} indicates the anisotropic tensor to represent the rotation of yield surface.

The specific shape of the yield surface used in the present paper is of a generalized form of the Modified Cam clay model (ref.10). The yield surface which translates with expansion is illustrated in Fig.1 and is given by the following equation:

$$f = \bar{J}_2/r(\bar{\theta})^2 + m\bar{I}_1^2 + k\bar{I}_1 = 0 \tag{2}$$

where m is a material constant and the invariants are

$$\bar{I}_1 = \bar{\sigma}_{kk}, \ \bar{J}_2 = \bar{s}_{ij}\bar{s}_{ij}/2, \ \bar{J}_3 = \bar{s}_{ij}\bar{s}_{jk}\bar{s}_{ki}/3, \ \cos(3\bar{\theta}) = -3\cdot 3^{1/2}\bar{J}_3/(2\bar{J}_2^{3/2}) \tag{3}$$

where

$$\bar{\sigma}_{ij} = \sigma_{ij} - \alpha_{ij} - \beta_{ij}I_1/3, \quad \bar{s}_{ij} = s_{ij} - \alpha'_{ij} - \beta'_{ij}I_1/3$$

$$\alpha'_{ij} = \alpha_{ij} - \alpha_{kk}\delta_{ij}/3, \quad \beta'_{ij} = \beta_{ij} - \beta_{kk}\delta_{ij}/3 \tag{4}$$

The variable $r(\bar{\theta})$ in eq. (2) has been given by Willam and Warnke (ref.11) and is expressed as

$$r(\bar{\theta}) = [2(1-\bar{r}^2)\cos\bar{\theta} + (2\bar{r}-1)\{4(1-\bar{r}^2)\cos^2\bar{\theta} + 5\bar{r}^2 - 4\bar{r}\}^{1/2}] / [4(1-\bar{r}^2)\cos^2\bar{\theta} + (1-2\bar{r})^2] \tag{5}$$

Since the tensor α_{ij} represents translation of yield surface, the yield function given by eq. (2) can expand with isotropic hardening and simultaneously translate with kinematic hardening in stress space, as shown in Fig.1.

The direction of incremental plastic strain is defined by specifying the shape of the plastic potential surface. As it has been suggested that plastic potential surface is usally different from yield surface, the non-associative flow rule is essentially to be employed in the incremental plasticity theory. The plastic potential used in the present paper is a generalized form of the Cam clay model (ref.12) and given by the following form:

$$g = J_2^{1/2}/(R(\theta)I_1) - \eta \ln(I_1/h_1) = 0 \tag{6}$$

where η is a material constant and h_1 is unnecessary to define explicitly. The parameter $R(\theta)$ is given by the form similar to the parameter $r(\bar{\theta})$ in eq. (2).

HARDENING FUNCTIONS AND STRESS-STRAIN RELATIONSHIPS

It is generally considered that the hardening process is composed of isotropic and kinematic hardening due to plastic deformation. It is necessary to provide the evolutional rules of each hardening. If the isotropic hardening may depend on not only the plastic work related with the change in shape but also that related with the change in volume, the evolution of isotropic hardening is expressed in the following forms (refs.5-9):

$$\dot{k} = \phi(\sigma_{ii}\dot{e}_{jj}^{(p)}/3 + ns_{ij}\dot{e}'^{(p)}_{ij}) \tag{7}$$

where ϕ and n denote material constants and $\dot{e}_{ij}^{(p)}$ and $\dot{e}'^{(p)}_{ij}$ are the plastic strain rate and the deviatoric plastic strain rate respectively.

The kinematic hardening used in the present paper is modified form of Ziegler's law (ref.13). This evolution is expressed as

$$\dot{\alpha}_{ij} = C\bar{s}_{ij}\dot{e}_p \tag{8}$$

where

$$C = D J_2^{1/2} r(\bar{\theta})^2/(2J_2(1+(W(\theta)/R(\theta))^2)^{1/2}) \tag{9}$$

where D is a material constant, $W(\theta) = \partial R(\theta)/\partial\theta$ and

$$\dot{e}_p = (\dot{e}'^{(p)}_{mn}\dot{e}'^{(p)}_{mn})^{1/2} \tag{10}$$

A general elastoplastic formulation is based on the assumption that the total strain response consists of elastic and plastic strain responses. This is represented as follows:

$$\dot{e}_{ij} = \dot{e}_{ij}^{(e)} + \dot{e}_{ij}^{(p)} \tag{11}$$

where e_{ij} and $e_{ij}^{(e)}$ the total strain and elastic strain respectively.

The plastic strain rate is expressed by the following form:

$$\dot{e}_{ij}^{(p)} = \Lambda \ \partial g/\partial \sigma_{ij} \tag{12}$$

where Λ is a scalar function dependent on stress and strain, and g is the plastic potential.

The function Λ is determined by use of the consistency condition which is expressed as

$$\partial f/\partial\sigma_{ij}\,\dot{\sigma}_{ij} + \partial f/\partial \alpha_{ij}\,\dot{\alpha}_{ij} + \partial f/\partial k\,\dot{k} = 0 \tag{13}$$

The constitutive relation in elastic range is given by the following equation:

$$\dot{e}_{ij}^{(e)} = (\dot{\sigma}_{ij} - \dot{p}\,\delta_{ij})/(2\,\mu p) + 0.435C_s\,\dot{p}\,\delta_{ij}/\{3(1+e_0)p\} \tag{14}$$

where μ is an elastic material constant, e_0 is the initial void ratio, C_s is the rebound compressibility of sand, $p=I_1/3$ and μ is given as follows:

$$\mu = \{3(1-2\,\nu)(1+e_0)\}/\{2(1+\nu)\ 0.435C_s\} \tag{15}$$

where ν is the Poisson's ratio.

SIMULATION OF UNDRAINED TRIAXIAL TESTS OF ANISOTROPICALLY CONSOLIDATED CLAY

In order to investigate the strength and deformation of saturated clay, undrained shear tests have been carried out under anisotropic consolidation or K_0-consolidation. The proposed model is applied to the simulation of mechanical behaviour of anisotropically consolidated clay. The predicted results are compared with experimental data reported in ref. 14. The physical property and material parameters used in prediction of undrained triaxial tests are shown in TABLE 1. The stresses employed in triaxial tests are effective mean principal stress p and stress difference q.

Figs. 2-6 show the comparison between simulation and experimental results given by Stipho (ref.14) for effective stress paths, stress-strain curves and pore water pressure responses when a sample is subjected to isotropic consolidation. It is found that the model simulation is capable of describing experimental results of triaxial compression and extension tests for every value of OCR. The prediction performed by Banerjee et al. seems to be not close to experimental data of triaxial compression test for OCR=1.0. This may be because the associative flow rule employed by Banerjee et al.(ref.3) is not adequate for the model prediction. On the other hand, the present simulation using the non-associative flow rule is successful in predicting experimental data of triaxial compression tests for OCR=1.0.

Figs. 7- 9 show the comparison between model simulation and experimental results in compression and extension tests for anisotropically consolidated clay ($N=\sigma_1/\sigma_3$). The proposed model is capable of predicting the effective stress path, stress-strain curve and pore water pressure obtained in experiments.

The effective stress path is shown in Fig.10 for anisotropic consolidation N=1.50. Figs.11-13 show the effective stress paths, stress-strain curves and pore water pressures for anisotropic consolidation N=1.75. Although the pore water pressure of experimental data increases with increment of axial strain for the value greater than OCR=2.0, the model simulation tends to increase in the lower range of axial strain and decrease in the higher range of axial strain.

REFERENCES

1 ST. Pietruszczak and Z. Mroz, On hardening anisotropy of K_0-consolidated clays, Int.J.Num.Anal.Meth.Geomech. 7, 1983, 19-38.

2 M.J. Kavvadas, A constitutive model for clays based on non-associated anisotropic elasto-plasticity, Proceeding of the International Conference on Constitutive Laws for Engineering Materials - Theory and Applications, Tucson, 1983, 263-270.

3 P.K. Banerjee, A.S. Stipho and N.B. Yousif, A theoretical and experimental investigation of the behavior of anisotropically consolidated clays, Development in Soil Mechanics and Foundation Engineering, II, P.K. Banerjee and R. Butterfield, Eds., Elsevier Applied Science, 1984, 1-41.

4 A. Anandarajah and Y.F.Dafalias, Bounding surface plasticity. III:application to anisotropic cohesive soils, J.Eng.Mech. ASCE, 112(12), 1986, 1292-1318.

5 H. Hirai, A combined hardening model of plasticity for sands, Constitutive Laws for Engineering Materials - Theory and Applications C.S. Desai et al.(Eds.), Elsevier, 1987, 557-564.

6 H. Hirai, An anisotropic hardening model for sand subjected to cyclic loading, Developments in Geotechnical Engineering, 42, A.S.Cakmak (Eds.), Elsevier and Computational mechanics publications, 1987, 53-67.

7 H. Hirai, Modelling of cyclic behaviour of sand with combined hardening, Soils and Foundations, 27(2), 1987, 1-11.

8 H. Hirai, An isotropic hardening model for cyclic plasticity of sand, Proceedings of Japan Society of Civil Engineers, 382(III-7), 1987, 217-225.

9 H. Hirai, An elastoplastic constitutive model for cyclic behaviour of sands, Int. J. Numer. Anal. Meth. Geomech. 11, 1987, 503-520.

10 K.H. Roscoe and J.B. Burland, On the generalized stress-strain behaviour of "wet" clay, Engineering Plasticity, J. Heyman and F.A. Leckie(Eds.), Cambridge University Press, 1968, 535-610.

11 K.J. Willam and E.P. Warnke, Constitutive models for the triaxial behavior of conrete, Int.Assoc.Bridge Struct.Eng., 19, 1975, 1-30.

12 A.N. Schofield and C.P. Wroth, Critical State Soil Mechanics, McGraw-Hill, 1968.

13 H. Ziegler, A modification of Prager's hardening rule, Quart. Appl. Math. 17, 1959, 55-65.

14 A.S. Stipho, Theoretical and experimental investigation of anisotropically consolidated Kaolin, Ph.D. Thesis, University College, Cardiff, 1978.

TABLE 1. Physical properties and material parameters used for clay tests.

Number of figure	2-13
Elastic parameters	
e_0	1.0
ν	0.3
C_s	0.115
Yield parameters	
m	0.0133
$\bar{r}$	0.95
Plastic potential parameters	
η	-0.204
$\bar{R}$	0.72
Hardening parameters	
n	1.0
ϕ	-0.57
D	342

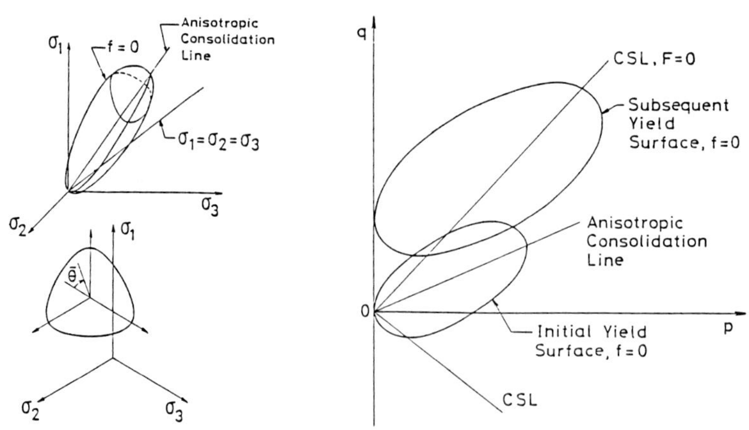

Fig.1. A combined hardening model of anisotropically consolidated clay.

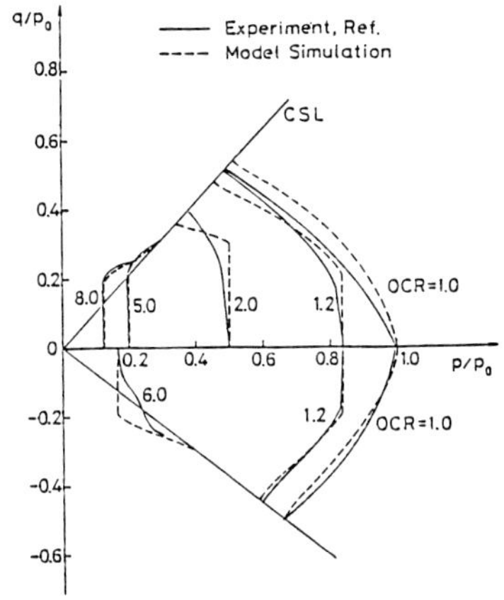

Fig.2. Comparison of model simulation with triaxial compression and extension tests of isotropically consolidated clay.

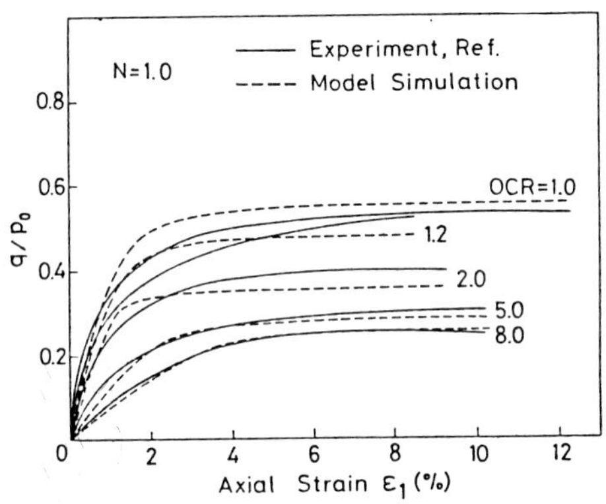

Fig.3. Relationships between stress and strain in triaxial compression tests.

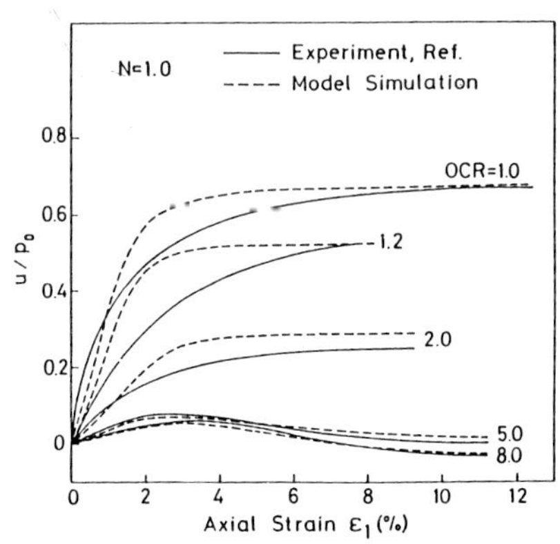

Fig.4. Relationships between strain and pore water pressure in triaxial compression tests.

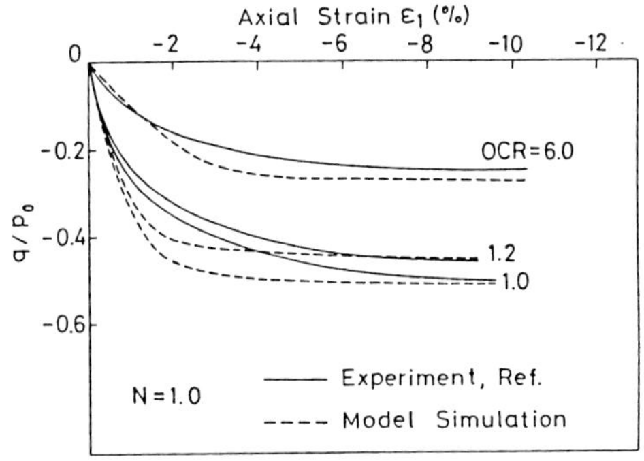

Fig.5. Relationships between stress and strain in triaxial extension tests.

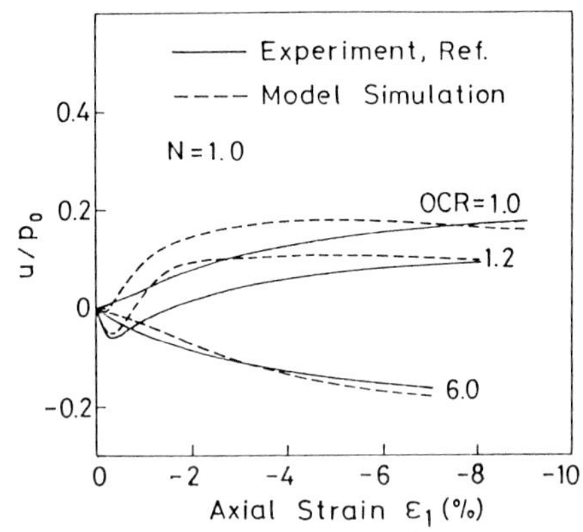

Fig.6. Relationships between strain and pore water pressure in triaxial extension tests.

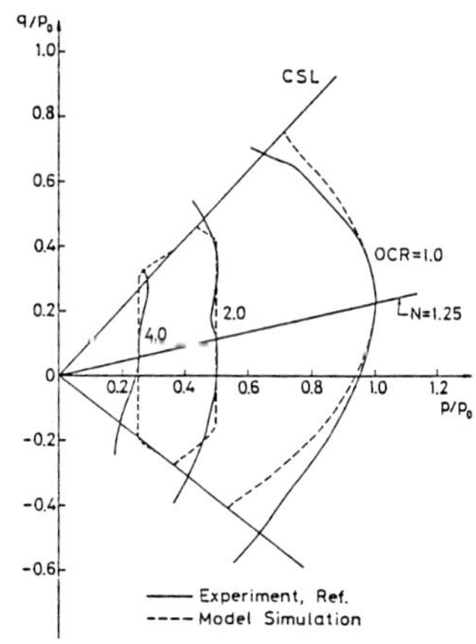

Fig.7. Effective stress paths in undrained test of anisotropically consolidated clay.

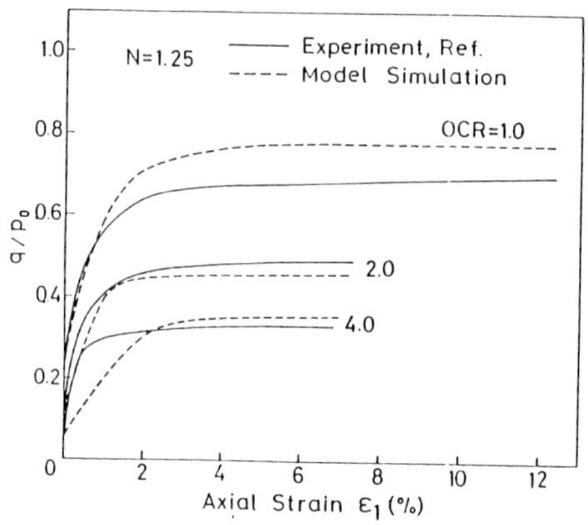

Fig.8. Stress-strain relationships of anisotropically consolidated clay.

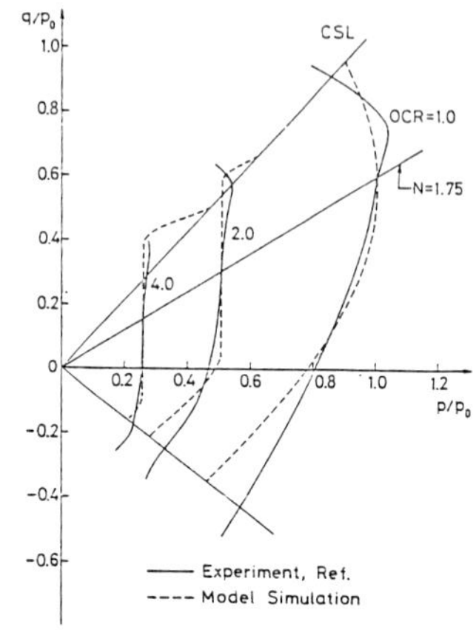

Fig.11. Effective stress paths in undrained test of anisotropically consolidated clay.

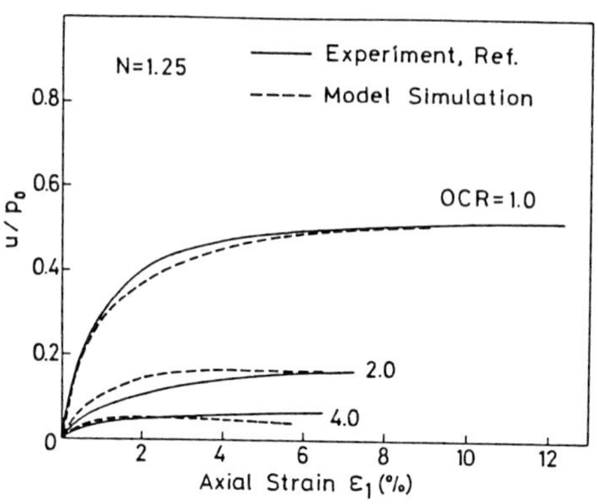

Fig.9. Pore water pressure responses of anisotropically consolidated clay.

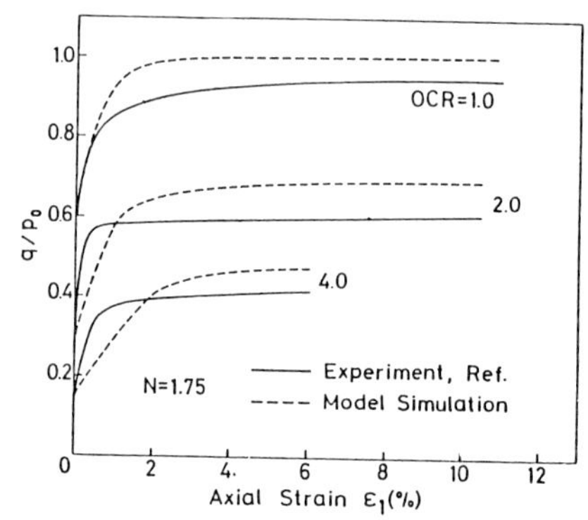

Fig.12. Stress-strain relationships of anisotropically consolidated clay.

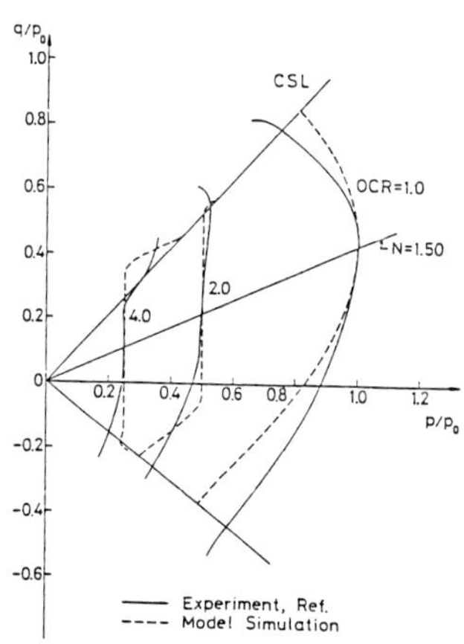

Fig.10. Effective stress paths in undrained test of anisotropically consolidated clay.

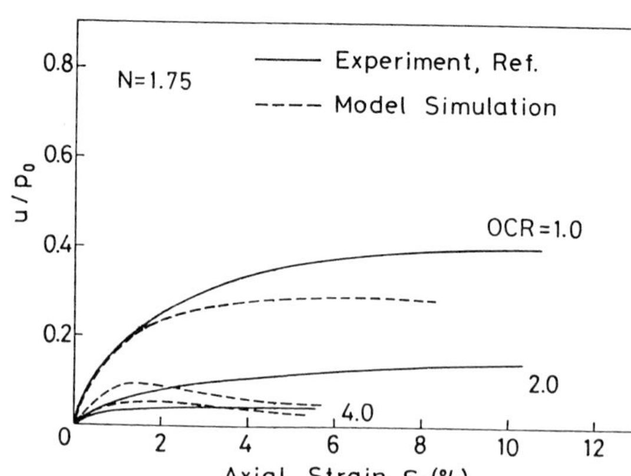

Fig.13. Pore water pressure responses of anisotropically consolidated clay.

Micromechanics of Granular Materials, edited by M. Satake and J.T. Jenkins
Elsevier Science Publishers B.V., Amsterdam, 1988 — Printed in The Netherlands

STABILITY OF A RADIAL FLOW OF GRANULAR MATERIAL

A. J. M. SPENCER

Department of Theoretical Mechanics, University of Nottingham, Nottingham, NG7 2RD, England

ABSTRACT

We consider small perturbations of the radial flow of a granular material, deforming according to the double-shearing theory, between smooth walls. It is shown that admissible perturbed solutions exist, and that under some conditions they may grow exponentially with time.

INTRODUCTION

In an analysis [1] of steady shearing flow of granular materials using the 'double-shearing' theory of flow of granular materials [2], it was found that, under the velocity boundary conditions appropriate to steady uniform shearing between parallel planes, the associated constant stress field is unstable, in the sense that the principal stress axes always rotate away from the directions they have in the steady stress state. Eventually this rotation seems to lead to the formation of localised shear bands, which is consistent with many observations.

The question naturally arises as to whether similar instabilities may occur in other steady flows. A few analytical solutions for steady flows using the double-shearing theory are known; some are described in [2]. In particular, flows in hoppers and bins are of practical as well as theoretical interest. There is a wealth of experimental results on bin and hopper flows; an excellent account of experimental methods and results has been given by Tüzün, Houlsby, Nedderman and Savage [3]. A wide variety of behaviours is reported, depending on material, hopper geometry, wall friction conditions, and so on. In many cases the flow is, to a reasonable approximation, steady and symmetrical. However this is not always the case and, for example, Lee, Cowin and Templeton [4] describe hopper flows which are cyclic and asymmetric. An objective of this paper is to consider whether the double-shearing theory (which is one of a number of constitutive theories which have been proposed) is capable of describing this type of behaviour. To this end we consider small perturbations of a very simple but inhomogeneous steady flow, namely radial flow in a smooth-walled wedge-shaped channel. We find that perturbed asymmetrical and unsteady solutions exist, and that in some circumstances the perturbations may, initially, grow exponentially with time.

NOTATION AND BASIC THEORY

We refer all variables to plane polar coordinates r, θ. Stress components are denoted by σ_{rr}, $\sigma_{\theta\theta}$, $\sigma_{r\theta}$, and velocity components by v_r, v_θ. Then, as in [2], we may write

$$\sigma_{rr} = -p + q \cos 2\psi , \qquad \sigma_{\theta\theta} = -p - q \cos 2\psi , \qquad \sigma_{r\theta} = q \sin 2\psi , \tag{1}$$

where p is the pressure, q is the maximum shear stress, and ψ is the angle the major principal stress axis makes with the radial direction. The Coulomb–Mohr yield condition can be expressed as

$$q = p \sin\varphi + c \cos\varphi \ , \tag{2}$$

where the cohesion c and the internal friction angle φ are assumed constant. With no body forces, the equilibrium equations are

$$\frac{\partial\sigma_{rr}}{\partial r} + \frac{1}{r}\frac{\partial\sigma_{r\theta}}{\partial\theta} + \frac{\sigma_{rr}-\sigma_{\theta\theta}}{r} = 0 \ , \qquad \frac{\partial\sigma_{r\theta}}{\partial r} + \frac{1}{r}\frac{\partial\sigma_{\theta\theta}}{\partial\theta} + \frac{2\sigma_{r\theta}}{r} = 0 \ . \tag{3}$$

For the velocity field, we adopt the 'double-shearing' constitutive model described in [2], in which the velocity components are governed by the equations

$$\frac{\partial v_r}{\partial r} + \frac{v_r}{r} + \frac{1}{r}\frac{\partial v_\theta}{\partial\theta} = 0 \ , \tag{4}$$

$$\left[\frac{1}{r}\frac{\partial v_r}{\partial\theta} + \frac{\partial v_\theta}{\partial r} - \frac{v_\theta}{r}\right]\cos 2\psi - \left[\frac{\partial v_r}{\partial r} - \frac{v_r}{r} - \frac{1}{r}\frac{\partial v_\theta}{\partial\theta}\right]\sin 2\psi$$
$$+ \sin\varphi\left[\frac{1}{r}\frac{\partial v_r}{\partial\theta} - \frac{\partial v_\theta}{\partial r} - \frac{v_\theta}{r} + 2\Omega\right] = 0 \ , \tag{5}$$

where

$$\Omega = \dot{\psi} = \frac{\partial\psi}{\partial t} + v_r\frac{\partial\psi}{\partial r} + v_\theta\frac{1}{r}\frac{\partial\psi}{\partial\theta} \ , \tag{6}$$

is the convected rotation-rate of the principal stress axes, and t is not necessarily physical time but may be any parameter which determines the sequence of events. Equation (4) is the incompressibility equation and (5) expresses the double-shearing deformation mechanism.

INITIAL SOLUTION

We consider solutions with $\psi = 0$ and $p = p(r)$. Then it is easily found that a solution of (1) – (3) is

$$q = q_0 = \frac{c\cos\varphi}{1-\sin\varphi}\left[\frac{r}{a}\right]^n \ , \qquad p = p_0 = c\cot\varphi\left\{\frac{1}{1-\sin\varphi}\left[\frac{r}{a}\right]^n - 1\right\} \ , \qquad \psi = 0 \ , \tag{7}$$

where

$$n = 2\sin\varphi/(1-\sin\varphi) \ . \tag{8}$$

The corresponding stress is

$$\sigma_{rr} = c\cot\varphi\left\{1-\left[\frac{r}{a}\right]^n\right\} \ , \qquad \sigma_{\theta\theta} = c\cot\varphi\left\{1-\frac{1+\sin\varphi}{1-\sin\varphi}\left[\frac{r}{a}\right]^n\right\} \ , \qquad \sigma_{r\theta} = 0 \ , \tag{9}$$

and the surface $r = a$ is traction-free.

Also, a solution of (4) – (6), with $\psi = 0$, is the radial flow solution

$$v_r = -\frac{Q}{r} \ , \qquad v_\theta = 0 \ , \tag{10}$$

where Q is constant (or a function of t) and, for positive plastic work-rate, $Q \geqslant 0$. This solution represents converging radial flow in a smooth-walled wedge-shaped channel of arbitrary angle 2α, as illustrated in Fig. 1. We consider the region $r > a$. The flow rate is $2\alpha Q$ per unit length normal to the flow plane.

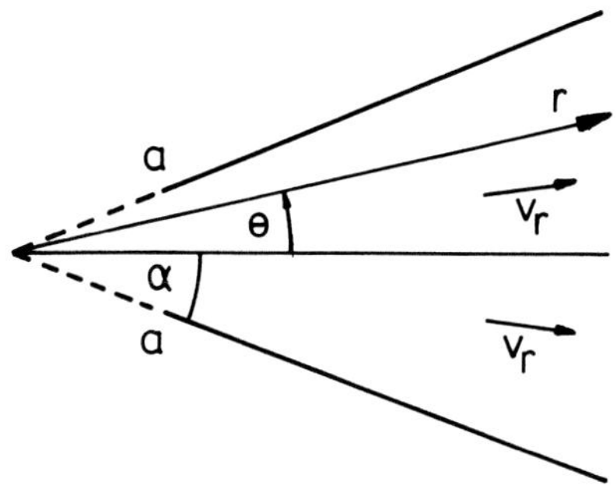

Fig. 1

PERTURBED SOLUTION

We consider small perturbations of the above solution. Thus, for the stress field, we seek solutions of (1) – (3) of the form

$$p = p_0 + \epsilon p' , \qquad q = q_0 + \epsilon q' , \qquad \psi = \epsilon \psi' , \tag{11}$$

where ϵ is a small parameter. The corresponding increment in stress is, to order ϵ, given as

$$\epsilon \sigma'_{rr} = \epsilon(-p'+q') , \qquad \epsilon \sigma'_{\theta\theta} = -\epsilon(p'+q') , \qquad \epsilon \sigma'_{r\theta} = 2\epsilon q_0 \psi' , \tag{12}$$

with $q' = p' \sin \varphi$. By substituting (12) into the equilibrium equations, there follows

$$- (1 - \sin \varphi)\frac{\partial p'}{\partial r} + \frac{2}{r}\frac{\partial}{\partial \theta}(q_0 \psi') + \frac{2 \sin \varphi}{r} p' = 0 ,$$

$$2 \frac{\partial}{\partial r}(q_0 \psi') - (1 + \sin \varphi)\frac{\partial p'}{\partial \theta} + \frac{4}{r} q_0 \psi' = 0 . \tag{13}$$

Solutions of these equations are

$$p' = \left[\frac{r}{a}\right]^{\beta}\left[\left\{1 + \gamma(1 - \sin \varphi)\right\}C_1(t)\left[\frac{r}{a}\right]^{\gamma} + \left\{1 - \gamma(1 - \sin \varphi)\right\}C_2(t)\left[\frac{r}{a}\right]^{-\gamma}\right]\sin k\theta ,$$

$$2q_0\psi' = k \cos^2\varphi \left[\frac{r}{a}\right]^{\beta}\left\{C_1(t)\left[\frac{r}{a}\right]^{\gamma} + C_2(t)\left[\frac{r}{a}\right]^{-\gamma}\right\}\cos k\theta , \tag{14}$$

where

$$\beta = (-1 + 2 \sin \varphi)/(1 - \sin \varphi) , \qquad \gamma = (1 - k^2 \cos^2 \varphi)^{\frac{1}{2}}/(1 - \sin \varphi) . \tag{15}$$

This solution applies when $k \cos \varphi < 1$, and γ is real. When $k \cos \varphi > 1$, the solution (14) still applies if $C_1(t)$ and $C_2(t)$ are regarded as complex functions of t, and p' and $2q_0\psi'$ are interpreted as the real parts of the expressions on right-hand sides of (14). The resulting expressions for p' and ψ' are then, in explicit form

$$p' = \left[\frac{r}{a}\right]^{\beta}\Big[\{C_1'(t)+|\gamma|(1 - \sin\varphi)C_2'(t)\}\cos\Big\{|\gamma|\ln\left[\frac{r}{a}\right]\Big\}$$

$$- \{C_2'(t)+|\gamma|(1 - \sin\varphi)C_1'(t)\}\sin\{|\gamma|\ln\left[\frac{r}{a}\right]\Big\}\Big]\sin k\theta \ ,$$

$$2q_0\psi' = k\cos^2\varphi\ \left[\frac{r}{a}\right]^{\beta}\Big[C_1'(t)\cos\Big\{|\gamma|\ln\left[\frac{r}{a}\right]\Big\} + C_2'(t)\sin\Big\{|\gamma|\ln\left[\frac{r}{a}\right]\Big\}\Big]\cos k\theta \ ,$$

where

$$C_1'(t) = \mathrm{Re}\{C_1(t)+C_2(t)\} \ , \qquad C_2'(t) = \mathrm{Im}\{-C_1(t)+C_2(t)\} \ .$$

There also exist similar solutions with p' and ψ' proportional to $\cos k\theta$ and $\sin k\theta$ respectively, but (14) seems to be the case of greater interest.

For a smooth-walled channel we require $\psi' = 0$ at the channel wall. Hence, for a channel with wedge angle α, we have

$$k = (i+\tfrac{1}{2})\pi/\alpha \ , \qquad i = 0,1,2,\dots \ . \tag{16}$$

VELOCITY FIELD

For the velocity field we seek perturbed solutions such that, in addition to (11),

$$v_r = -\frac{Q}{r} + \epsilon v_r' \ , \qquad v_\theta = \epsilon v_\theta' \ . \tag{17}$$

Then the incompressibility condition (4) gives

$$\frac{\partial v_r'}{\partial r} + \frac{v_r'}{r} + \frac{1}{r}\frac{\partial v_\theta'}{\partial\theta} = 0 \ . \tag{18}$$

It is convenient to introduce a stream function χ such that

$$v_r' = \frac{1}{r}\frac{\partial\chi}{\partial\theta} \ , \qquad v_\theta' = -\frac{\partial\chi}{\partial r} \ , \tag{19}$$

and then (18) is identically satisfied. By substituting (11) and (17) into (5) and (6), discarding terms of order ϵ^2, and introducing (19) we obtain

$$(1 - \sin\varphi)\frac{\partial^2\chi}{\partial r^2} - (1 + \sin\varphi)\left\{\frac{1}{r}\frac{\partial\chi}{\partial r} + \frac{1}{r^2}\frac{\partial^2\chi}{\partial\theta^2}\right\} = -\frac{4Q}{r^2}\psi' + 2\Omega'\sin\varphi \ , \tag{20}$$

$$\Omega' = \frac{\partial\psi'}{\partial t} - \frac{Q}{r}\frac{\partial\psi'}{\partial r} \ . \tag{21}$$

With ψ' given by (14), equation (20) has solutions of the form

$$\chi = h(r,t)\cos k\theta \ . \tag{22}$$

By substituting (7), (14) and (22) into (20), there follows

$$(1 - \sin\varphi)\frac{\partial^2 h}{\partial r^2} - (1 + \sin\varphi)\left\{\frac{1}{r}\frac{\partial h}{\partial r} - \frac{k^2 h}{r^2}\right\} =$$

$$\frac{k(1 - \sin\varphi)\cos\varphi}{c}\left[\left\{\dot{C}_1(t)\left[\frac{r}{a}\right]^{-\delta+\gamma} + \dot{C}_2(t)\left[\frac{r}{a}\right]^{-\delta-\gamma}\right\}\sin\varphi\right.$$

$$\left. - \frac{Q}{a^2}\left[\{2-(\delta-\gamma)\sin\varphi\}C_1(t)\left[\frac{r}{a}\right]^{-\delta+\gamma+2} + \{2-(\delta+\gamma)\sin\varphi\}C_2(t)\left[\frac{r}{a}\right]^{-\delta-\gamma-2}\right]\right] , \quad (23)$$

where

$$\delta = n - \beta = 1/(1 - \sin\varphi) . \quad (24)$$

The general solution of (23) is

$$h = D_1(t)\left[\frac{r}{a}\right]^{\delta+\gamma} + D_2(t)\left[\frac{r}{a}\right]^{\delta-\gamma} + e_1\dot{C}_1(t)\left[\frac{r}{a}\right]^{-\delta+\gamma+2} + e_2\dot{C}_2(t)\left[\frac{r}{a}\right]^{-\delta-\gamma+2}$$

$$+ f_1 C_1(t)\left[\frac{r}{a}\right]^{-\delta+\gamma} + f_2 C_2(t)\left[\frac{r}{a}\right]^{-\delta-\gamma} , \quad (25)$$

where

$$e_1 = \frac{ka^2\cos\varphi}{c\{(2\delta-\gamma-2)^2-\gamma^2\}} , \qquad f_1 = -\frac{Qk\cos\varphi\{2-(\delta-\gamma)\sin\varphi\}}{c\{(2\delta-\gamma)^2-\gamma^2\}} , \quad (26)$$

and e_2 and f_2 are obtained by replacing γ by $-\gamma$ in the expressions for e_1 and f_1 respectively.

If γ is purely imaginary, then $C_1(t)$, $C_2(t)$, $D_1(t)$ and $D_2(t)$ are taken to be complex quantities and h is interpreted as the real part of the expression on the right-hand side of (25).

UNSTEADY PERTURBED VELOCITY FIELD

If the traction is prescribed to be that which corresponds to that of the initial solution on any surface r = constant (for example, if zero traction is prescribed on $r = a$), then we require $\psi' = 0$, $p' = 0$ at that surface, and it follows from (14) that ψ' and p' are zero everywhere. Nevertheless the theory admits the possibility of unsteady perturbed velocity fields, for with $C_1(t) = 0$, $C_2(t) = 0$, equation (23) has the solution

$$h = D_1(t)(r/a)^{\delta+\gamma} + D_2(t)(r/a)^{\delta-\gamma} ,$$

with corresponding velocity components

$$v_r' = -k\{D_1(t)(r/a)^{\delta+\gamma-1} + D_2(t)(r/a)^{\delta-\gamma-1}\}\sin k\theta ,$$
$$v_\theta' = -\{(\delta+\gamma)D_1(t)(r/a)^{\delta+\gamma-1} + (\delta-\gamma)D_2(t)(r/a)^{\delta-\gamma-1}\}\cos k\theta , \quad (27)$$

for the case in which γ is real. When γ is imaginary, we have

$$h = \left[\frac{r}{a}\right]^{\delta}\left[D_1'(t)\cos\left\{|\gamma|\ln\left[\frac{r}{a}\right]\right\} + D_2'(t)\sin\left\{|\gamma|\ln\left[\frac{r}{a}\right]\right\}\right] ,$$

where

$$D_1'(t) = \mathrm{Re}\{D_1(t)+D_2(t)\} , \qquad D_2'(t) = \mathrm{Im}\{-D_1(t)+D_2(t)\} ,$$

and, correspondingly,

$$v_r' = -\frac{k}{a}\left[\frac{r}{a}\right]^{\delta-1}\left[D_1'(t)\cos\left\{|\gamma|\ln\left[\frac{r}{a}\right]\right\} + D_2'(t)\sin\left\{|\gamma|\ln\left[\frac{r}{a}\right]\right\}\right]\sin k\theta \ ,$$

$$v_\theta' = -\frac{1}{a}\left[\frac{r}{a}\right]^{\delta-1}\left[\{\delta D_1'(t)+|\gamma|D_2'(t)\}\cos\left\{|\gamma|\ln\left[\frac{r}{a}\right]\right\}\right.$$

$$\left. + \{\delta D_2'(t)-|\gamma|D_1'(t)\}\sin\left\{|\gamma|\ln\left[\frac{r}{a}\right]\right\}\right]\cos k\theta \ .$$

In either case, when k is given by (16), we have $v_\theta = 0$ at $\theta = \pm\alpha$, as required. Of course, the solution is required to be compatible with any other specified velocity boundary conditions. In particular, periodic velocity solutions arise if $D_1(t)$ and $D_2(t)$ are periodic.

UNSTEADY PERTURBED STRESS AND VELOCITY FIELDS

Of more interest is the possibility of non-trivial perturbed solutions arising under homogeneous boundary conditions. We consider the case in which the velocity at two surfaces $r = a$ and $r = b$ is prescribed to be that of the initial solution. Thus

$$v_r' = 0 \ , \qquad v_\theta' = 0 \qquad \text{at } r = a \text{ and } r = b \ .$$

It follows from (19) and (22) that

$$h = 0 \ , \qquad \partial h/\partial r = 0 \qquad \text{at } r = a \text{ and } r = b \ . \tag{28}$$

We seek solutions of the form

$$\{D_1(t),\ D_2(t),\ C_1(t),\ C_2(t)\} = \{d_1,\ d_2,\ c_1,\ c_2\}e^{\mu t} \ ,$$

where d_1, d_2, c_1, c_2 and μ are constants. Then (28) gives

$$d_1 + d_2 + (e_1\mu+f_1)c_1 + (e_2\mu+f_2)c_2 = 0 \ ,$$

$$d_1\rho^{\delta+\gamma} + d_2\rho^{\delta-\gamma} + (e_1\mu\rho^2+f_1)\rho^{-\delta+\gamma}c_1 + (e_2\mu\rho^2+f_2)\rho^{-\delta-\gamma}c_2 = 0 \ ,$$

$$(\delta+\gamma)d_1 + (\delta-\gamma)d_2 + \{(-\delta+\gamma+2)e_1\mu + (-\delta+\gamma)f_1\}c_1$$
$$+ \{(-\delta-\gamma+2)e_2\mu + (-\delta-\gamma)f_2\}c_2 = 0 \ ,$$

$$(\delta+\gamma)\rho^{\delta+\gamma}d_1 + (\delta-\gamma)\rho^{\delta-\gamma}d_2 + \{(-\delta+\gamma+2)e_1\mu\rho^2 + (-\delta+\gamma)f_1\}\rho^{-\delta+\gamma}c_1$$
$$+ \{(-\delta-\gamma+2)e_2\mu\rho^2 + (-\delta-\gamma)f_2\}\rho^{-\delta-\gamma}c_2 = 0 \ , \tag{29}$$

where $\rho = b/a$.

For non-trivial solutions the determinant of coefficients of d_1, d_2, c_1 and c_2 in (28) has to be zero, thus

$$\begin{vmatrix} 1 & 1 & e_1\mu+f_1 & e_2\mu+f_2 \\ \rho^{\delta+\gamma} & \rho^{\delta-\gamma} & (e_1\mu\rho^2+f_1)\rho^{-\delta+\gamma} & e_2\mu\rho^2+f_2)\rho^{-\delta-\gamma} \\ \delta+\gamma & \delta-\gamma & \begin{matrix}(-\delta+\gamma+2)e_1\mu \\ +(-\delta+\gamma)f_1\end{matrix} & \begin{matrix}(-\delta-\gamma+2)e_2\mu \\ -(\delta+\gamma)f_2\end{matrix} \\ (\delta+\gamma)\rho^{\delta+\gamma} & (\delta-\gamma)\rho^{\delta-\gamma} & \begin{matrix}\{(-\delta+\gamma+2)e_1\mu\rho^2 \\ +(-\delta+\gamma)f_1\}\rho^{-\delta+\gamma}\end{matrix} & \begin{matrix}\{(-\delta-\gamma+2)e_2\mu\rho^2 \\ -(\delta+\gamma)f_2\}\rho^{-\delta-\gamma}\end{matrix} \end{vmatrix} = 0 \ . \tag{30}$$

When expanded, (30) is a quadratic equation for μ. If there exists at least one root of this equation with positive real part, then, at least initially, small perturbations of the initial solution which conform to the velocity boundary conditions will grow, indicating lack of stability of the initial steady state solution.

It is difficult to draw general conclusions about the roots of (30) by analysis. However there is some simplification in the case $\rho = (b/a) \gg 1$. In this case, noting that $\delta > 1$ and $\delta > \gamma > 0$, the leading terms in the expansion of (30) give

$$\gamma e_1 e_2 \mu^2 + \mu\{e_2 f_1(\gamma-1)+e_1 f_2(\gamma+1)\}+ \gamma f_1 f_2 = 0 \ . \tag{31}$$

A full investigation of the nature of the roots of (31) has not been undertaken, but a numerical exploration shows that (31) has at least one positive real root over a range of physically realistic values of φ and α. The boundaries of this range have not been determined.

DISCUSSION

The problem we have considered is a simple illustrative one, and caution should be exercised in drawing conclusions. Among the important factors which are not taken into account are gravity, inertia, dilatation and wall friction; all of these have an important influence in real hopper and bin flows. However the analysis does show that the double-shearing theory admits a variety of mechanical behaviours, at least in small perturbations from the steady state, and that these can include asymmetric and cyclic flows.

A similar approach can be applied to other problems for which an initial steady-state solution is known analytically, but such solutions are few. One which would be of interest is that of radial flow between rough walls, given by Pemberton [5] and outlined in [2], but this is much more complicated than the problem considered here.

The question of stability appears to depend critically upon boundary conditions. If the boundary conditions are such that ψ is fixed at the boundaries, then the time-independent stress field seems to be stable. If ψ is not fixed (as when velocity boundary conditions are given) then there appears to be a possibility of exponential growth of perturbations of the stress field, and consequent loss of stability.

We note that stability of granular flows has been considered in a more general way by Schaeffer [6].

REFERENCES

1 A. J. M. Spencer, *Acta Mechanica* **64** (1986) 77–87.

2 A. J. M. Spencer, in H. G. Hopkins and M. J. Sewell (Eds.), *Mechanics of Solids*, Pergamon, Oxford, 1982, pp. 607–652

3 U. Tüzün, G. T. Houlsby, R. M. Nedderman and S. B. Savage, *Chem. Engng. Sci.*, **37** (1982) 1691–1709.

4 Jannan Lee, S. C. Cowin and J. S. Templeton III, *Trans. Soc. Rheol.*, **18** (1974) 247–269.

5 C. S. Pemberton, *J. Mech. Phys. Solids*, **13** (1965) 351–360.

6 D. G. Schaeffer, *J. Diff. Eq.* To appear.

Micromechanics of Granular Materials, edited by M. Satake and J.T. Jenkins
Elsevier Science Publishers B.V., Amsterdam, 1988 — Printed in The Netherlands

KINETIC THEORETICAL ANALYSIS OF A RAPIDLY DEFORMING DISK ASSEMBLY

M. NAKAGAWA

Dept. of Civil Engineering, Tohoku University, Aoba, Sendai 980 (Japan)

SUMMARY

Kinetic theoretical approach is employed in analyzing the flow behavior of a rapidly deforming granular disk assembly. Binary collisions are assumed and two types of energy dissipation mechanisms are carefully investigated. The roughness coefficient that relates the relative velocities at contact before and after a collision is assumed to depend upon the orientation of the colliding disks. The present theory takes account of the geometrical dependence of a collision in the averaging scheme and estimates the constitutive relations accordingly.

INTRODUCTION

Employing the most careful kinetic theoretical approach, Jenkins and Richman (ref. 1) considered plane flows of a dense gas of identical, rough inelastic circular disks. Assuming the perturbation from the equilibrium is small, they determined the single particle velocity distribution function as an approximate solution to its equation of evolution. Inspite of their devotion to the use of statistical averaging method, less attentions were given to the dynamics of binary collision mechanisms. They recognized the fact that the change in the tangential component of the relative velocity of the contact point depends on the geometry of the collisions, however, the roughness coefficient was left as a constant parameter in their analysis. Here the collisional orientation of two disks is effectively expressed through the collision angle γ in the averaging, and constitutive quantities are evaluated based on the above scheme.

THEORY

Exact Binary Collision Model

Consider two-dimensional motion of identical, rough, inelastic circular disks. The mass of a disk is m, the diameter is σ, and the moment of inertia about its axis of spin is I. Greek subscripts denote components of vectors and tensors with respect to the orthogonal unit vectors $\underline{e}_1$ and $\underline{e}_2$ in the plane, and prime denotes their component values after a collision. In a collision, the balance of linear momentum requires

$$c'_{1\alpha} = c_{1\alpha} - \frac{1}{m} J_\alpha \tag{1}$$

$$c'_{2\alpha} = c_{2\alpha} + \frac{1}{m} J_\alpha \tag{2}$$

where $c_{1\alpha}$ and $c_{2\alpha}$ are the components of the velocities of the centers of disk 1 and 2, respectively, and J_α are the components of the impulsive force exerted by disk 1 upon disk 2. The balance of moment of momentum relates the spins, ω_1 and ω_2 before a collision, to those ω'_1 and ω'_2, after a collision by

$$\omega'_1 = \omega_1 - \frac{\sigma}{2I} \varepsilon_{\beta\alpha} k_\beta J_\alpha \tag{3}$$

and

$$\omega'_2 = \omega_2 - \frac{\sigma}{2I} \varepsilon_{\beta\alpha} k_\beta J_\alpha \tag{4}$$

where k_α are the components of the unit vector directed from the center of disk 1 to that of disk 2 at impact. The symbol $\varepsilon_{\alpha\beta}$ represents two-dimensional sign permutations, $\varepsilon_{12} = -\varepsilon_{21} = 1$ and $\varepsilon_{11} = \varepsilon_{22} = 0$. When Greek subscripts are repeated, they are to be summed from one to two.

Immediately prior to a collision the components v_α of the velocity of the contact point on the first disk relative to that on the second is given by

$$v_\alpha = g_\alpha + \frac{\sigma}{2} \varepsilon_{\beta\alpha} k_\beta (\omega_1 + \omega_2) \tag{5}$$

where $g_\alpha \equiv c_{1\alpha} - c_{2\alpha}$ is the relative velocity. A similar expression for the post-collisional components v'_α and g'_α is easily obtained.

The normal and tangential components of v_α before and after a collision are assumed to be related through the coefficient of restitution e and the roughness coefficient β as

$$k_\alpha v'_\alpha = -e\,(k_\alpha v_\alpha), \tag{6}$$

and

$$\varepsilon_{\alpha\beta} k_\alpha v'_\beta = -\beta\,(\varepsilon_{\alpha\beta} k_\alpha v_\beta), \tag{7}$$

respectively. These coefficients characterize the modes of energy dissipation. The usual coefficient of restitution is e, and the coefficient β is assumed to vary between -1 and 1. For values between 0 and 1, β may be interpreted as the coefficient of restitution for the tangential deformation of rough disks. Values of β between - 1to 0 can be thought of the coefficient of friction between rigid rough disks that slip during a collision. When $\beta = 0$, surface friction and inelasticity are sufficient to give zero tangential component of the relative velocities. Upon solving eqns.(1)-(4), (6) and (7), the impulse J_α may be expressed as

$$J_\alpha = amv_\alpha + bm(k_\beta v_\beta)k_\alpha , \tag{8}$$

where

$$a \equiv \kappa(1+\beta)/2(1+\kappa) \tag{9}$$

with

$$\kappa \equiv 4I/m\sigma^2 , \quad b = r - a \quad \text{and} \quad \gamma \equiv (1+e)/2 . \tag{10}$$

In order to relate β to the coefficient of friction, μ, Coulomb's law of friction is assumed to apply when the disks slip during the collision. Then

$$|\varepsilon_{\beta\alpha} k_\beta J_\alpha| = \mu(k_\phi J_\phi) . \tag{11}$$

When eqns.(8), (9) and (10) are used in eqn.(11), it provides an explicit expression for β,

$$\beta = -1 + \mu(1+e)(1+\kappa)|\cot\gamma|/\kappa \tag{12}$$

in which γ is the collision angle between v_α and k_α. It is clear that β always equals to -1 for glancing collisions ($\gamma = \pi/2$) or for smooth disks ($\mu = 0$).

With eqn.(8), the expressions for the post-collisional translational velocities of each disk are

$$c'_{1\alpha} = c_{1\alpha} - av_\alpha - b(k_\beta v_\beta)k_\alpha \tag{13}$$

and

$$c'_{2\alpha} = c'_{2\alpha} + av_\alpha + b(k_\beta v_\beta)k_\alpha . \tag{14}$$

Also, the post-collisional spins are

$$\omega'_1 = \omega_1 - \frac{2a}{\sigma\kappa}\varepsilon_{\beta\alpha}k_\beta v_\alpha \tag{15}$$

and

$$\omega_2' = \omega_2 - \frac{2a}{\sigma\kappa}\varepsilon_{\beta\alpha}k_\beta v_\alpha \quad . \tag{16}$$

In these, the coefficients a and b are functions of both the material properties of disks and the collision angle γ.

Statistical Consideration

The mean of a disk property $\psi(\underline{c})$ is defined as

$$\langle\langle \psi(\underline{c}) \rangle\rangle \equiv \frac{1}{n}\int \psi(\underline{c})\, f(\underline{c}, \underline{r}, t)\, d\underline{c} \quad , \tag{17}$$

in which $f(\underline{c},\underline{r},t)$ is the single particle velocity distribution function, and $n(\underline{r}, t)$ is the local number density of disks. When the spin velocity s is defined by $s \equiv \omega\sqrt{I/m}$, mean velocity of $\underline{u}$ and mean spin velocity w are given by $\langle\langle \underline{c} \rangle\rangle$ and $\langle\langle s \rangle\rangle$, respecitvely. Then, based on the definition of the mean above, the transport equation for the mean amount of any particle property of $\underline{c}$, $\psi(\underline{c})$, is expressed as

$$\frac{\partial}{\partial t}\langle\langle \rho\psi \rangle\rangle = \langle\langle n\underline{F}\cdot\frac{\partial\psi}{\partial\underline{c}} \rangle\rangle - \nabla\cdot\langle\langle \rho\underline{c}\,\psi \rangle\rangle + ¢(m\psi) \tag{18}$$

in which $\underline{F}$ is the generalized force that involves both external force and couple, and the last term, $¢(m\psi)$, is the rate of change of ψ due to collisions. The collision term involves the pair distribution function, and assuming Enskog's notion of molecular chaos of a dense gas it is expressed as the product of the radial distribution function g and two single-particle velocity distribution functions;

$$f(\underline{c}_1, \underline{r}_1, \underline{c}_2, \underline{r}_2) = g(|\underline{r}_2 - \underline{r}_1|/2)\, f(\underline{c}_1, \underline{r}_1)\, f(\underline{c}_2, \underline{r}_2) \; . \tag{19}$$

The collisional contribution to the rate of change of the mean can be decomposed into a collisional source $\chi(m\psi)$ and a collisional flux $\theta_\alpha(m\psi)$ as

$$¢(m\psi) = \chi(m\psi) - \frac{\partial}{\partial r_\alpha}\theta_\alpha(m\psi) \, . \tag{20}$$

The integral expressions for the source and flux terms are obtained by examining the dynamics of a binary collision and characterizing certain symmetries on the likelihood of such collisions. One of the most detailed decompositions that lead to the integral expressions for the source and flux above is done in (ref. 2). There the continuum mechanical requirement of the balance of total angular momentum, $\sqrt{I/m}\;\chi(m S) = -\varepsilon_{\alpha\beta}\theta_\alpha(m C_\beta)$, is identically satisfied, and the invariant and anti-invariant nature of collisional properties are expressed in a binary collision. $\underline{C}$ and S are the fluctuations in velocity, $\underline{c}-\underline{u}$, and spin, s-w, respectively.

The development of eqns.(17), (18), (19) and (20) is completely general so that these expressions are hold regardless the form of the velocity distribution function. Here, assuming that the system of interest does not deviate far from the equilibrium and that mean spins and the translational flux of spin are small, the linear expression of the perturbed Maxwellian single velocity distribution function is written as

$$f(\underline{C}, s, \underline{r}, t) = [1 + \frac{a_{\alpha\beta} C_\alpha C_\beta}{2\bar{T}^2} - \frac{a_{\alpha\beta\beta} C_\alpha}{8\bar{T}^2}\{4 - \frac{C_\gamma C_\gamma}{\bar{T}}\} - \frac{a_{\alpha 33} C_\alpha}{2\tilde{T}\bar{T}}\{1 - \frac{s^2}{\tilde{T}}\}]\ F(\underline{C}, s, \underline{r}, t) \tag{21}$$

where

$$F(\underline{C}, s, \underline{r}, t) \equiv \frac{n}{(2\pi\bar{T})(2\pi\tilde{T})^{\frac{1}{2}}} \exp[\frac{-C_\alpha C_\alpha}{2\bar{T}} - \frac{s^2}{2\tilde{T}}] \quad . \tag{22}$$

In the above translational and spin granular temperatures are defined

$$\bar{T} = 1/2 \langle\langle C_\zeta C_\zeta \rangle\rangle \qquad \text{and} \qquad \tilde{T} = \langle\langle S^2 \rangle\rangle \quad . \tag{23}$$

Also in finding eqn.(21) only the transport coefficients that have immediate physical significance are retained. The deviatoric part of the second moment $a_{\alpha\beta} = \langle\langle C_\alpha C_\beta \rangle\rangle$, the contracted third moments, $a_{\alpha\beta\beta} = \langle\langle C_\alpha C^2 \rangle\rangle$ and $a_{\alpha\beta\beta} = \langle\langle C_\alpha S^2 \rangle\rangle$ are determined as approximate solutions to the higher moment equations obtained from the expression (18) when it is casted in terms of the fluctuation velocity and spin. Detailed calculations can be found elsewhere (ref. 1), however, it must be mentioned that to the leading order approximation, these coefficients in (ref. 1) agree with those obtained in the present analysis,even though the collision dynamics in the averaging scheme is refined.

CONSTITUTIVE RELATIONS

It is the main objective here to investigate whether the more exact treatment of the dynamics of a binary collision affects the constitutive relations. The constitutive relations of special interest are the fluxes and sources that appear in the field equations for mass, linear momentum, translational energy and spin energy. As examples, the lowest order constitutive relations for the sources of spin and translational energies are:

$$\chi(\tfrac{m}{2}s^2) = \frac{3\sqrt{\pi} g_0 n^2 \sigma m}{\kappa} \bar{T}^{3/2} \Big[\frac{1}{6}\mu^2(1+e)^2(2-3\sin\gamma_0 + \sin^3\gamma_0)$$

$$+ \frac{1}{6}\frac{\kappa}{(1+\kappa)^2}\sin^3\gamma_0 - \Big\{\frac{1}{3}\mu(1+e)\cos^3\gamma_0 + \frac{1}{3}\frac{\kappa}{1+\kappa}\sin^3\gamma_0$$

$$+ \frac{1}{12} \frac{\mu^2(1+e)^2}{\kappa} (2-3\sin\gamma_0 + 3\sin^3\gamma_0) + \frac{1}{12} \frac{\kappa}{(1+\kappa)^2} \sin^3\gamma_0 \tag{24}$$

$$- \frac{5}{4} \frac{\mu^2(1+e)^2}{\kappa} \left(\frac{\pi}{4} - \frac{\gamma_0}{2} - \frac{\sin^2\gamma_0}{4}\right)$$

$$- \frac{5}{4} \frac{\kappa}{(1+\kappa)^2} \left(\frac{\gamma_0}{2} - \frac{\sin^2\gamma_0}{4}\right)\Big\} \tilde{T}/\bar{T}\Big]$$

and

$$\chi(\tfrac{m}{2}C^2) = -18\sqrt{\pi}g_0 n^2 \sigma m \bar{T}^{3/2} \mu \ln \mu \tag{25}$$

At leading order the expressions for fluxes of linear momentum, translationsal energy and spin energy are identical to those obtained in (ref. 1).

The ratio of the temperatures and then the translational temperature are found by specializing the balance equations for a steady, homogeneous, rectilinear flow with a strain rate of . For such a flow the spin energy equation reduces to the requirement that the source of spin energy vanishes, $\chi(\frac{m}{2}S^2) = 0$. This then determines the lowest order temperature ratio

$$\tilde{T}/\bar{T} = 2\mu + \phi_1 \tag{26}$$

where ϕ_1 is the function that is proportional to μ^2, $(1-e)^2$ or $\mu(1-e)$. The assumption that μ is small corresponds to Jenkins and Richman's assumption in (ref. 1) that $(1+\beta)$ is small. Similarly, when the translational energy equation is specialized for a simple shear flow on x-y plane, x being the direction of the flow and y being perpendicular direction to it, it reduces to the condition requiring that

$$2\lambda P_{xy} = \chi(\tfrac{m}{2}C^2) \tag{27}$$

in which Pxy is the off-diagonal components of the pressure tensor:

$$P_{\alpha\beta} = \rho(1+2G)\bar{T}\delta_{\alpha\beta} - \frac{16m\nu G\bar{T}^{1/2}}{\sigma\pi^{3/2}} \left\{1+\frac{\pi}{8}\left[\frac{1}{G^2} + \frac{2}{G} + 1\right]\right\} \tag{28}$$

with $G = \nu g$. The collisional source of translational energy, $\chi(\frac{m}{2}C^2)$, has a $\mu \ln \mu$ term at its leading order. Consequently, in satisfying the translational energy equation, the dimensionless parameter $R \equiv 2\sigma\lambda / \bar{T}^{1/2}$ is found to be

$$R = A\mu \ln \mu + \phi_2 \tag{29}$$

where $A \equiv 64G/[\sigma\pi(1+G)+8G]$. The function ϕ_2 is the error introduced in determining R. It may contains terms proportional to μ or 1-e.

The above expression indicates that if σ/L, L being a typical length scale of the flow, is assumed to be small and the first order as in (ref. 1), then the present theory is applicable to disks with a friction coefficient μ only if $\mu \ell n \mu$ is second order in σ/L. For example, if $\sigma/L = 0.1$, then the value of μ that satisfies this would be $\mu = 0.0015$. Consequently, a theory of the type proposed by Jenkins and Richman that is restricted to dissipation of the order of (σ/L) is unrealistic.

DISCUSSION

The fact that collisions between disks with such a minute friction coefficient produces unexpectedly high translational energy dissipation may be understood by studying the correlationship between the coefficient of friction and the collision angle γ in the expression for β.

Recalling the eqn.(12) it is noticed that slip occurs throughout those collisions with a collision angle γ such that:

$$\cot \gamma < \frac{\kappa}{\mu(1+e)(1+\kappa)} \tag{30}$$

The collision angle γ^* separating a collision terminating with slip from one terminating without slip is defined as

$$|\cot \gamma^*| = \kappa / \mu (1+e)(1+\kappa) . \tag{31}$$

The angle γ^* increases in the range between 0 and $\pi/2$ as either μ or e increases. That is, as either the friction coefficient or coefficient of restitution increases, the number of collisions ending with slip becomes smaller. This is shown in (Fig.1) as a plot of β against γ for different values of μ. If, for example, $\mu = 0.5$, β ranges between -1 and 0 only for collision angles between 71° and 90°. For all other collisions β is 0 and the energy is dissipated most there. Supposing that all collision angles between 0° and 90° are equally likely, only a rather small percentage of disks leave collisions with slip.

Fig. 2 shows the fraction of $\beta = 0$ collisions for different values of μ. It is seen that for μ larger than 0.5 over 80% of the collisions take place in such a way that $\beta = 0$.

From these observations it is reasonable to conclude that, as far as the geometrical dependence of the roughness coefficient is concerned, setting $\beta = 0$ may not result in a substantial error in characterizing the flow behavior.

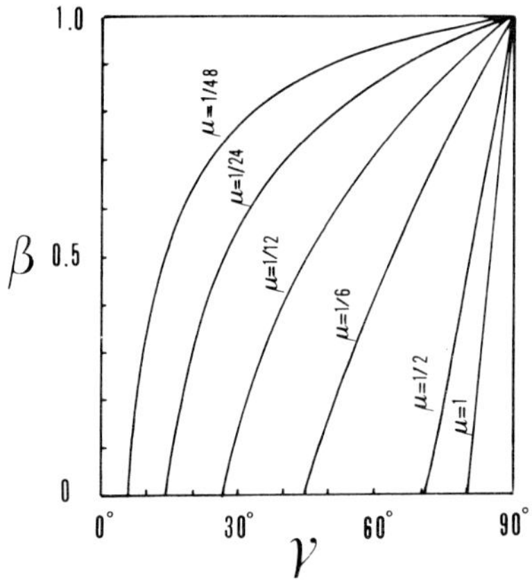

Fig. 1. The variation of roughness coefficient, β, with collision angle, γ, for various values of friction coefficient, μ. Here e = 1 , κ = 1/2 and γ max = π/2.

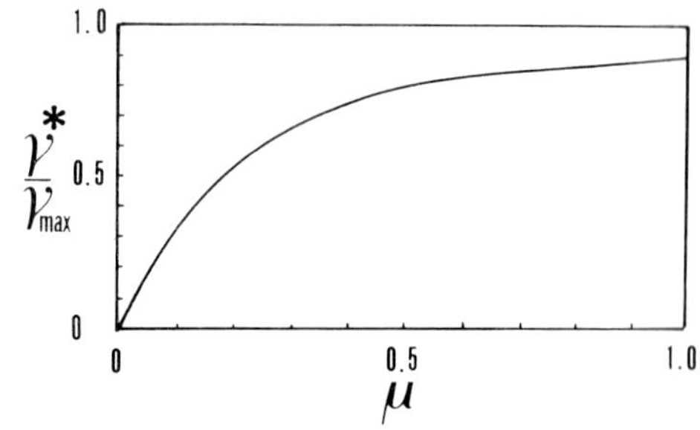

Fig. 2. The variation of the probability of β = 0 collisions with friction coefficient, μ. Here e = 1, κ = 1/2 and γ max = π/2.

ACKNOWLEDGEMENT

The author wishes to thank Professors J. T. Jenkins at Cornell University and M. W. Richman at Worcester Polytechnical Institute for their sincere advice.

REFERENCES

1 J. T. Jenkins and M. W. Richman, Kinetic theory for plane flows of a dense gas of identical, rough, inelastic, circular disks, Phys. Fluids 28(12) (1985)
2 M. Nakagawa, Ph.D. thesis, Cornell University

Micromechanics of Granular Materials, edited by M. Satake and J.T. Jenkins
Elsevier Science Publishers B.V., Amsterdam, 1988 — Printed in The Netherlands

EXPERIMENTS ON CHUTE FLOWS OF GRANULAR MATERIALS

H. AHN, C.E. BRENNEN, and R.H. SABERSKY
California Institute of Technology, Pasadena, CA 91125 (U.S.A.)

ABSTRACT

Experiments on continuous, steady flows of granular materials down an inclined channel or chute were made with the object of acquiring information on the rheological properties of the granular material flow and the nature of the boundary condition on the base of the channel. Specifically measurements were made of the mean material velocities and velocity profiles on all boundaries of the flow using cross-correlation of two neighboring fibre-optic displacement probes. The output from these probes was used to obtain (1) the unsteady or random component of the particle velocity in the longitudinal direction and (2) a measure of the volume fraction of the flow in contact with the base by counting the frequency of passage of the particles. Measurement was also made of the depth of the flow, the mass flow rate and the shear stress on the base. The latter employed a strain-gauged shear force plate built into the base.

The experiments are currently in progress and so further data will be presented at a later date. Nevertheless the preliminary data have yielded a number of interesting features.

INTRODUCTION

Recent research has added greatly to our knowledge of the rheology of rapid granular flows. In particular the theoretical models (for example Jenkins and Savage [1983] and Lun et al [1984]) have led to a comprehension of how the stresses and volume fraction in a granular flow are related to the velocity gradients and to the kinetic energy associated with the random motions of the particles (the so-called granular temperature). These advances have been greatly aided by the computer simulations (e.g. Campbell and Brennen (1985a and b), Walton (1984)) which have been able to provide qualitative and quantitative information in situations in which the complexity is too great to allow evaluation with the current analytical methods. Progress in experimental methods for granular materials has been much slower, hindered by the obvious difficulties involved in trying to make point measurement of velocity, volume fraction, or granular temperature in the interior of a granular material flow.

The present paper reports on an ongoing experimental program studying the continuous, steady flows of granular materials down an inclined channel or chute. One of the objectives of this program is to acquire information on the rheological properties of the granular material flow and the nature of the boundary condition at the kind of smooth, solid walls which would be common in engineering practice. Since the program is ongoing further data will be presented at a later time. Despite its incompletness the data presented here has some interesting features which we thought would be of interest to this workshop.

The present state of the experimental information on granular material flows consists of a number of Couette flow studies (e.g., Savage and McKeown [1983], Hanes [1985], Craig et al [1986] and several studies of flows down inclined chutes (e.g. Bailard [1978], Augenstein and Hogg [1978], Sayed and Savage [1983]). The understandable, initial objective of some of the Couette flow experiments (such as those of Savage and McKeown) was to produce a simple shear flow with a uniform velocity gradient, uniform volume fraction and, hopefully, uniform granular temperature. Toward this end the solid containing surfaces were roughened to produce a no-slip condition. Practical engineering circumstances require us to learn how to model the conditions at smooth walls at which slip occurs. This presents some difficulties because the boundary conditions on the velocity and granular temperature at smooth walls are far from clear (see, for example Campbell and Gong [1987]). The computer simulations of chute flows with speculative "smooth wall" conditions (Campbell and Brennen [1985]) indicate a boundary layer next to the wall which has a lower volume fraction and a higher granular temperature than the material further from the wall. They also indicate a "conduction" of granular heat from this layer toward the bulk.

EXPERIMENTAL MEASUREMENTS

The experiments were carried out in a 7.62 wide 120 cm. long rectangular aluminum channel or "chute" installed in a continuous flow, granular material facility previously described (Patton et al [1986]). In this paper, results are described for the flow of glass beads with a diameter, d, of 0.126 cm. The flow into the channel is regulated by a vertical gate, the opening between this and the channel base being referred to as the gate height, h_o. The channel is inclined at different angles, θ, to the horizontal and measurements were only taken after a steady state flow had been established.

The following measurements were taken for a range of h_o and θ. Point probes were used to record the depth, h, of the flow at several longitudinal locations in the channel and the mass flow rate was obtained by timed collection of material discharging from the chute. To measure the shear stress between the flowing material and the chute base, a rectangular hole, 11.4 cm long and 3.8 cm wide was cut into the chute base and replaced by a plate supported by strain-gauged flexures sensitive to the shearing force applied to the plate. Calibration of this balance was achieved by placing weights on the plate with the channel set at various inclinations. The clearance between the plate and the rest of the chute base was adjusted to be about 0.02 cm, much smaller than the particle size. Nevertheless, dirt particles would occasionally get trapped in the gap and this necessitated cleaning of the gap prior to each measurement.

In addition to the above, a fibre-optic probe similar to that originally devised by Savage [1979] was developed to measure particle velocities on the free surface, the sidewalls and the channel base. Several of these were fabricated. They each consisted of two MTI fibre-optic displacement probes set with their faces flush in an aluminum plug which was, in turn, either set flush in the channel base or sidewalls or held close to the free surface of the flowing granular material. The probe faces were 0.16 cm in diameter and of the type in which one semi-circle of the face consisted of transmitting fibres and the other of receiving fibres. The specific geometry is shown in figure 1. These probes were calibrated by placing them close to a revolving drum to which particles had been glued and measuring both the probe output (see below) and the drum peripheral velocity.

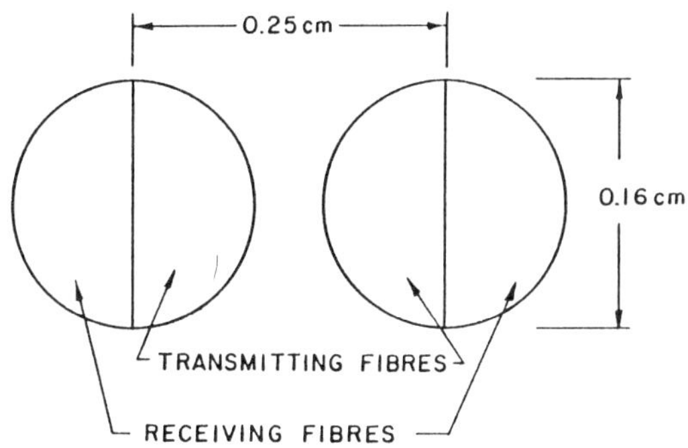

Figure 1. Geometry of the faces of the two displacement probes used for velocity measurements with the 0.126 cm diameter glass beads.

The output from each of these velocity measuring devices was processed in the following way. First the signals from each of the two displacement probes were simultaneously digitized and stored using a data acquisition system recording at 3×10^4 samples/sec. Except for very slow flows, the record time was about 0.5 secs; typically each record detected the passage of 300-600 particles. The two records were digitally cross-correlated over the entire record in order to obtain the mean particle velocity, $\bar{u}$. This information was then used to identify the peaks on the two records corresponding to the passage of a particular particle. Where no such correspondence could be firmly established or where the peak was below a certain threshold, the data was discarded for the purposes of this second part of the analysis. However, where positive identification was made, the velocity of that individual particle was obtained from the time interval between the peaks it generated on the two records. In this way, a set of instantaneous particle velocities were obtained and ensemble-averaging was used to obtain both the mean velocity, $\bar{u}$ and the root-mean-square of the fluctuating velocity, u'. Though the latter represents only one component of the fluctuating kinetic energy, it should nevertheless be some measure of the granular temperature. Finally, the number of particles per unit time detected by the probe, $\dot{n}$, was divided by the velocity, $\bar{u}$, to obtain a linear concentration, ν^*, of particles per unit length.

Though other measurements are being carried out as a part of the on-going program (including measurements of the heat transfer rates) this paper will be confined to presentation of some of the results obtained with the above equipment.

PRELIMINARY OBSERVATIONS

Early in the present program it was found that the data were quite sensitive to the cleanliness of the base and the sides of the channel. After running for some time, a thin layer of grime would be deposited on the chute by the particles and would apparently alter the surface boundary conditions. The obvious solution was regular cleaning and some of the data was obtained with this precaution. But a virtue was also made of this difficulty and data was taken with a dirty chute in order to determine those characteristics of the rheology which were sensitive to the boundary condition at the wall and those which are not sensitive. Thus, as well as presenting data for three gate openings, h_o, of 1.59 cm, 2.54 cm, and 5.08 cm and various chute inclination ranging from $\theta = 15°$ to $37°$, data is also classified by whether the chute was "clean", "moderately dirty" or "dirty". Though these characterizations are clearly subjective, the differences and/or similarities in the data are, we believe, of significance.

Fibre-optic probe measurements were made at three lateral locations on the channel base and at four lateral locations on the free surface and at approximately 12 vertical positions on a sidewall. For brevity, details of these profiles will not be included here but will be available later. It is sufficient to note the following characteristics of the profiles. The profiles on the free surface at all speeds and on the base at higher speeds were always close to uniform. Only on the base at lower speeds was the nonuniformity due to the sidewalls significant. Under these conditions the lateral profiles appeared to be roughly parabolic with a side-wall velocity of not less than 80% of the velocity at the channel center.

The velocity profiles at the sidewall were examined with greater resolution since the gradients in velocity and solid fraction were expected to be greater in the direction normal to the base. These sidewall profiles were used as an indication of the vertical profile elsewhere in the flow, though we do not have evidence for the validity of these indications. The profiles of fluctuating velocity and linear solid fraction indicated surprisingly small variation in these quantities over the depth of the flow except within a distance of about one particle diameter from the base. Furthermore, the mean velocity gradient normal to the base, $d\bar{u}/dy$, was similarly constant outside this smaller-than-expected boundary layer. As a result of these observations, we shall assume in the remainder of this presentation that the solid fraction is characterized by the mean (or overall) solid fraction, ν_m (see below) and that the shear rate, $d\bar{u}/dy$, can be characterized by the difference between the base and free surface velocities, Δu, divided by the depth, h, of the flow.

The differences between all the velocity measurements were sufficiently small so that an average mean velocity, $\bar{u}_M$, for any of the flows could be fairly accurately obtained by taking an appropriate mean. From $\bar{u}_M$ and the mass flow rate $\dot{m}$, the local flow depth measurement, h, and the particle density, ρ, a mean solid fraction, ν_M could then be obtained as $\nu_M = \dot{m}/\rho h b \bar{u}_M$ where b is the channel width. From previous experiments (Savage [1979], Sayed and Savage [1983], Bailard [1978]) and from computer simulations (Campbell and Brennen [1986b]) there are indications that the solid fraction, ν, is smaller in a layer next to the base than in the bulk of the material which appears to ride on this boundary layer. The sidewall measurements mentioned above revealed a similar deficiency in the linear solid fraction near the base; however, the boundary layer seemed to be only about one particle diameter thick.

Another perspective on this boundary layer can be gained by comparing the linear concentration of particles per meter, ν^*, detected by the fibre-optic probes with the mean solid fraction, ν_m. If the former is roughly converted to a wall solid fraction, ν_w, by assuming that the probe only detects a single particle layer so that $\nu_w = \pi(d\nu^*)^3/6$ it transpires that ν_w is close to ν_m The largest deviation occurs for low volume fraction flows (as one would expect) where ν_w may be as little as one-half of ν_m.

STRESSES AND RHEOLOGICAL BEHAVIOR

Neglecting the effects of the side wall in these relatively shallow and wide flows the normal stress, τ_N, on the channel base is calculated as $\rho\nu_M g h$; as previously discussed the shear stress, τ_S, is measured directly. The analytical work of Jenkins and Savage [1983] and Lun et al [1984] suggests that the rheological behavior of rapidly-flowing granular materials in a shearing flow such as occurs in the chute is of the form

$$\tau_S \propto g_S(\nu)\rho d(\frac{d\bar{u}}{dy})c' \tag{1}$$

$$\tau_N \propto g_N(\nu)\rho(c')^2 \tag{2}$$

where $g_S(\nu)$, $g_N(\nu)$ are functions of the solid fraction (which differ somewhat depending on the analytical approach), c' is the root mean square of the velocity fluctuations and $d\bar{u}/dy$ is the mean velocity gradient. Here, an attempt will be made to compare these relations with the experimental measurements. Though there are indications that the random fluctuations may be nonisotropic, particularly at denser packings (Campbell and Brennen [1985a]) we shall assume isotropy and therefore that $(c')^2 = 3(u')^2$.

First, in figure 2, we plot the experimental values of

$$g_S(\nu) = \tau_S/\rho d\frac{\Delta u}{h}(u)' \tag{3}$$

versus the mean solid fraction, ν_m. As can be observed, all of the data obtained to date appears to conform to a single function $g_s(\nu)$ over a wide range of velocities (inclinations)

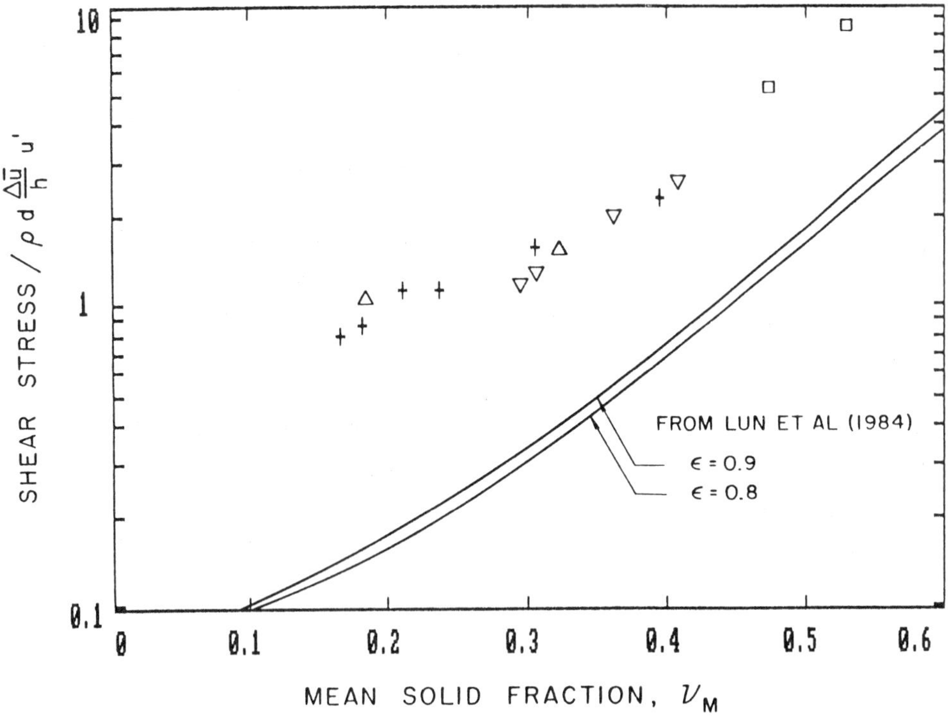

Figure 2. The shear stress function, $h\tau_S/\rho d\Delta\bar{u}u'$, against mean solid fraction, ν_M. Data points h_o = 1.59 cm and moderately dirty surface: $+$, h_o = 1.59 cm and clean surface: $\triangle$, h_o = 2.54 cm and moderately dirty surface: ∇, h_o = 5.08 cm and moderately dirty surface: $\square$. Also shown are the analytical results of Lun et al [1984] for coefficients of restitution, ϵ, of 0.8 and 0.9.

and flow depths (gate height). It should be noted that if, instead of $\Delta u/h$, the mean bulk shear rate, $d\bar{u}/dy$, from the sidewall velocity profiles were employed in evaluating $g_S(\nu)$ then the ordinates in figure 2 would only be altered in a minor way and the correlation would still hold.

The theoretical prediction from Lun et al (1984) for a *simple* shear flow is also presented in figure 2; though Lun et al present results for different coefficients of restitution, ϵ between the particles the variation of the results with this parameter is small for the expected range of ϵ relevant to the experiments. The discrepancy between the theoretical and experimental results of figure 2 is substantial. Part of the reason for this may be that the channel flow is far from being a simple shear flow even though, according to the sidewall velocity profiles, the shear rate is relatively constant over the bulk of the depth.

The corresponding normal stress function $g_N(\nu) = \tau_N/\rho(u')^2$ is presented in figure 3. Unlike the shear stress data it appears to be less well correlated; in fact, one can distinguish different lines for different gate heights, h_o, and surface frictional qualities.

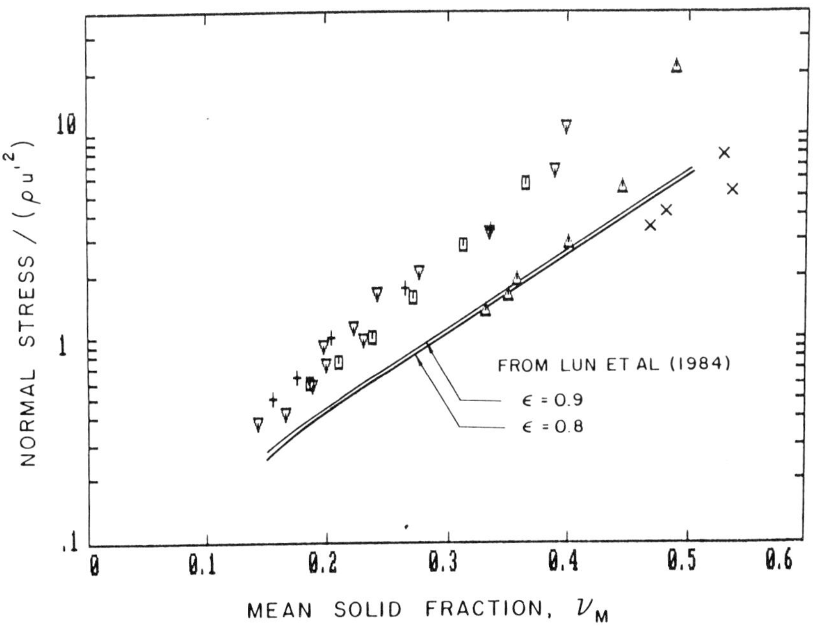

Figure 3. Normal stress function of $\tau_N/\rho(u')^2$ plotted against mean solid fraction, ν_M Data points h_o = 1.59 cm and clean surface:□, h_o = 2.54cm and clean surface: △, h_o = 1.59 cm and moderately dirty surface: ∇, h_o = 5.08 cm and moderately dirty surface: ×, h_o = 1.59 cm and dirty surface: +. Also shown are the analytical results derived from Lun et al [1984] for coefficients of restitution, ϵ of 0.8 and 0.9.

On the other hand it more closely agrees with the theoretical predictions of Lun et al, especially at the lower solid fractions where the theory is expected to be more accurate and the components of the fluctuations more isotropic.

The friction coefficient or ratio of τ_S/τ_N is shown in figure 4 and is remarkably constant and Coulomb-like. This implies that a Coulomb frictional condition is appropriate at the wall and therefore that the relations (1) and (2) cannot both hold at the solid boundary.

It is therefore important to examine one of the most important constituents of the theories, namely the ratio of the fluctuating velocity to the shear rate, $d\bar{u}/dy$. From the current experiments we estimate this parameter as $S = d\Delta u/hu'$ and values of this quantity are plotted against mean solids fraction in figure 5.

Note that the data for various depths and surface qualities do not reduce to a single correlation, $S(\nu_M)$. The analysis of Lun et al [1984] predicts that, in a simple shear flow with a given coefficient of restitution, S is a function only of ν. The fact that the data do not reduce to a single function $S(\nu)$ implies that the channel flow is not a simple

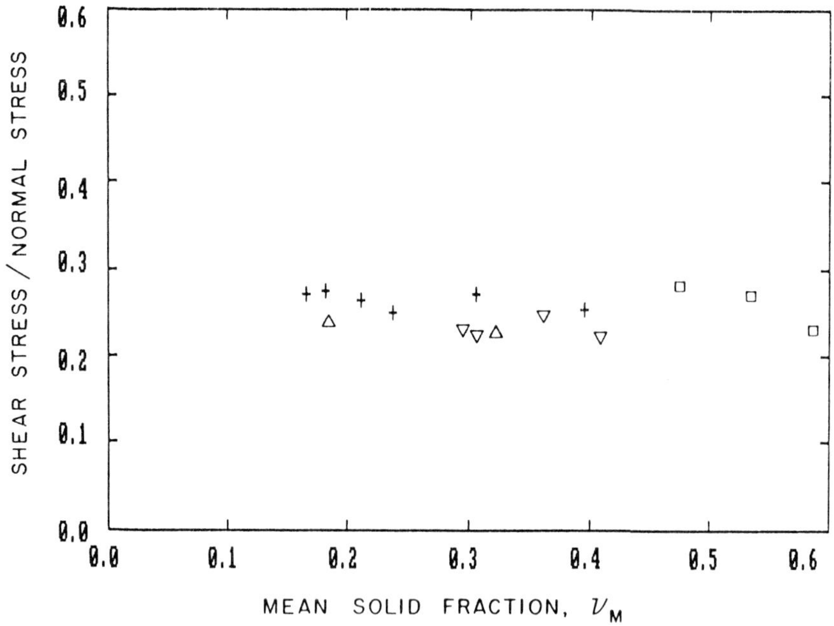

Figure 4. The friction coefficient of ratio. τ_S/τ_N as a function of mean solid fraction, ν_M. Data as in Figure 2.

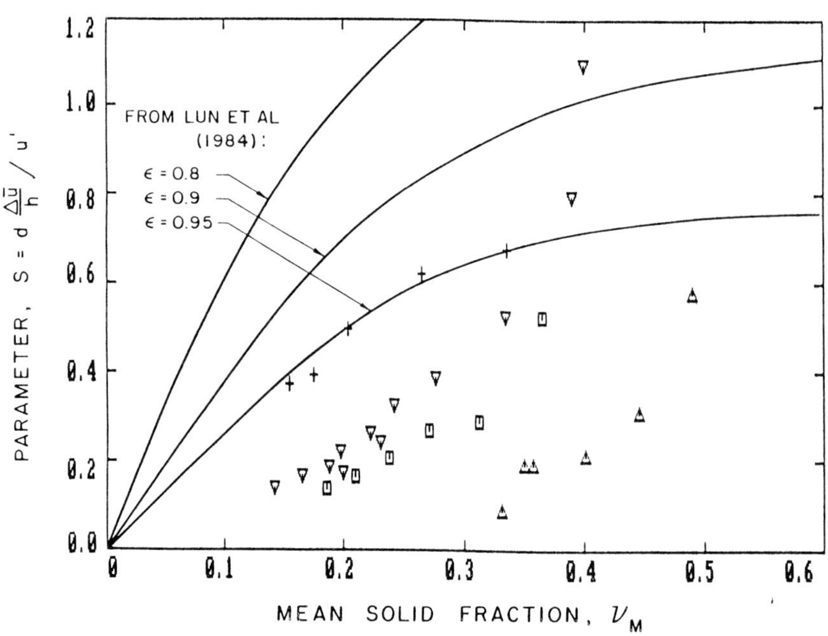

Figure 5. The parameter, $S = d\Delta\bar{u}/hu'$, plotted against mean solid fraction, ν_M. Point notation as in Figure 3. Also shown are analytical results derived from Lun et al [1984] for coefficients of restitution, ϵ, of 0.8, 0.9 and 0.95.

shear flow. It would appear that the imposition of a Colomb boundary condition caused substantial departure from the simple shear flow relations in the neighborhood of the solid surface. This may involve significant anisotropy and/or conduction of granular heat in the boundary layer close to the wall.

CONCLUDING REMARKS

Experiments in inclined channels were designed to measure the granular temperature as well as the normal and shear stresses in a series of rapid flows of granular materials. The objective was to evaluate some features of existing rheological models for granular materials. The results of Lun et al (1984) were chosen as representatives of the latter.

Preliminary results indicate that the shear stresses on the base of the channel when normalized by dividing by the density, the mean shear rate and the fluctuating velocity do appear to reduce to a single function of the solid fraction. This function is, however, significantly larger than predicted by the theory. On the other hand, the normal stress when normalized by $\rho(u')^2$ seems to yield values which depend not only on the solid fraction but also on the base frictional quality and other factors. The same is true of the parameter S or ratio of the shear velocity to the fluctuating velocity. The friction coefficient is quite constant and independent of the solid fraction. Thus the flow near the solid surface is not a simple shear flow but departs from this state to satisfy a Coulomb frictional condition. Experiments are continuing in an attempt to clarify these relationships.

ACKNOWLEDGEMENT

The authors are most grateful for the help given by S. Wakayama, D. Pillsbury, and C. Bertani during the conduct of this research. This work was supported by the National Science Foundation under Grant No. MEA 8400322. Some support from Union Carbide is also gratefully acknowledged.

REFERENCES

1 Augenstein, D.A., and Hogg, R., 1978, "An Experimental Study of the Flow of Dry Powders on Inclined Surface," Powder Technology, **19**, p.205.

2 Bailard, J., 1978, "Experimental Study of Granular - Fluid Flow, Ph.D Thesis, University of California, San Diego.

3 Campbell, C.S. and Brennen, C.E., 1985a, Computer simulation of granular shear flows, J. Fluid Mech. Vol. 151, p. 167.

4 Campbell, C.S. and Brennen, C.E., 1985b, Chute flows in a two-dimensional granular shear flow, J. Fluid Mech, Vol. 52, p. 172.

5 Campbell, C.S. and Gong, A., 1986, The stress tensor in a two-dimensional granular shear flow. J. Fluid Mech. Vol. 164, p. 107.

6 Craig, K., Buckholz, R.H., and Domoto, G., 1986, "An Experimental Study of the Rapid Flow of Dry Cohension Less Metal Powder," J. Appl. Mech., Vol. 53, p. 935.

7 Hanes, D.M., and Inman, D.L., 1985, "Observations of Rapidly Flowing Granular Fluid Materials," J. Mech., Vol. 150, p. 357.

8 Jenkins, J.T. and Savage, S.B., 1983, A theory for the rapid flow of identical smooth nearly elastic particles, J. Fluid Mech. Vol. 130, p. 1987.

9 Lun, C.K.K., Savage, S.B., Jeffrey, D.J. and Chepurniy, N. 1984, Kinetic theories for granular flow: inelastic particles in Couette flow and slightly inelastic particles in a general flowfield, J. Fluid Mech. Vol. 140, p. 223.

10 Patton, J.S., Sabersky, R.H., and Brennen, C.E. 1986 Convective heat transfer to rapidly flowing granular materials. Int. J. Heat Mass Transfer, Vol. 29, No. 8, p. 1263.

11 Sayed, M., and Savage, S.B., 1983, "Rapid Gravity Flow of Cohesionless Granular Materials down Inclined Chutes," J. Appl. Math. Phys., Vol 34, p. 84.

12 Savage, S.B. 1979 Gravity flow of cohesionless granular materials in chutes and channels, J. Fluid Mech., Vol.92, p. 53.

13 Savage, S.B. and McKeown, S., 1983, "Shear Stress Developed During Rapid Shear of Dense Concentrations of Large Spherical Particles between Concentric Cylinders," J. Fluid Mech., Vol. 127, p. 453.

14 Walton, O.R., 1984, Computer simulation of particulate flow. Energy and Tech. Rev., Lawrence Livermore Lab., May 1984, p. 24.

Micromechanics of Granular Materials, edited by M. Satake and J.T. Jenkins
Elsevier Science Publishers B.V., Amsterdam, 1988 — Printed in The Netherlands

A MONTE CARLO SIMULATION OF A SIMPLE SHEAR FLOW OF GRANULAR MATERIALS

MARK A. HOPKINS[1] and HAYLEY H. SHEN[2]

[1]Thayer School of Engineering, Dartmouth College, Hanover, New Hampshire 03755

[2]Civil and Environmental Engineering Department, Clarkson University, Potsdam, New York 13676

SUMMARY

A Monte Carlo simulation method is developed to produce the stresses generated in simple shear flow of circular, spherical and polygonal grains. In contrast to existing analytical methods, this method creates the velocity distribution exactly, instead of deriving it approximately with various perturbation techniques. The construction of this method and its computational simplicity are demonstrated. Excellent agreement is observed between the results obtained with this method and the full dynamic simulations.

INTRODUCTION

The definition of a "rapid" flow of granular materials here is a flow in which grain interactions are dominated by binary collisions. Extensive theoretical study of such flows has been made recently. These theories enable the determination of the constitutive laws and the boundary conditions necessary to mathematically model such a granular flow within boundaries of simple geometry (refs. 1-5). These theoretical results generally agree with data from physical experiments (refs. 6,7) and computer simulations (refs. 8,9).

Current theoretical analyses adopt the framework of the gas kinetic theory (ref. 10). For example, the transport of a quantity ψ for a system of identical disks of diameter σ is formulated as

$$\frac{1}{2}\chi \iiint (\psi_1^* \psi_1) f^{(1)}(\underline{c}_1, \underline{r} + \frac{1}{2}\sigma \underline{k}) f^{(1)}(\underline{c}_2, \underline{r} - \frac{1}{2}\sigma k)(g \cdot k)\, \sigma \, dk d\underline{c}_1 d\underline{c}_2 \qquad (1)$$

where $\chi f^{(1)} f^{(1)}$ is the pair distribution function of the velocities $\underline{c}_1, \underline{c}_2$ for the two grains involved in a binary collision, $\psi_1^* - \psi_1$ is the difference of ψ in grain 1 after and before the collision, $\underline{k}$ is a unit vector along the direction from center of grain 2 to center of grain 1, $\sigma \underline{k} = \underline{r}_1 - \underline{r}_2$, and $\underline{g} = \underline{c}_1 - \underline{c}_2$ is the velocity of grain 1 relative to grain 2. The above is integrated over those $\underline{c}_1$ and $\underline{c}_2$ where $\underline{g} \cdot \underline{k} > 0$. The central problem of formulating any transport parameters in such granular flows is the determination of the velocity distribution function $f^{(1)}$.

Early theories assumed the form of $f^{(1)}$. Those that followed the kinetic theory assumed that $f^{(1)}$ is Maxwellian (ref. 11). Another crude approach, on the other hand, assumed a Delta functional form for $f^{(1)}$ (refs. 12-14).

Sophisticated theories have been developed to modify these assumptions. Some of these theories are based on heuristic arguments (ref. 15). Some are based on rigorous derivations using perturbation methods (ref. 16). In either case, one still is not sure if an accurate description of the velocity distribution is indeed arrived at.

In this paper, an alternative theoretical method is presented. We adopt a Monte Carlo simulation procedure to create the velocity distribution. This procedure enables us to avoid lengthy perturbation methods in searching for the velocity distribution. This method is not restricted to the considerations of smooth, circular and nearly elastic grains. The only assumption is a state of "molecular chaos" amongst colliding grains. That is, the velocity distributions for any two points in the flow field are independent.

An illustration of this method and a comparison of the results with the existing data will be provided for a steady homogeneous planar simple shear flow of disks and regular polygonal shapes.

THEORETICAL BACKGROUND

Consider a homogeneous simple shear flow of disks with constant solid fraction ν. The corresponding number density n is defined in terms of the velocity distribution function $f^{(1)}$ as

$$n = n(r) = \int f^{(1)}(\underline{c},\underline{r})d\underline{c} \tag{2}$$

where $f^{(1)}(\underline{c},\underline{r})d\underline{c}$ represents the number of grains per unit volume at $\underline{r}$ which possess a velocity of $\underline{c}$ in $d\underline{c}$. In this homogeneous flow, $f^{(1)}$ is independent of $\underline{r}$ except for its mean value $u(\underline{r}) = \int \underline{c}\ f^{(1)}(\underline{c},\underline{r})d\underline{c}$, which is spatially varying according to the given shear flow.

As shown in ref. 17, for a homogeneous flow field, the collision frequency per unit area F is a constant, where

$$F = \iiint \chi(\nu)f^{(1)}(\underline{c}_2,\underline{r}_2)f^{(1)}(\underline{c}_1,\underline{r}_2 + \sigma\underline{k})(\underline{g}\cdot\underline{k})\sigma d\underline{k}d\underline{c}_1 d\underline{c}_2 \tag{3}$$

in which $\underline{r}_2$ is any point in the flow field and $\chi(\nu)$ is the value of the radial distribution function at grain contacts (ref. 10). There are many studies quantifying $\chi(\nu)$. We will use the following formula given in ref. 16.

$$\chi(\nu) = \frac{1}{2(1-\nu)} + \frac{1}{2(1-\nu)^2} + \frac{\nu}{16(1-\nu)^2} \tag{4}$$

We define the collisional pair configuration distribution as the following normalized function, which represents the probability of a collision between grains 1 and 2 with a configuration $(\underline{c}_1,\underline{c}_2,\underline{k})$ such that $\underline{g}\cdot\underline{k}>0$,

$$f_c^{(2)}(\underline{r}_2,\underline{k},\underline{c}_1,\underline{c}_2) = \frac{f^{(1)}(\underline{c}_2,\underline{r}_2)f^{(1)}(\underline{c}_1,\underline{r}_2+\sigma\underline{k})\ (\underline{g}\cdot\underline{k})\sigma}{\iiint f^{(1)}(\underline{c}_2,\underline{r}_2)f^{(1)}(\underline{c}_1,\underline{r}_2+\sigma\underline{k})(\underline{g}\cdot\underline{k})\sigma d\underline{k}d\underline{c}_1 d\underline{c}_2} \tag{5}$$

In real terms, we may find the distribution of collisional configurations by recording the configuration $(\underline{c}_1,\underline{c}_2,\underline{k})$ of all colliding pairs in the flow with one grain in $d\underline{r}_2$ centered at $\underline{r}_2$.

The collisional stress defined as the rate of momentum transfer across a unit line in the flow field is (ref. 10)

$$\frac{1}{2}\iiint \chi(\nu)f^{(1)}(\underline{c}_1,\underline{r}_2+\frac{1}{2}\sigma\underline{k})f^{(1)}(\underline{c}_2,\underline{r}_2-\frac{1}{2}\sigma\underline{k})m\underline{k}(\underline{c}_2^*-\underline{c}_2)\sigma^2(\underline{g}\cdot\underline{k})d\underline{k}d\underline{c}_1 d\underline{c}_2 \tag{6}$$

where m is the mass of a single grain. The kinetic stress is (ref.10)

$$\int f^{(1)}(\underline{c}_1)m(\underline{c}_1-\underline{u})(\underline{c}_1-\underline{u})d\underline{c}_1 \tag{7}$$

MONTE CARLO SIMULATION

We established a sample space of size n equal to the number of disks per unit area of flow. The reference frame is chosen to move with the mean velocity of this point in the flow. Therefore, the sample velocities are the fluctuation components. In this space, each sample consists of the three velocity components (v^1,v^2,ω), where $\underline{C} = v^1\underline{i}+v^2\underline{j}$ and $\underline{\omega} = \omega(\underline{i}\times\underline{j})$ represent the translational and rotational velocities of a disk relative to a particular point in the flow field. To initialize the simulation, all values are set to zero.

The statistics of the sample space is described by $f^{(1)}(\underline{C})$ where $\underline{C}$ includes both the translational and rotational velocities unless otherwise specified. The purpose here is to produce the correct $f^{(1)}(\underline{C})$ which in reality is created by the shearing flow and depends on the solid fraction ν, the deformation rate $\partial u_1/\partial x_2$ and the physical properties of the grains. The physical properties of the grains are the disk diameter σ, density ρ_s, the coefficient of restitution ε and their frictional coefficient μ.

The Monte Carlo simulation starts by choosing θ uniformly at random from 0 to 2π. This θ, as shown in Fig. 1, defines the line of center $\underline{k}$ in a collision. A random pair of fluctuation velocities (v_1^1,v_1^2,ω_1) and (v_2^1,v_2^2,ω_2) are then uniformly and independently chosen at random from the current sample space. Let

$$\underline{C}_1 = v_1^1\underline{i} + v_1^2\underline{j} \quad \text{and} \quad \underline{C}_2 = v_2^1\underline{i} + v_2^2\underline{j} \tag{8}$$

be the two fluctuation velocities.

$$\underline{c}_1 = \underline{C}_1 + \frac{1}{2}\sigma\frac{\partial u_1}{\partial x_2}\sin\theta\underline{i} \quad \text{and} \quad \underline{c}_2 = \underline{C}_2 - \frac{1}{2}\sigma\frac{\partial u_1}{\partial x_2}\sin\theta\underline{i} \tag{9}$$

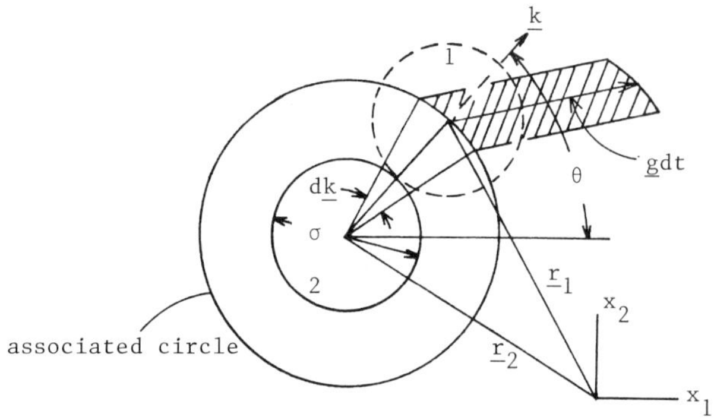

Fig. 1. Collisional configuration of disks.

then represent the two possible velocities of a pair of disks in contact measured in a reference frame moving with the contact point. A collision is impending if $\underline{g}\cdot\underline{k}>0$ where $\underline{g} = \underline{c}_1-\underline{c}_2$. Post collision velocities $\underline{c}_1^*$, $\underline{c}_2^*$ and ω_1^*, ω_2^* are obtained using rigid body dynamics described in ref. 18.

As discussed in deriving (5), the probability of collision with configuration $(\underline{c}_1,\underline{c}_2,\underline{k})$ arbitrarily chosen as shown in (9) is proportional to $f^{(1)}(\underline{c}_2, -\frac{1}{2}\sigma\underline{k})$. These samples must be weighted according to $\underline{g}\cdot\underline{k}$ to form the collisional configuration distribution $f_c^{(2)}$ defined in (5). The collisional configuration distribution will be used to create the single grain velocity distribution $f^{(1)}(\underline{C})$. A filter is formally defined by the constraint

$$\underline{g}\cdot\underline{k} - KR > 0 \qquad (10)$$

in which, K is a positive number and R is a uniformly distributed random number between 0 and 1 chosen anew for each configuration. Both the accuracy and the computational time of the Monte Carlo simulation increase with increasing K. Only when (10) is satisfied is the randomly chosen configuration $(\underline{c}_1,\underline{c}_2,\underline{k})$ used to produce the updated $f^{(1)}(\underline{C})$ samples. That is, when (10) is satisfied the corresponding post collision translational fluctuations

$$\underline{C}_1^* = \underline{c}_1^* - \frac{1}{2}\sigma\frac{\partial u_1}{\partial x_2}\sin\theta\underline{i} \quad \text{and} \quad \underline{C}_2^* = \underline{c}_2^* + \frac{1}{2}\sigma\frac{\partial u_1}{\partial x_2}\sin\theta\underline{i} \qquad (11)$$

together with the rotational components ω_1^* and ω_2^* are returned to the sample space to replace the original pre-collision velocities $\underline{C}_1,\omega_1$ and $\underline{C}_2,\omega_2$.

After each Δt, where $\Delta t = 1/F$, a collision among these disks will take place because Δt represents the average amount of time between two consecutive collisions. The collision frequency F derived in (3) is computed as

$$F = \frac{\pi}{2} n^2\sigma\chi \frac{\Sigma\,\underline{g}\cdot\underline{k}}{\Sigma 1} \qquad (12)$$

where χ is defined in (4) and the summation is over all cases where $\underline{g}\cdot\underline{k} > 0$.

In this time Δt between two consecutive collisions, all disks represented by the velocity space have moved to a new location and thus have new fluctuation velocities relative to their new locations. Therefore, we update all velocities in the sample space by

$$\underline{C}_{new} = \underline{C}_{old} - \frac{\partial u_1}{\partial x_2} \Delta t(\underline{C}_{old} \cdot \underline{j})\underline{i} \tag{13}$$

This process which was begun by choosing the collisional configuration $(\underline{c}_1,\underline{c}_2,\underline{k})$ is repeated many times until the velocities thus produced in the process become statistically stationary. The velocities created comprise the velocity distribution function $f^{(1)}(\underline{C})$ that we need for evaluating all the transport parameters. Using the definitions given in (6) and (7), we then compute the collisional stress $\underset{\sim}{\tau}^c$ and the kinetic stress $\underset{\sim}{\tau}^k$ in the following way.

$$\underset{\sim}{\tau}^c = \frac{\pi}{2}\,\frac{\sigma^2 n^2 \chi \Sigma m\underline{k}(\underline{c}_1^* - \underline{c}_1)(\underline{g}\cdot\underline{k})}{\Sigma 1} \tag{14}$$

where the summation is over all configurations $(\underline{C}_1,\underline{C}_2,\underline{k})$ such that $\underline{g}\cdot\underline{k} > 0$;

$$\underset{\sim}{\tau}^k = mn\,\frac{\Sigma\,\underline{C}\,\underline{C}}{\Sigma 1} \tag{15}$$

where the summation is over many iterations for all values of $\underline{C}$ in the sample space.

RESULTS FOR SIMPLE SHEAR FLOW OF DISKS

The nondimensional collisional plus kinetic stresses are given in Figs. 2a and 2b where $\tilde{\tau}_{ij} = \tau_{ij}/\rho_s\sigma^2(\partial u_1/\partial x_2)^2$. Comparisons of these results with the computer simulated disk flows show excellent agreement.

SIMPLE SHEAR FLOW OF POLYGONAL GRAINS

We now extend the Monte Carlo method to study a simple shear flow of uniform regular polygonal grains. The definition sketch for a pair of colliding polygons is given in Fig. 3.

Let Ns denote the number of sides of the regular polygons. Each polygon has an orientation defined by the angle θ from the x_1 direction to the direction of a reference line shown in Fig. 3.

Similar to but slightly more complicated than the circular case, the polygonal velocity distribution is $f^{(1)}(\underline{c},\underline{r},\theta)$ where

$$n = n(\underline{r}) = \int_{\theta=0}^{2\pi} \int_{\underline{c}} f^{(1)}(\underline{c},\underline{r},\theta)\,d\underline{c}\,d\theta \tag{16}$$

is the number density of polygons

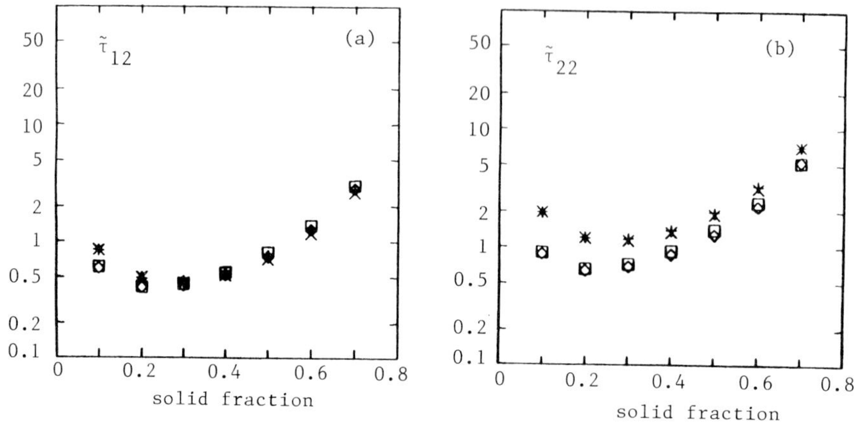

Fig. 2. Non-dimensional shear and normal stresses for disks with ε=0.8. +(μ=0.0), □ (μ=0.5): Monte Carlo; x(μ=0.0), ◇ (μ=0.5): Full Simulation (ref. 18).

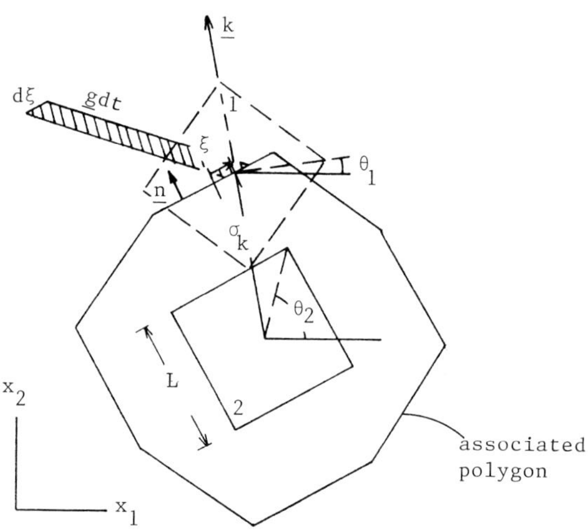

Fig. 3. Collisional configuration of polygons.

As shown in Fig. 3, when a polygon denoted by 2 is chosen to have a configuration $(\underline{c}_2,\theta_2)$, the polygon denoted by 1 will collide with it with a given configuration $(\underline{c}_1,\theta_1)$ if its center lies on the associated polygon. Note that the number of sides on this associated polygon is 2Ns.

We mark a length $d\xi$ on the associated polygon. Within a time duration dt, those polygons that will arrive at $d\xi$ with a configuration $(\underline{c}_1, \theta_1)$ must have their centers lie in the shaded parallelogram as shown in Fig. 3, where $\underline{g} = \underline{c}_1 - \underline{c}_2$. The area of this parallelogram is $\underline{g}\cdot\underline{n} d\xi dt$. Therefore, the probable number of polygons in the shaded parallelogram as $dt \to 0$ is

$$\sum_{i=1}^{2N_s} \chi_p(\nu) f^{(1)}(\underline{c}_1, \theta_1, \underline{r}_2 + \sigma_k \underline{k}) |\underline{g}\cdot\underline{n}_i| d\xi \, d\underline{c}_1 d\theta_1 \tag{17}$$

The factor $\chi_p(\nu)$ is the value of the radial distribution function at contact for polygons. Its value is estimated by a computer simulation (ref. 19). The results are given in Table 1.

	χ_3	χ_4	χ_5	χ_8	χ_∞
ν=0.1	1.30	1.35	1.28	1.25	1.18
ν=0.4	2.90	2.60	2.50	2.40	2.24

Table 1. Radial distribution function at contact.

The collision frequency per unit area of flow is therefore

$$F_p = \frac{1}{2} \sum_{i=1}^{2N_s} \iiiint \chi_p \, f^{(1)}(\underline{c}_2, \theta_2, \underline{r}_2) f^{(1)}(\underline{c}_1, \theta_1, \underline{r}_2 + \sigma_k \underline{k}) |\underline{g}\cdot\underline{n}_i| d\xi \, d\underline{c}_1 d\theta_1 d\underline{c}_2 d\theta_2 \tag{18}$$

The above integration is over $\underline{g}_c\cdot\underline{n}_i > 0$ where $\underline{g}_c$ is the relative velocity at the contact point defined as

$$\underline{g}_c = \underline{c}_1 - \underline{c}_2 + \underline{r}_1 \times \underline{\omega}_1 - \underline{r}_2 \times \underline{\omega}_2 \tag{19}$$

The absolute value $|\underline{g}\cdot\underline{n}_i|$ is used in (18) because the possibility of having collisions when $g_c \cdot n_i > 0$ with $\underline{g}\cdot\underline{n}_i < 0$.

Similar to the case of circular grains, we define the collisional pair configuration distribution as follows,

$$f_c^{(2)}(\underline{r}_2, \underline{k}, \underline{c}_1, \theta_1, \underline{c}_2, \theta_2) = \frac{f^{(1)}(\underline{c}_2, \theta_2, \underline{r}_2) f^{(1)}(\underline{c}_1, \theta_1, \underline{r}_2 + \sigma_k \underline{k}) |\underline{g}\cdot\underline{n}_i|}{\sum_{i=1}^{2N_s} \iiiint f^{(1)}(\underline{c}_1, \theta_2, \underline{r}_2) f^{(1)}(\underline{c}_1, \theta_1, \underline{r}_2 + \sigma_k \underline{k}) |\underline{g}\cdot\underline{n}_i| d\xi \, d\underline{c}_1 d\underline{c}_2 d\theta_1 d\theta_2} \tag{20}$$

where the integral is over all $\underline{g}_c\cdot\underline{n}_i > 0$. The above represents the probability of the collisional configuration $(\underline{c}_1, \theta_1, \underline{c}_2, \theta_2, \underline{k})$.

The Monte Carlo procedure is designed to produce the single grain velocity and orientation distribution $f^{(1)}(\underline{C}, \omega, \theta)$ where $\underline{C}$ is the translational fluctuation velocity and ω is the rotational velocity. This distribution is uniform in space because of the homogeneity in the flow field.

In the sample space of size n, where n is the number of polygons per unit area, each sample contains two translational velocity and one rotational velocity plus the orientation indicated by θ. Thus, $(v^1, v^2, \omega, \theta)$ together represent a sample grain that may be found at a given time and a location in the flow field. The toality of this sample space consists of a set of possible velocities that may be found in a unit area of the flow field in a given instance. Initially, all velocities are set to zero and the orientation θ is randomly assigned from 1 to 2π.

Two samples $(\underline{C}_1, \omega_1, \theta_1)$ and $(\underline{C}_2, \omega_2, \theta_2)$ are independently chosen at random from this sample space. The center of polygon 1 is randomly and uniformly chosen to lie on side i, $1 \leq i \leq 2Ns$, with location ξ on side i, $-\frac{\ell}{2} \leq \xi \leq \frac{\ell}{2}$. As shown in Fig. 3, the centerline direction $\underline{k}$ is defined once i and ξ are chosen. The velocities of two polygons relative to the contact point are

$$\underline{c}_1 = \underline{C}_1 + (\nabla\underline{u})\cdot\underline{r}_1 \quad \text{and} \quad \underline{c}_2 = \underline{C}_2 + (\nabla\underline{u})\cdot\underline{r}_2 \tag{21}$$

whether a collision is allowed to take place depends on two conditions. The first condition is $\underline{g}_c \cdot \underline{n}_i > 0$, where $\underline{g}_c$ is the relative velocity at the contact point defined in (19). The second condition is $|\underline{g}\cdot\underline{n}_i| - RK > 0$, where $0 \leq R \leq 1$ is a uniformly distributed random number chosen anew for each configuration and K is a positive constant. As discussed before, the second condition is the filter designed to produce $f_c^{(2)}$ defined in (20). The post collision velocities $\underline{c}_1^*, \omega_1^*$ and $\underline{c}_2^*, \omega_2^*$ are obtained by using rigid body dynamics described in ref. 18. The new fluctuations

$$\underline{C}_1^* = \underline{c}_1^* - (\nabla\underline{u})\cdot\underline{r}_1 \quad \text{and} \quad \underline{C}_2^* = \underline{c}_2^* - (\nabla\underline{u})\cdot\underline{r}_2 \tag{22}$$

and the new rotations ω_1^* and ω_2^* then replace the original samples. The orientations are unchanged.

These collisions are allowed to occur every Δt in time where Δt is the reciprocal of the collisional frequency defined in (18). After each collision described above, all samples in the sample space move Δt in time to a new location. Because of the mean shear flow, their x_1 fluctuation velocity component will change as given in (13). Because of their rotation, their new orientations become $\theta_{new} = \theta_{old} + \omega\Delta t$. After updating all values in the sample space, we repeat the process starting with choosing the collisional configurations.

When the statistics of the velocities thus produced stabilizes, we have arrived at the required single grain velocity distribution $f^{(1)}(\underline{C}, \underline{\omega}, \theta)$. The collisional stress $\underset{\sim}{\tau}^c$ and kinetic stress $\underset{\sim}{\tau}^k$ are computed as

$$\underset{\sim}{\tau}^c = \frac{1}{2}(2N_s L)n^2\chi_p \frac{\Sigma|\underline{g}\cdot\underline{n}|m\underline{k}(\underline{c}_1^* - \underline{c}_1)}{\Sigma 1} \tag{23}$$

where L is the length of a side of the polygon and Σ is over all cases where

$\underline{g}_c \cdot \underline{n}_i > 0;$

$$\underset{\sim}{\tau}^k = \frac{mn\Sigma \underline{C}\,\underline{C}}{\Sigma 1} \tag{24}$$

where the Σ is over many iterations for all samples in the sample space.

The results for shear and normal stresses are given in Fig. 4, in which $\tilde{\tau}_{ij} = \pi\tau_{ij}/4\rho_s A(\partial u_1/\partial x_2)$ and A is the polygon area. As Ns $\rightarrow \infty$, the results reduce to that of the circular flows. One important thing to note here is that because of the possibility of off-center collisions, polygons can transfer linear momentum to angular momentum in the absence of friction. This is the main difference between circular and polygonal cases. Because of this, the linear momentum transfer is generally reduced in a collision and consequently the stresses are expected to be reduced also.

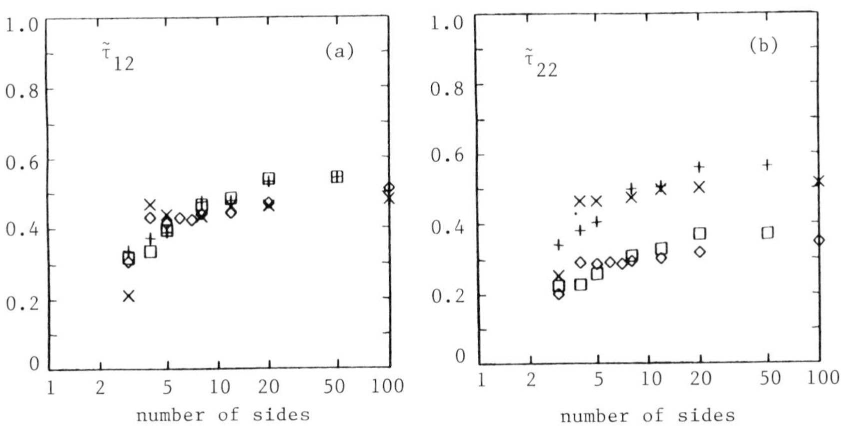

Fig. 4. Non-dimensional shear and normal stresses for polygons with ν=0.4 and ε=0.8. +(μ=0.0), □(μ=0.0): Monte Carlo; x(μ=0.0), ◇ (μ=0.5): Full Simulation (ref. 18).

CONCLUSION

An alternative to the well-established theoretical method for studying rapid simple shear flow of granular material is introduced in this paper. This Monte Carlo method is computationally efficient and may be implemented using a small computer. Here we demonstrate its utility in producing simple shear flows of circular material, obtaining excellent agreement with full dynamic simulation results. Extension of this method to study polygonal material is also presented with a few results given.

ACKNOWLEDGEMENTS

This study was partially supported by the grant number MSM-8419416 of the

National Science Foundation.

REFERENCES

1 K. Hui, P.K. Haff, J.E. Ungar and R. Jackson, Boundary conditions for high-shear grain flows, J. Fluid Mech., 145 (1984) 223-233.

2 J.T. Jenkins and M.W. Richman, Boundary conditions for plane flows of smooth nearly elastic, circular disks, J. Fluid Mech., 171 (1986) 53-69.

3 G. Gutt, Boundary conditions on continuum theories of granular flow, 6th Eng. Mech. Specialty Conf., ASCE/EMD, Buffalo, NY, May 20-22, 1987, p. 137.

4 G.C. Pasquarell and N.L. Ackermann, Boundary value problems for two-dimensional, rectilinear, rapidly sheared granular flows, 6th Eng. Mech. Specialty Conf., ASCE/EMD, Buffalo, NY, May 20-22, 1987, p. 138.

5 G.C. Pasquarell, Boundary value problems for planar, smooth, rapidly sheared granular flows including phase changes, Ph.D. Dissertation, Clarkson University, 1987, 250 p.

6 B. Sanders, Experimental study of two dimensional gravity driven granular flow, M.S. Thesis, Clarkson University, 1986, 161 p.

7 B. Sanders and N.L. Ackermann, Two dimensional gravity driven granular flow, 23rd Annual Meeting Soc. Eng. Sci., Buffalo, NY, Aug. 25-27, 1986 p. 512.

8 S.C. Campbell and A. Gong, The stress tensor in a two-dimensional granular shear flow, J. Fluid Mech., 164 (1986) 107-125.

9 O.R. Walton, Lawrence Livermore National Lab. Granular Solids Flow Project-Quarterly Reports, 1985.

10 S. Chapman and T.G. Cowling, The mathematical theory of non-uniform gases, Cambridge Press, 1970.

11 J.T. Jenkins and S.B. Savage, A theory for the rapid flow of identical, smooth, nearly elastic, spherical particles, J. Fluid Mech., 130 (1983) 187-202.

12 H.H. Shen and N.L. Ackermann, Constitutive relationships for fluid-solid mixtures, J. Mech. Div., ASCE 108(EM5) (1982) 748-763.

13 H.H. Shen and N.L. Ackermann, Constitutive equations for a simple shear flow of a disk shaped granular mixture, Int. J. Eng. Sci., 22(7) (1984) 829-843.

14 G.C. Pasquarell, N.L. Ackermann, H.H. Shen and M.A. Hopkins, Collisional stress in a granular flow: Bagnold revisited, J. Eng. Mech., ASCE (in press).

15 C.K.K. Lun, S.B. Savage, D.J. Jeffrey and N. Chepurniy, Kinetic theories for granular flow: Inelastic particles in Couette flow and slightly inelastic particles in a general flowfield, J. Fluid Mech., ASCE, 140 (1984) 223-256.

16 J.T. Jenkins and M.W. Richman, Kinetic theory for plane flows of a dense gas of identical, rough, inelastic, circular disks, Phys. Fluids, 28(12) (1985) 3485-3494.

17 J.H. Jeans, An introduction to the kinetic theory of gases, Cambridge Press, 1948.

18 M.A. Hopkins, Granular flow simulations, Civil and Environmental Engineering Research Report 87-7, Clarkson University, 1987.

19 M.A. Hopkins, Constitutive relations for rapidly sheared granular flows: A Monte Carlo form based on the kinetic theory of dense gases, Ph.D. Dissertation, Clarkson University, 1987.

Micromechanics of Granular Materials, edited by M. Satake and J.T. Jenkins
Elsevier Science Publishers B.V., Amsterdam, 1988 — Printed in The Netherlands

PHYSICAL EXPERIMENTS AND NUMERICAL SIMULATION OF TWO DIMENSIONAL CHUTE FLOW

BRIAN E. SANDERS[1], MARK A. HOPKINS[2], AND NORBERT L. ACKERMANN[3]

[1]Civil and Environmental Engineering Department, Clarkson University, Potsdam, New York 13676, U.S.A.

[2]Mark A. Hopkins, Thayer School of Engineering, Dartmouth College, Hanover, New Hampshire 03755, U.S.A.

[3]Civil and Environmental Engineering Department, Clarkson University, Potsdam, New York 13676, U.S.A.

SYNOPSIS

Numerical simulations and physical experiments were used to study the two dimensional flow of disks in a chute. For a chute with a fixed geometry, inclination, and material properties the physical experiments and their computer simulations produced a series of steady, uniform flows over a range of bulk solids concentrations from approximately $\bar{C}$=0.1 to $\bar{C}$=0.7. These results were in contrast to findings obtained from the numerical integration of the governing equations that described the corresponding boundary value problem. General agreement was observed between computer simulated values of velocity and solids concentration and those dtermined from physical experiments.

INTRODUCTION

The governing equations and boundary conditions for the steady, two dimensional, rapidly sheared, granular flow in a chute were shown by Pasquarell (ref. 2) to provide a unique solution for the local velocity, concentration and granular temperature. Physical experiments and the corresponding numerical simulations however do not demonstrate such a unique relationship. Instead a continuous range of steady, uniform flows were found for a given chute geometry, angle of inclination and granular material. For these flows the concentration of solids, when averaged across the channel width, varied continuously between approximately $\bar{C}$=0.1 to $\bar{C}$=0.7.

Numerically simulated flows, having the identical width-averaged bulk concentrations as those calculated by Pasquarell produced simulated velocity and concentration distributions that yielded a volume flow rate which was approximately equal to that predicted by Pasquarell's unique solution. The procedure for the numerical simulation was developed by Hopkins (ref. 1) while the results from physical experiments are obtained from ongoing tests which are partially reported by Sanders (ref. 3).

A close agreement was found between the numerically simulated and laboratory measurements for the velocity and solids concentration distributions.

LABORATORY EXPERIMENTS

Air table and granular material

The granular material used in the chute flow experiments consisted of identical plastic disks 0.25 inch (6.3mm) thick and 1.75 inches (44.5mm) in diameter having an average mass of 11.18 g. The disks were driven by gravity down the surface of an air table which had a length of 26 feet (7.9 meters) and whose slope was maintained at inclinations of approximately 1 degree. The width of the chute could be changed and had a maximum value of 10.3 disk diameters. The dimensions of the disk and channel (air table) are shown in Fig. 1.

The table consisted of a long hollow chamber or box filled with compressed air. The top surface of this box formed the chute or channel surface along which the disks moved. Compressed air escaped from the chamber through small holes drilled through the table surface. The holes were spaced so that a disk placed anywhere on the table would be supported by a cushion of air produced by the air jets from at least four holes.

The flow of disks down the chute was confined between two straight parallel walls. The walls were made bumpy with half-disks fastened to the surface as shown in Fig. 1. The interaction between the flowing disks and wall half-disks were considered to provide the only boundary resistance to the gravity forces acting on the flowing disks.

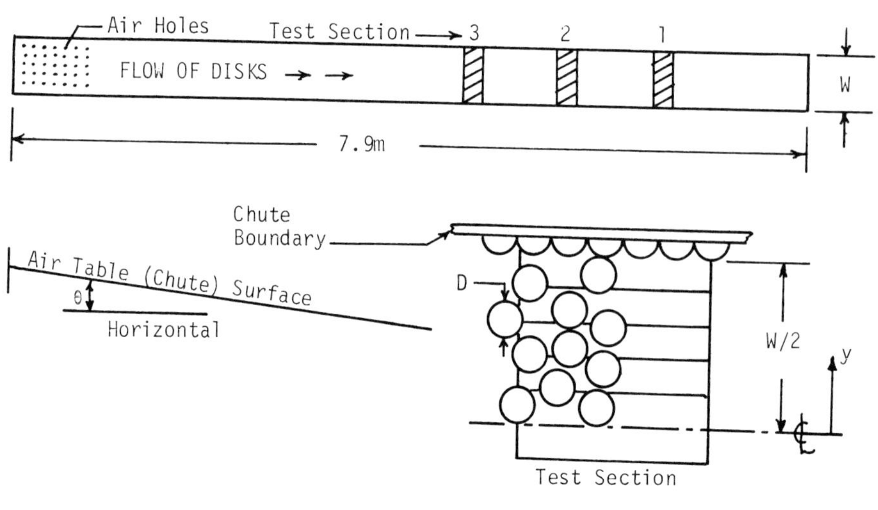

Fig. 1. Air Table and Test Section.

Figure 1 also shows the position of test sections along the length of the chute. Each test section was divided into longitudinal strips parallel to the boundary. Each strip, one disk diameter wide, was identified by lines painted on the floor of the chute. Video cameras with built-in timers were used to determine the time and coordinates of each disk traversing each of three separate test sections. Average disk velocities and solids concentrations were determined from that information.

A 16 mm high speed motion picture camera was also used to determine the pre and post collision linear and angular velocity of disks in experiments to determine the coefficient of restitution and the coefficient of friction. The average coefficient of restitution, e_D, for the flowing disks was e_D=0.952, the average coefficient of restitution, e_W, for a flowing disk and a fixed boundary half-disk was e_W=0.661 and the average coefficient of kinetic friction, μ, between the disks was μ=0.118.

Experimental procedure

Prior to each test the entire channel surface was covered uniformly with disks. The initial number of disks was varied for different tests. A test was initiated by providing a supply of compressed air to the table which then produced a cushion of air under each disk. Approximately 20 seconds were required for all of the disks to flow over the end of the table when the channel width was 10.3 disk diameters and the channel was on a slope of 64 minutes.

Steady, uniform flow

Steady, uniform flow corresponded to a condition in which the gravitational force acting on the disks was balanced by the boundary resistance. Steady, uniform flow was considered to be obtained when velocity and concentration profiles of the flowing disks remained constant with time and were identical at adjacent test sections.

In order to obtain an indication when steady, uniform flow occurred, the centerline velocity and concentration were first determined thoughout an entire test. This information is presented in Fig. 2 for a test that had 1400 disks on the table surface at the initiation of the test run. As shown in Fig. 2 a period of steady, uniform flow existed after approximately 8 seconds from the time the air supply to the table was introduced.

PRESENTATION OF RESULTS

Laboratory experiments and numerical simulations with a fixed chute geometry, angle of inclination and material properties produced numerous flow conditions that appeared to be both steady and uniform. Width integrated

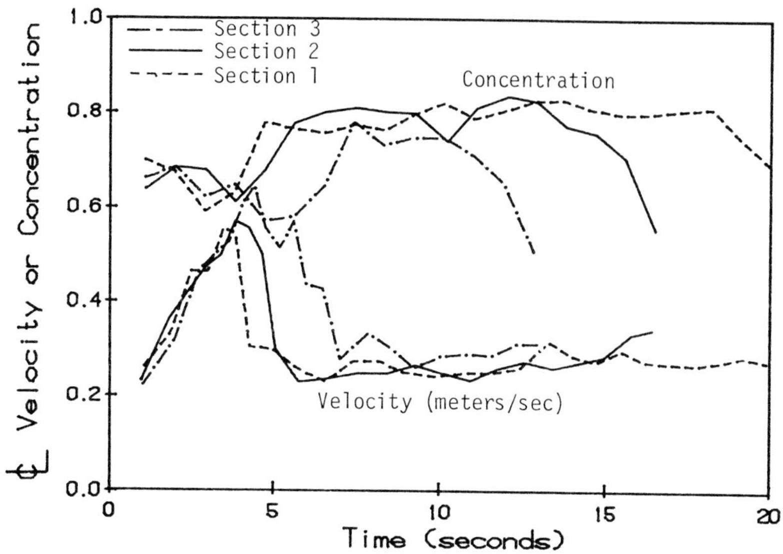

Fig. 2. Velocity, Concentration, Time Relationship.

values of local average velocity, u(y) and concentration, c(y) for a flow that was steady and uniform could then be described by the following relationship

$$f[Q,\overline{C},W/D,e_D,e_W,\mu,h/D] = 0 \tag{1}$$

where $$Q = \frac{\int_0^W u\ c\ dy}{W(Dg\sin\theta)^{1/2}}, \qquad \overline{C} = \frac{\int_0^W c\ dy}{W} \tag{2,3}$$

In the experimental results which are reported only the terms Q, $\overline{C}$ and W/D were varied.

As shown in Figs. 3 through 7, the numerical simulation is sometimes found to overestimate the dimensionless discharge rates Q determined from physical measurements. In general, however, the agreement between these two results were favorable. The unique solution obtained from the numerical integration of the governing equations is also presented on Figs. 3 through 7.

Fig. 8 presents a superposition of the results for the computer simulated data which are described in more detail in Figs. 3 through 7. Fig. 9 shows a typical velocity and solids concentration distribution across the width of the chute as determined from the physical tests and numerical simulation.

CONCLUSIONS

Agreement was found between measurements from physical experiments and corresponding information from computer simulated two-dimensional flows of disk shaped solids on an inclined chute. It was found that in a chute of fixed

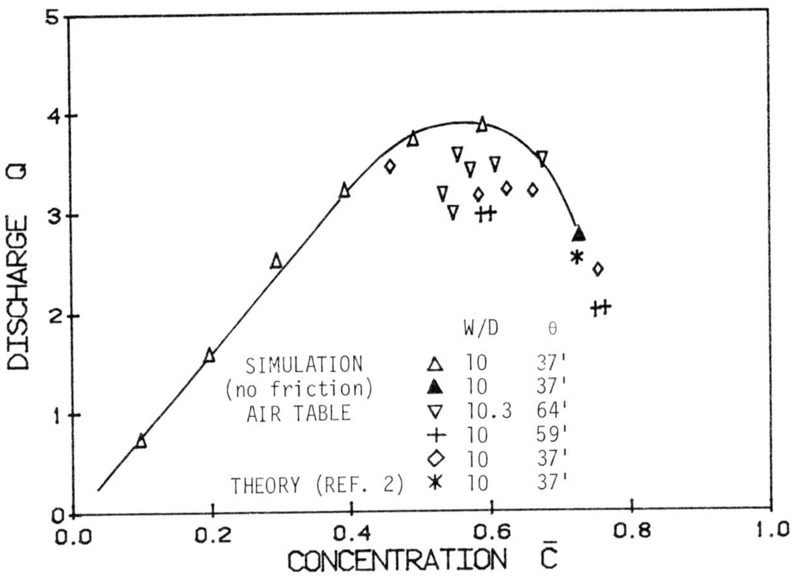

Fig. 3. Discharge, Concentration Relationship.

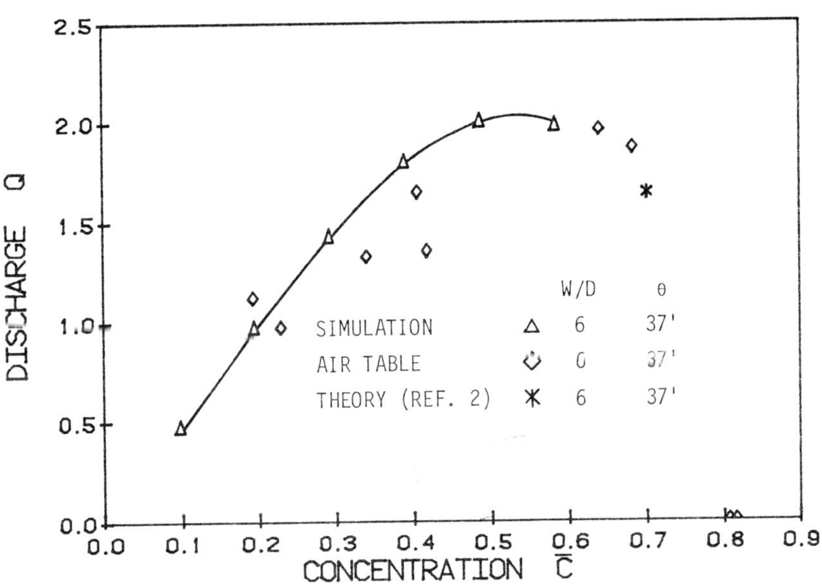

Fig. 4. Discharge, Concentration Relationship.

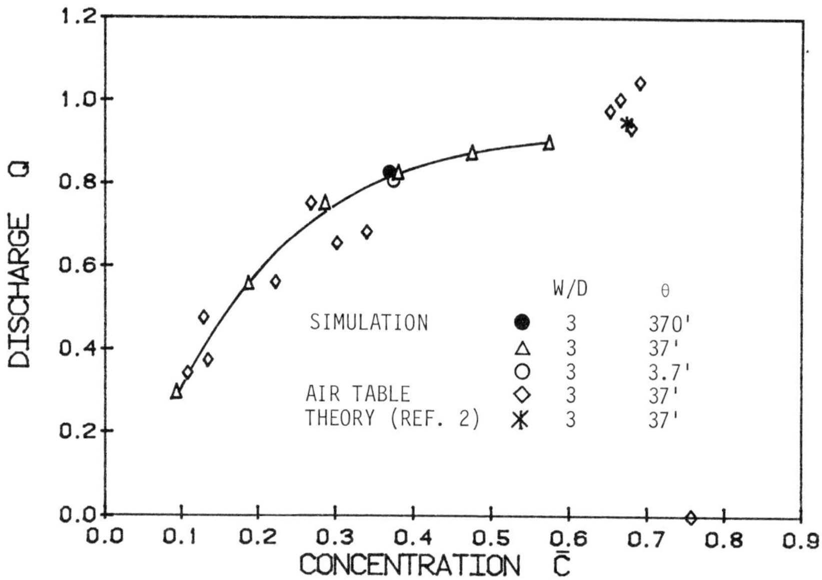

Fig. 5. Discharge, Concentration Relationship.

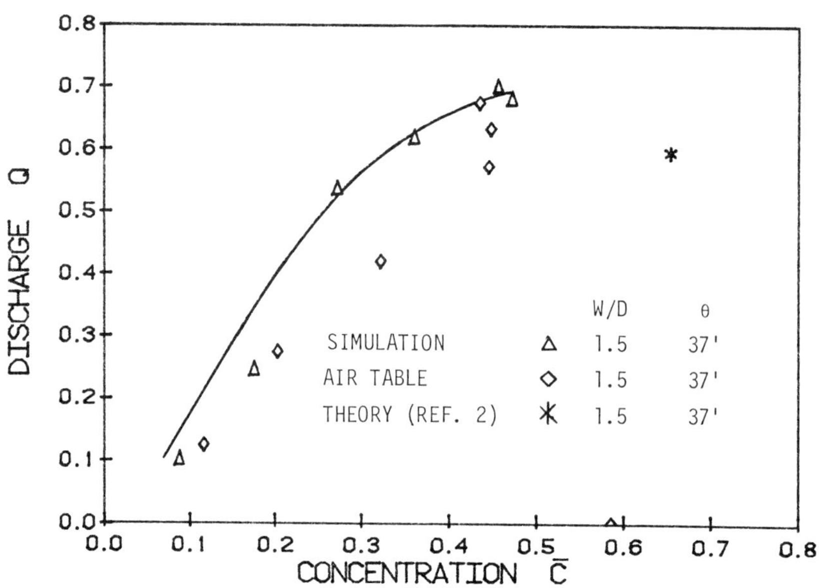

Fig. 6. Discharge, Concentration Relationship.

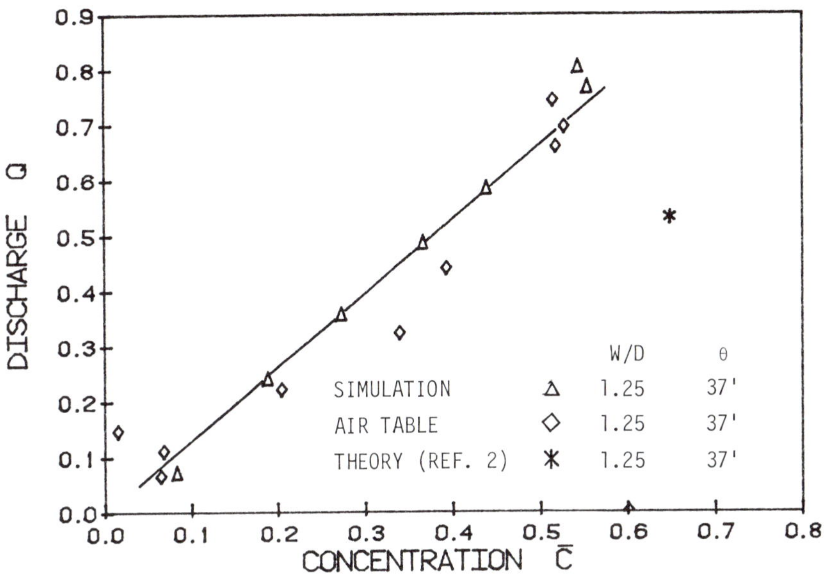

Fig. 7. Discharge, Concentration Relationship.

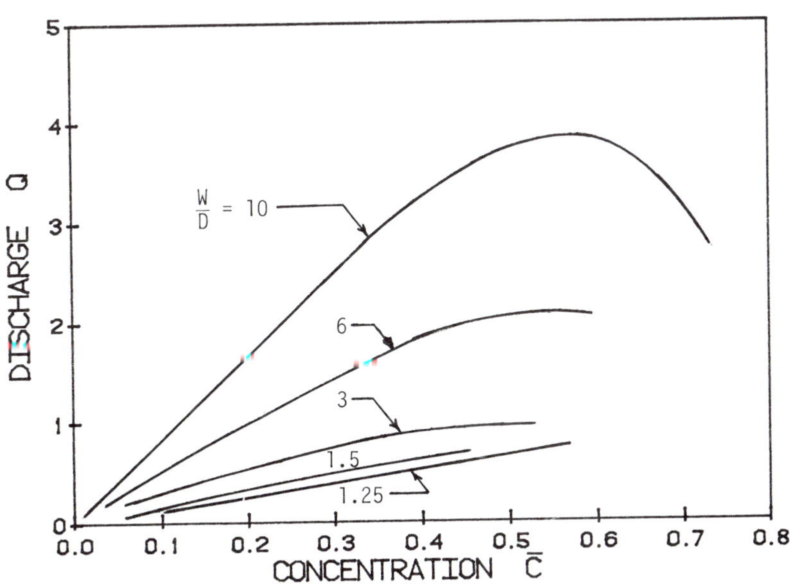

Fig. 8. Simulated Results for Discharge Concentration Relationship.

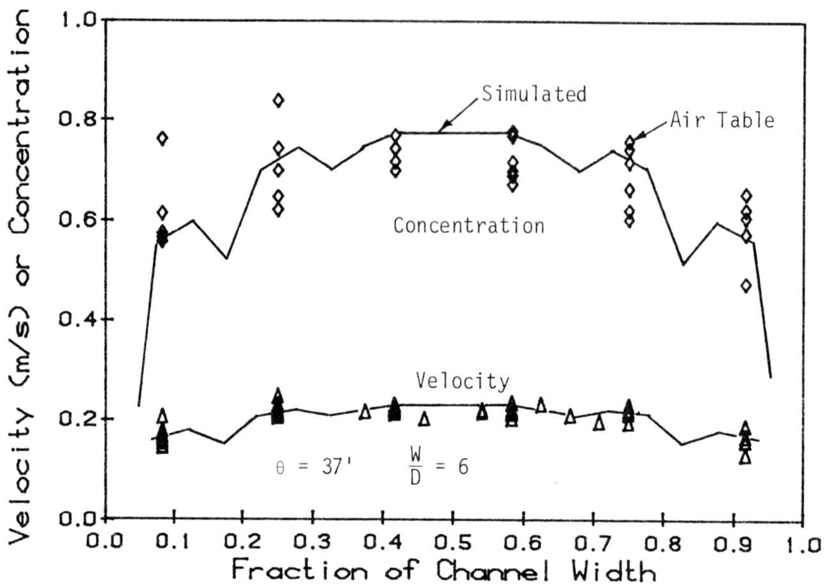

Fig. 9. Velocity and Concentration Distribution.

geometry and material properties a continuous range of steady, uniform flows could be produced where the average solids concentration varied from approximately $\overline{C}$=0.1 to $\overline{C}$=0.7. These results were in contrast to results from the numerical integration of the governing equations which predicted only a single steady, uniform flow condition for a given channel geometry, inclination and granular material.

ACKNOWLEDGEMENTS

This work was supported by NSF under Grant No. MSM-8419416.

REFERENCES

1 Hopkins, M.A., Granular Flow Simulations, Civil and Environmental Engineering Research Report 87-7, Clarkson University, Potsdam, NY, 13676, 1987.
2 Pasquarell, G.C., Boundary Value Problems for Planar, Smooth, Rapidly Sheared Granular Flows Including Phase Changes, Thesis for Doctor of Philosophy, Clarkson University, Potsdam, NY, 13676, 1987.
3 Sanders, B.E., Experimental Study of Two-Dimensional Gravity Driven Granular Flow, Thesis for Master of Science, Clarkson University, Potsdam, NY, 13676, 1986.